The Oxford Anthology of
Indian Wildlife
Volume I
Hunting and Shooting

The Oxford Anthology of
Indian Wildlife

Volume I
Hunting and Shooting

edited by
MAHESH RANGARAJAN

UNIVERSITY PRESS

OXFORD
UNIVERSITY PRESS

YMCA Library Building, Jai Singh Road, New Delhi 110 001

Oxford University Press is a department of the University of Oxford. It furthers the
University's objective of excellence in research, scholarship, and education
by publishing worldwide in

Oxford New York

Athens Auckland Bangkok Bogota Buenos Aires Calcutta
Cape Town Chennai Dar es Salaam Delhi Florence Hong Kong Istanbul
Karachi Kuala Lumpur Madrid Melbourne Mexico City Mumbai
Nairobi Paris Sao Paolo Singapore Taipei Tokyo Toronto Warsaw

with associated companies in Berlin Ibadan

Oxford is a registered trade mark of Oxford University Press
in the UK and in certain other countries

Published in India
By Oxford University Press, New Delhi

© Oxford University Press 1999

The moral rights of the author have been asserted
Database right Oxford University Press (maker)
First published 1999

The previously published material included in this anthology is reprinted
with the permission of the original copyright holders. A list of those who have
granted permission is included at the end of this book.

All rights reserved. No part of this publication may be reproduced,
stored in a retrieval system, or transmitted, in any form or by any means,
without the prior permission in writing of Oxford University Press,
or as expressly permitted by law, or under terms agreed with the appropriate
reprographics rights organization. Enquiries concerning reproduction
outside the scope of the above should be sent to the Rights Department,
Oxford University Press, at the address above

You must not circulate this book in any other binding or cover and you must
impose this same condition on any acquirer

ISBN 019 564 5928

Typeset in Garamond
by Excellent Typesetters. Pitampura, Delhi 110 034
Printed by Pauls Press, New Delhi 110 020]
Published by Manzar Khan, Oxford University Press
YMCA Library Building, Jai Singh Road, New Delhi 110 001

*In memory of
Kailash Sankhala
and
Oona Mansingh*

Acknowledgements

Like the tail of Hanuman at the court of Ravana, the list of those who helped me in compiling this set of articles is an ever-growing one. Naming them all would be impossible but all such help is gratefully acknowledged. The sense of *esprit de corps* in the wildlife and ecology fraternity is strong enough to cut across barriers of age, discipline and nationality. Often, an urgent phone call or post card was enough for a friend to flood me with a list of photocopies of articles worthy of inclusion. This collection is only intended to give a glimpse into a genre of writing about nature, not only about the large and spectacular animals of India, but its smaller, equally fascinating furred and feathered inhabitants. In recent years, our own perspectives have widened to include trees and flowering shrubs, reptiles and amphibians, many of which are relatively easy to observe and write about. If a century ago it was the landed gentry, the princes and civil and military officials who recorded their experiences, today it is a mix of professional wildlife biologists, self-taught naturalist-writers or photographers and film-makers and occasional foresters whose works make up the corpus of popular writings. Newspapers such as *The Statesman* and *The Hindu*, the now deceased *Science Today* and *The Illustrated Weekly of India* and smaller, privately published periodicals like *The Newsletter for Bird Watchers* from Bangalore and *The Hamadryad* (Chennai) and *The World Wide Fund for Nature Quarterly* have now joined the ranks with the much older *Journal of the Bombay Natural History Society*. The *Sanctuary Asia* magazine is perhaps unique in being a commercial venture that combines colour pictures and nature writing; *The India Magazine*, with a wider compass also has some excellent accounts of our wildlife.

I would be failing in my duties if I did not thank at least those individuals whose private collections I ransacked and whom I plagued with queries: R. R. Chari, Rom Whitaker, Janaki Lenin, Indraneil Das and S. Theodore Baskaran (in Chennai), D. K. Lahiri

Chaudhury in Calcutta and Valmik Thapar, Divyabhanusinh, Pallava Bagla, Mahendra Vyas, Tariq Aziz and Pradeep Sankhala in Delhi. I am also grateful to Rukun Advani and Anuradha Roy for suggesting I put together this collection and to Rimli Borooah and several others at OUP for all kinds of help. The Director and Fellows of the Nehru Memorial Museum and Library provided an intellectual base and support for which I am grateful. And may those not named here forgive me!

My family has both tolerated my obsessions and encouraged my interests. My mother and my wife Geetha Venkataraman helped in different ways. My daughter Uttara is too young to read the pieces, though she can growl 'like a tiger' when she sees a photo of one. I am especially grateful to them.

It is a matter of deep sadness that two remarkable individuals with whom I shared an interest in wildlife and who played a major role in my formative years are not here today. Both were neighbours in Kaka Nagar, a leafy housing colony in New Delhi but there the similarity ends. Kailash Sankhala was the Director of Project Tiger in 1974 when I first met him. It is typical of him that he treated me (then ten years of age) as an equal once he was convinced we had a shared interest in wildlife. He was never too busy to answer a query, always generous with rare books from his collection and willing to sit and go through drafts of my articles written for newspapers which rarely got published (for no fault of his). When we last met in 1993, he encouraged me, as a historian, to delve into the archives to find out more about the distribution and decline of India's fauna. He died a little more than a year later.

If Kailash Sankhala was a veteran, Oona Mansingh was a peer, scarcely a year older, and like me, an active member of an environmental group, Kalpavriksh. Oona's interests were in the outdoors: she was a member of three varsity teams, shooting, tennis and swimming. After a degree in rural management, she had settled in Kumaon, working to revive community management of forests. Her tragic and sudden death in 1996 due to food poisoning, which also claimed the life of her young daughter, has left a void in the lives of all who knew her. I humbly dedicate this collection to the memory of Kailash Sankhala and Oona Mansingh.

Contents

Introduction: Cheetah Hunts, Tiger Tracks 1

PART I: SPORTSMAN'S PARADISE 7

A Lion Shoot in Haryana 9
G. C. Mundy

Hunting Large Animals 13
William Rice

Sport in Lower Bengal 24
Edward Braddon

Game Animals and Birds of the Plains 36
James Forsyth

Rhinos, Javan and Indian 54
Frank B. Simson

Hunting the Houbara 61
J. H. Baldwin

Shooting Elephants 78
P. D. Stracey

Sport in Upper Burma: Sumatran Rhino 99
F. T. Pollock and W. S. Thom

A Wild Buffalo Hunt in Bastar 112
R. P. Noronha

PART II: ON HILL AND MOUNTAIN 129

Himalayan Black Bears 131
Isabel Savory

The Nilgiri Wild Goat 138
F. W. F. Fletcher

The Kashmir Stag	155
C. H. Stockley	
Stalking the Gaur	168
James Forsyth	

PART III: PRINCES AND SAHIBS 181

Govindgurh	183
Louis Roussellet	
Panther Shoots with Ranji	196
C. B. Fry	
A Viceroy's Game Diary	207
Lord Hardinge	
A 'Criminal' Lion in Kathiawar	214
Guy Fleetwood Wilson	
Visitors and Big Game	218
J. W. Best	

PART IV: TRAPPERS AND TRACKERS 235

Elephant-Catching in Mysore	237
C. P. Sanderson	
My Shikaree Friends	260
A. Mervyn Smith	
How to Trap and Train a Cheetah	268
Divyabhanusinh	
Cheetah in the Kolhapur Deccan	280
Suydam Cutting	
Following the Lion's Trail: The Lion Trackers of Mytiala	286
R. S. Dharmakumarsinh	
Gulam Hussain Baazdaar: The Falconer	292
R. S. Dharmakumarsinh	
Aboriginal Methods Employed in Killing and Capuring Game	297
K. M. Kirkpatrick	

A Novel Method of Destroying Man-eaters
and Cattle-lifters 302
S. R. Daver

PART V: MAN-EATERS AND ROGUES 321
The Panar Man-eater 323
Jim Corbett

The Man-eating Leopard of Rudraprayag 340
Jim Corbett

The Man-Eater of Segur 355
Kenneth Anderson

The Crossed Tusker of Gerheti 375
Kenneth Anderson

Tigers in Bastar 392
R. P. Noronha

On Corbett's Trail: Tracking the Man-eater 420
A. J. T. Johnsingh and G. S. Rawat

Abbreviations 429

Glossary 431

Bibliography 437

Copyright Statement 441

Introduction
Cheetah Hunts, Tiger Tracks

Hunting anecdotes and essays may seem out of place in an age of ecological awakening. Tales of the hunt seem part of a dim, best-forgotten past. This feeling deepens if we tot up the 'scores' of big-game hunters—for example, the Maharaja of Sarguja, Ramanuj Saran Singh Deo, who was hunting till well into the 1950s and killed over 1100 tigers, 2000 leopards, 2 lions and even a cheetah. Nobody else ever managed to inflict such a high rate of mortality on the animal world, but several did their best to emulate if not equal his exploits.

The world of the hunt pervaded not only the drawing room or verandah with its draped tiger skins and mounted stags' heads; it also entered the dining room—you often ate what you had shot. Here possibilities ranged from the houbara bustard in the western grasslands to ducks and geese from marshes and ponds. Small game was just as important as venison or boar and the cooking of game was as much an art as taxonomy, the preparation of trophies. It is only natural that at the dawn of the twenty-first century, we are trying to transcend such dubious and troubling legacies.

But there was much more to hunting than the blood and gore, the recovery of trophies and the totting up of records. For well over a century and a half at the pinnacle of the Raj, those who set out to track, trap or shoot game, big and small, also left behind written accounts of what the land looked and felt like. This collection is not so much a celebration of their deeds as a compendium of memorable impressions of the natural world. It is not motivated by a sense of nostalgia but out of the need to learn creatively from the past. It is certainly often about a world we have lost. But it hopes to remind all who care for the wild today that we now have only fragments of our natural heritage left. Given the incredible diversity of terrain and habitats in India, hunting can now be seen—ironically—as one way of knowing the country. There were indeed many insensate killers. At the same time, some had a keen sense of natural history, an eye for detail, and even an appreciation of the wild.

In the days before the motor car, hunting often required arduous journeys on foot, mule, horse or elephant-back. Over the years, advances in technology, especially of weapons and the increasing mobility of the hunter began to tilt the scales against the quarry. But the very fact that such a range of wildlife did exist and that seeking it out meant hard travel and ingenuity, makes the records of hunts simultaneously an eyewitness account of the natural wealth and beauty of the land.

There was the second heaviest land animal on earth, the Asian elephant, an animal to be trapped alive rather than shot at. In the mountainous regions was quarry reminiscent of alpine and temperate areas in Europe—black and brown bears, wild goats and sheep, and a local variant of the red deer, the Kashmir stag. Of all big game, the tiger was the most eagerly sought after, but it was by no means the only one. The smaller leopard was always more widespread, though much more elusive. It was normal for big-game hunters in India to try and match the 'big five' of Africa—the lion, leopard, elephant, rhino and wild buffalo—all mainly shot in the plains of east and southern Africa. India could throw up not one but many such 'lists'—the tiger, the leopard, the gaur, wild buffalo and sambhar were an obvious quintet. And by moving even a few hundred miles, the hunter could be in strikingly different terrain. The imperial capital of Delhi was at the doorstep of antelope country; there were tigers in the Aravalli hills to the west as in the foot-hills of the Himalaya and in the reed beds along the Ganga at Garhmukteswar, where the river flows closest to Delhi.

Some animals the hunters pursued now exist only in the small patches that remain of their once widespread domains. So, today when lions only survive in the dry thorn and teak forests of the Gir forest, it is the lion-hunter's tales that remind us they once roared and reigned over the savannah in north India not far from Delhi. Similarly, the mangroves of the Gangetic delta were once home to the Javan rhino. Frank Simson, one of the few people to have hunted both the great Indian one-horned rhino and the Javan rhinoceros, writes of the remarkable difference between the habits and habitats of the two. Sanderson recalls the *keddahs* or elephant round-ups in Mysore, and provides an early account of the behaviour of the elephant in the wild. James Forsyth's essay on the wild animals and bird-life of the central Indian plains is rich with insights into the ecology and human life-styles in the region.

Of course, hunters themselves can be further grouped into two broad sets. There was the gentrified lot—the British military men and officers, the civil servants and boxwallahs, the Indian princes, aristocrats and landed gentry. They hunted 'for glory', for proving their prowess at arms, their skills of marksmanship, for leisure or pleasure. Over time, their ventures into the jungles and hills became more of a ritual, with elaborate preparations to ensure 'a kill'.

Such *shikar*, as the ceremonial hunt was often called, had long Indian antecedents but it reached a climax under the British standard. The Indian princes played a key role by acting as hosts to big-game shoots. The Nawab of Junagarh called in a chosen few to 'take' a lion. A neighbouring ruler, Ranjitsinhji, the renowned batsman, could not summon lions for his British guests, but had mongrels tied up as bait for the leopards they would shoot. Far more elaborate were the arrangements for the Viceroy and for British royalty.

The second set of hunters were more humble, pursuing their quarry under far more adverse conditions. Many local hunters were tribals, hunting for food or to trade in skins, hides, tusks or meat. But they were also critical to the successes of princely and *sahib-log* hunts when they took part in them. Though there are accounts of their range of skills and their myriad methods of catching wild animals, this is perhaps the first time a hunting anthology gives them the attention they deserve.

Local hunters were usually seen by upper-class sportsmen as rivals for game. They were often stereotyped for their supposed 'cruelty', 'waste' and 'recklessness'; but at times, another picture does emerge. Mervyn Smith provides a sketch of two Indian 'native shikaris' from the Chotanagpur plateau, including one from a hereditary caste of Bagmaris or 'tiger-slayers'. Daver pays tribute to the courage and skill of tribal hunters who kill big cats without recourse to fire arms. Much more riveting are the accounts of the means used to live-trap and train the cheetah. From Bhavnagar, a small princely state in western India, come two first-rate accounts by one of India's best ornithologists, Dharmakumarsinh. One recounts the dexterity of the *pagis* or lion-trackers and guides; another is a life-sketch of a great falconer.

Man-eaters and rogue animals must occur in any book of hunting stories. Ever since the publication of Jim Corbett's *The Man Eaters of Kumaon* nearly half a century ago, the thrill and danger of the man-eater's trail has never ceased to hold the reader spellbound. It is not widely appreciated that Corbett was not alone or exceptional in shooting man-eaters. Anderson's stories are set in the south, Noronha's tiger tales are from Bastar, central India. Such ace hunters who had to slay the rare animal dangerous to humans were at times fine naturalists as well. Corbett's vivid descriptions of landscapes remain appealing enough for two present-day biologists, A. J. T. Johnsingh and G. S. Rawat, to walk his trails. They discuss how the land has changed since Corbett's time.

In a sense, this volume is a record of ways of life, of times long past and of the world of the forest. The writings span more than a century and a half and represent the 'pick' from over 250 books and innumerable journal and magazine articles on hunting and on the natural world. Many of the pieces reproduced here are now out of print and not easily accessible. But all are of wide interest being simultaneously informative and readable.

There is no doubt that the fortunes of India's wildlife have been transformed in the period between the publication of the earliest and most recent pieces included here. By the time Sudyam Cutting watched the tame cheetahs of the raja of Kolhapur race across the Deccan plains, he was struck by the fact that each of those cheetahs had been imported from Africa. They were already a rarity in the India of the 1930s and would soon be extinct. Ironically, many hunting journals and diaries attest to rapid ecological collapses and describe how once abundant species can 'crash' under sustained human pressure. There was a dark side to the thrill of the chase and sport-hunters played a role in delivering the *coup de grace* to many big-game animals. While they were never alone in doing so, it is impossible not to look with misgiving at the ethic of conquest that they embodied. A few had the sense to impose a regime of self-restraint. Junagarh's Nawabs gave the lions of Gir Forest a protective shield and the Mysore ruler protected tigers from all-out assault. Not that the means used, sometimes draconian and often highly coercive, made these popular ventures. As with shikar in general, the protection of game has left us a mixed legacy, one we cannot wish away but must transcend.

Perhaps the last drama of the shikar era was played out in the 1950s. An American television company staged a tiger-netting hunt in southern India with the help of an ex-shikari of a north Indian princely house. Tiger-netting used to be a means of ridding hillsides of cattle-lifting tigers in the days when fire-arms were confined to the elite. For the television shoot, hundreds of men armed with spears closed in on a huge tiger, encircling it with nets and eventually spearing it. The staged netting had elements of both tragedy and farce. What was death for one of the participants was all in a day's work of television shooting for the others. But the wheel was already turning. Less than two decades thereafter, tiger-hunting itself was outlawed. The night of the big-game hunter was over.

PART I

Sportsman's Paradise

> The boar, the mighty boar's my theme
> Whatever the wise may say!
> My morning thought, my midnight dream,
> My hope throughout the day
> Youth's daring spirit, Manhood's fire
> Firm hand and eagle eye
> Must they require,
> Who dare aspire
> To see the wild boar die
>
> J. A. STOCQUELLER, *c.* 1853

He emerged out of the tangled growth of bushes ten feet from His Highness and the Nawab (of Tonk) dropped him in his tracks with a bullet from his .465 bore India rifle planted at the base of his skull. Hardly had the loud report from His Highness' rifle died out than a full grown young cub, the size of a tiger came out roaring and the Nawab killed him too with another bullet from his rifle. '200 up, sir,' I whispered in joyful glee like a schoolboy who heralds the finish of a second century in a cricket match with those words.

ABDUL SHAKOOR KHAN, *c.* 1968

G. C. MUNDY

A Lion Shoot in Haryana

By 1888, there were no lions left in India outside the Kathiawar peninsula, Gujarat. But half a century earlier, the king of beasts had an extensive range across much of central and north India. Warren Hastings had hunted lions in Hissar, Haryana. Mundy's account of a similar lion hunt in the 1820s provides a look at a landscape that has been lost forever. Lions often lived in open country, partly living off livestock, prompting herders to help civil and military officers armed with modern weapons to locate and shoot the great cats. The scarcity of water made it easier to find and shoot them. The lion is more approachable than the tiger; it also roars often, advertising its presence and inviting its doom. The Gir forest where they still survive is strikingly different in terms of terrain from the tracts where Mundy and friends shot their lions. On this hunt Mundy was accompanying the Commander-in-Chief of the British troops in India. Of all his shikar escapades, this one shows us best how lions lived, not in forest but in scrub jungle and open savannah. Their reign was ending, even before their habitat vanished under the plough.

Oct. 31st.—Camp broke up from Patialah, and marched 14 miles to the village of Koralee—the face of the country still uninteresting, and devoid of large trees, but well cultivated.

Nov. 1st.—A heavy shower brought the thermometer down to 79°, relieving us of 13 degrees of heat,—an amelioration hailed by us as the precursor of the cool season.

This day's march brought us into the territories of the Khytul Rajah, which are divided by a small stream from the Patialah dominions. The effects of the efficient government of the latter province are plainly visible in the great superiority in cultivation possessed by it over the Khytul country, whose Rajah, a mere boy, is ruled entirely by his mother, who is, in her turn, swayed by a host of paramours and parasites.

Two marches brought us to Pewur, through a country which, with the exception of sufficient cultivation round the villages for the bare existence of the inhabitants, presents one vast sheet of wild jungle, abounding in game. Two mounted Shikkarees of the Rajah brought a couple of wild hogs into camp which they had killed, and three others arrived from Colonel Skinner, to assist us to find a lion between this place and Hansi. Of these animals there were formerly great numbers in the surrounding wilderness: but from the zeal of English sportsmen, and the price put upon their heads by Government, the royal race of the forest—like other Indian dynasties—is either totally extinct, or has been driven farther back into the desert. By *crack* sportsmen the lion is reputed to afford better sport than the tiger: his attack is more open and certain; a peculiarity arising either from the noble nature of the jungle king, or from the country which he haunts being less favourable for a retreat than the thick swampy morasses frequented by the tiger. Colonel Skinner relates many interesting anecdotes of lion-hunts, with the exploits and narrow escapes of the horsemen of his corps, who always accompanied the line of elephants into the jungle on these occasions. Major Fraser of the same regiment, is, however, the lion-queller par excellence.

A gentleman of our party had, perhaps, as perilous an adventure with one of these animals as any one of the former; he having enjoyed the singular distinction of lying for some moments in the very clutches of the royal quadruped. Though I have heard himself recount the incident more than once, and have myself sketched the scene, yet I am not sure that I relate it correctly. The main feature, however, of the anecdote, affording so striking an illustration of the sagacity of the elephant, may be strictly depended upon.

A lion had charged my hero's elephant, and he, having wounded him, was in the act of leaning forward in order to fire another shot, when the front of the howdah suddenly gave way, and he was precipitated over the head of the elephant into the very jaws of the furious beast. The lion, though severely hurt, immediately seized him, and would doubtless shortly have put a fatal termination to the conflict had not the elephant, urged by the mahout, stepped forward though greatly alarmed, and, grasping in her trunk the top of a young tree, bent it down across the loins of the lion, and thus forced the tortured animal to quit his hold! My friend's life was thus preserved, but his arm was broken in two places, and he was severely clawed on the breast and shoulders. The lion was afterwards slain by the other sportsmen who came up.

The village of Pewur towers boldly up from the midst of the jungle; and is situated on a small river, the banks of which are ornamented by several neat ghauts. Like the generality of towns in this part of the country, the houses are built of good brick; but, like the rest, the abodes of the present generation are stuck cheek by jowl with the ruins of their ancestors' habitations, no one troubling himself to clear away the rubbish of the fallen buildings, and few caring to finish an edifice commenced by their deceased predecessors. By this accumulation of material, however, an ancient village in course of time gains one advantage; namely, a site elevated above the surrounding flat country. I shot my way over to Pewur this morning, and got a good bag of partridges, hares, and wild fowl.

Nov. 5th.—A short march of 8 miles to the town of Khytul, which gives its name to the Rajah of the province. This prince came a mile or two out of the town to meet the Commander-in-Chief, and escorted him into camp. He is a fat, uninteresting, heavy-looking boy of 12 years, but is said to possess more brain than his appearance indicates. The town is irregularly built, but of very good brick; and from the kilns are produced large quantities of sal-ammoniac. The palace is a striking lofty building, and, as we approached, we had a very flattering view of it through a break in a fine grove of trees overhanging a spacious sheet of water. The Rajah is a relation of him of Patialah, and his revenue only amounts to 5 lacs of rupees, whilst that of his more powerful cousin is at least 5 times as great. The boy, however, being of the elder branch, and by inheritance a Goru or holy teacher, his gigantic relative cannot sit in his presence without special permission.

In the evening, the young king having signified his intention of visiting Lord Combermere, I was despatched, with the Persian interpreter, on an elephant, to marshal him into camp. We met him at the city gates, and he raised himself 100 per cent in my estimation by presenting us with a handsome matchlock and a couple of bows.

Nov. 6th.—The shikkarees having brought intelligence of two tigers in the jungle about 14 miles off, four of us started very early in search of them. The swamp where they had been seen proved the Slough of Despond to us, for we found nothing of them but their footmarks, those of a young female and a full-grown male—for the experienced natives can distinguish the prints. After a hot day's work we returned to Khytul, where Head-quarters had halted.

The day following we reached the hamlet of Kussaun, where the bush-jungle was so thick and extensive that we could scarcely find a spot to pitch the camp: and the morning after, we made a march of 14 miles to Shamdore. The morning was extremely cold, and the mid-day equally hot—therm. 82°, 2 P.M. In this province the cultivators adopt the following plan for scaring birds from their graincrops. Several platforms are raised in every large field, upon each of which a man stands, armed with a sling, formed (much like those used by mischievous boys at Eton to break windows and bargemen's heads) of two pieces of thong or cord, with a leathern receptacle for the missile, a pellet of hardened clay. This is thrown to the distance of several hundred yards, and each shot is followed by a loud crack—like that of a French postilion's whip—managed by the slinger when he lets fly the string of his weapon.

Excerpted from Pen and Pencil Sketches in India, Journal: A Tour in India *(London: John Murray, 1858).*

WILLIAM RICE

Hunting Large Animals

From 1850 onwards for four long years any chance of sport during the vacations saw William Rice and his friends set out with guns and camping gear from the military cantonment at Neemuch in central India. Though well-known as a slayer of tigers, Rice also pursued other large animals, such as the sambhar, India's largest deer, and the sloth bear. Such forays into the forest were common for British military men. When not subduing the human inhabitants of India, they hunted wild life, to hone their skills with horse and gun, for trophies and to add a variety of meats to their diet. Rice's records of wild dogs in the jungles of eastern Rajasthan are of interest, for these are not areas where they are now known, except for stray packs.

Continuing our operations, we set out on the 27th March to beat a ravine called 'Amko;' the men put out a bear, but we did not then see him. They pointed out the direction in which he had gone, and presently we saw him quietly climbing up a distant hill. Off we ran at our best speed, and luckily came upon him, lying up in some bamboo clumps or bushes. We instantly gave him three shots, which, luckily, finished his existence, fortunately without any charging taking place, for we had no spare guns up with us at the time, having left the men far behind in our race after the bear.

In many instances, when running after big animals in the hope of overtaking them, or of getting a parting shot as they made off, we used to find that we had far outstripped the men who carried our spare guns, and who were required, at least, to be able to keep up in a chase of this sort. This did not arise from our being able to compete with these men in the art of getting over broken ground, but solely from the fact that the gun-carriers had not the same interest in the chase. As it was quite a matter of indifference to them whether we killed or not, they took matters more easy, and lagged behind. This lazy conduct often left us an ugly chance of being overmatched on attacking the beast we had overtaken: a strong argument in favour of a repeating rifle.

Afterwards we looked for game in some other good covers about, but without success. Returning to our tent, I smashed the fore leg of a fine buck sambur, with splendid horns, but he managed to get off, for it was a shot at least 200 yards, as the noble looking animal stood watching us from a distant rocky peak. It was rather late in the season for him to wear his horns, for they generally shed them at the beginning of the hot weather. Provided we felt certain that there was little chance of getting a shot at any wild beast in the neighbourhood, we seldom neglected killing sambur, solely for the sake of their skins. The leather is beautifully soft and thick. From it we had shoes and gaiters made, which never, by any chance, hurt the feet, though ever so new or ill made. The skin easily stretches, being very pliable, and so smooth that it was quite unnecessary to wear socks with these boots. This leather, moreover, is the very best possible protection from thorns, and seldom ever tears. It makes also capital covers for saddles, which otherwise would soon be worn out and spoilt in such rough work as 'hog hunting,' or knocking about for months in the jungles. The flesh is coarse in the extreme, and very tasteless, but the marrow-bones are excellent.

On one occasion, I well remember, we were glad enough to live on sambur meat for more than a week, being out of supplies, and far away in the woods from where any better food could be procured. If sambur are started from any place, it argues but little chance of your finding either tiger or panther in that particular cover; but instances did occur in which this rule was upset, so we mostly allowed the sambur to escape, lest, by firing at them, a shot might be lost at more worthy game; otherwise, almost any number could have been slaughtered, as they dashed past us at only a few yards' distance, often in herds of even twenty together, in particular places where at all numerous. Where sambur abound, there will surely tigers be also found, but not living in the same patch of jungle or cover. The Bheels with us were always very desirous that we should shoot these animals, as well as the 'cheetul' or spotted deer ('axis'), for them to eat. They also contrive to kill them even with arrows, silently following for miles on the wounded deer's tracks. They have even a better chance of 'bagging' the sambur with an arrow than with the small bullet used with their matchlocks, for the long broad blade of the arrow, the points or heads of which serve them for all the purposes of a knife, by cutting deep at each stride of the poor wounded deer, soon causes him to stop and seek for safety, by endeavouring to hide itself in some dense cover, where, however, it is soon spied by such lynx-eyed gentry as the Bheels. It is perfectly astonishing how close deer will sometimes lie when thus hiding themselves. I have even fired into them quite near, but without their showing the slightest sign of having been struck, when, on attempting to lay hands on them, they will start off again seemingly none the worse.

The 'cheetul' are very handsome creatures (about the same size as fallow deer, and spotted in the same manner), and are in some jungles, where there is plenty of water and shade, very numerous—it being no uncommon sight to meet them in herds of even a hundred or more together; but the meat, like that of the sambur, is nearly as coarse and rank. Besides these deer, we often disturbed the little 'beadlah,' or 'beekree,' a small sort of four-horned deer, that somehow appears to have a touch of the sheep in his make. These deer are found singly, or at most in pairs; they lie during the heat of the day in thick high grass, or cool shady ravines, and start up even at the hunter's feet, bounding off in a series of quick high jumps, rendering it difficult to shoot them with ball, though so close. At evening they are often heard calling—a queer bleating sort of bellowing noise. The flesh is first-rate, far superior to any other kind of deer, and very tender. So, on returning from our day's sport to the tent, we generally looked out for a shot at these little creatures for the pot. The fact of their meat being so good, may be perhaps accounted for from their never exposing themselves to the sun. The tiny 'mouse,' or little hog deer (for they have tusks like a boar, but on a very small scale), we did not meet with in this part of the country, neither did we see many jungle fowl, so common in most large forests, but the little spur fowl, less even than the smallest bantam, were numerous enough.

Wild dogs, too, seemed to be utterly unknown in this country, which seemed rather strange considering their numbers in other wild extensive jungles over which we had hunted, where they were often met with in packs of even thirty or more together. The wild dog is about eighteen inches high, of a brick red or bright fox colour, with a thick bushy tail tipped with black, the ears are also black, pointed, and upright. They are the most determined enemies of the tiger, hunting him whenever they meet with him. I have been assured by Bheels that they have sometimes seen a tiger kept prisoner up a large tree with a pack of these dogs baying around him, when on no other occasion would a tiger attempt to save himself by climbing trees. On the approach of the men the wild dogs dispersed, when the tiger jumped down, and gladly made his escape. This I firmly believe to be a fact, for the story arose out of a casual remark one Bheel made to another in my hearing, as we were passing a certain large tree (with a straight stump about five yards high before the branches began, up which a tiger had jumped) in another part of the country over which I was hunting. Perhaps these dogs hunt the tiger for poaching on their deer, or it may be only the old antipathy between cat and dog, on a large scale.

A man came early on the 28th March to our tent, and told of a tiger he had seen four miles off, so we at once went to the spot, but could nowhere find this beast, although we looked carefully over every likely bit of ground for miles round. Did not get home till late. News of bullocks being killed at 'Doraee,' six miles off, also reached us at night, so we agreed to hunt there next day.

Early in the morning of the 29th March, having sent on our guns and men, we rode after them, and got to our ground by 9 A.M. Here we were delighted to hear that two tigers were known to live in a deep ravine in the hills near; so we at once, after examining the nature of the cover, decided on posting ourselves at the bottom of one end of the deep ravine, while the beaters went silently round to the other. Almost directly after they had commenced making a noise, we saw a fine tigress moving very cautiously towards us. We let her come close up, without seeing us, when we poured in a volley (one barrel missed fire, so she only got three instead of four bullets). These, however, brought her roaring and charging down past our hiding-place in beautiful style, but so quick that we had no time to exchange our empty rifles for loaded guns before she was out of sight in the high grass beyond. Scarcely had we quickly reloaded, when a magnificent tiger came walking in a most stately manner on to our front. At about twenty yards off he stopped still and stared steadily at us. Being quite ready for him, we at once agreed to open our fire, which was done with good effect; for the tiger began to bound about in all directions, giving us capital opportunities of finishing him, which we did far easier than we had expected, judging from his great size. He measured a few inches over eleven feet, and was exceedingly stout, with an extraordinary quantity of long hair about the face, and even had long curly locks down the back of his neck, much more so than in any we have before or since killed; this greatly added to his personal appearance.

Again loaded, and the beaters having now joined us, we proceeded to look for the first wounded tiger. For this purpose a procession is formed as follows. In front, stooping down between us, is our head 'shikarree,' or chief huntsman, who, by carefully observing each footprint or slightest drop of blood, points out the direction in which the wounded game has gone. Keeping guard over him with full cocked rifles, we lead the wedge-shaped procession. Immediately behind us follow our best or steadiest men bearing the spare loaded guns. Next comes the 'band,' which consists of four or five kettle-drums and one big drum, a man ringing a tremendous bell (novel method of 'belling the cat!'), with perhaps others, either blowing a large brass horn, or beating cymbals, besides two of our men constantly loading and firing blank shots from a pair of old horse-pistols. At either side of these are some men armed with drawn swords and two 'halberds,' or most formidable-looking spears, which serve to keep the beaters well together in passing through high thick grass or reeds, often high over head, for all can see their broad sharp glittering points. Last of all come a number of men engaged in constantly slinging and throwing large stones, which they either pick up as they advance, or take a supply with them before starting, according whether the ground is stony or not. These stones fall just in front, and on all sides of the whole party, often starting a wounded tiger that perhaps would otherwise allow us to pass, in spite even of the deafening noise so close to him. Overlooking all is a man up a tree, which he climbs from time to time as we pass them, keeping a good look-out on all sides for any large game moving ahead. The whole party, in a compact body, keep close together, move at a snail's pace, yell with their utmost power, and create what really is a most infernal din. No tiger will face such a mass of men and noise as this. They sometimes charge down within even a few yards, but then invariably turn off, mostly getting well 'peppered,' or are shot dead in the attempt.

With this system there is perfect safety to every one, no matter how dense the jungle may be; whereas solitary men or stragglers would be 'mauled' to a certainty by the wounded tiger should they attempt to invade the fastness he had sought refuge in.

While thus 'following up,' there is great difficulty in keeping the men well together, for they seem to consider that every wounded beast must be either already dead or just dying, and so get very careless, and are apt to lag behind unless well looked after. To avoid this noise the wounded tiger will go as far as he possibly can, bleeding often to death before his body is discovered. At other times he will decline moving an inch, roaring awfully at the men from some strong position where he cannot be seen; but a few shots fired at guess seldom fail to start him, so that he can be seen and shot. Sometimes this is easily done by climbing a tree overlooking his retreat. We never had any trouble in persuading men to accompany us after tigers, when they found there was so little danger in so doing; for they seemed to consider the whole business as a good joke. A body of men passing through a cover in this fashion will beat it far more effectually than if drawn out in a long line, for the tiger leaves directly he finds his retreat so disturbed without troubling you to beat every individual bush for him. In some very wide places, or in open plains of high grass, we divided the beaters into two bodies, and so beat over a vast extent of ground at once, each party moving parallel to the other.

In the present case we had not gone above a hundred yards through some very thick jungle, when we were called back by some men in the rear, who said we had passed the tiger. It took us some time to distinguish what they pointed out as the game. It was lying about twenty yards off. We fired together, and instantly down charged the tiger upon us. So thick was the cover that we could not see to fire our second barrels; we merely knew the brute was coming by its roars and the crashing of the bushes. She luckily pulled up just short of us, and again crouched down, being no doubt badly wounded. What she did in the 'charge,' however, caused a most laughable confusion, for nearly every man of us tumbled down over somebody else, as each, on hearing such roars coming on, involuntarily stepped back a pace. There was a temporary panic, but we soon re-formed and retreated, showing a front to the enemy. After a little delay we advanced again in a different direction, and soon saw the tiger lying down flat under a shady bush, evidently very much done. A few more shots now made the kill secure. We carried both the bodies to our tent, being obliged for some little distance to clear a space among the brushwood with axes, before we could take away this last killed brute, which proved to be a fine tigress. It was dark before we had finished skinning and pegging down the two skins.

Every one seemed much pleased with the day's sport. The large tiger had killed a camel the day before, and had completely gorged himself, but most ungallantly had refused to allow the lady tigress to dine with him, for she was quite empty, although by their footprints of yesterday's date we knew they must both have been together in the same ravine.

During the next three days we were engaged in fruitless searches for tigers reputed to have recently committed great ravages. Bear was all that we could find. On the 2nd April we met and shot a bear. On going up to her, discovered she had two cubs on her back, the usual method with bears of carrying their little ones when too young to save themselves by flight. The little brutes squalled lustily on our nearing them, and at length left their dead mother, whereupon we gave chase, and each caught one alive. Little got a severe bite from his cub in the hand. They were too old to bring up by hand or tame, so the Bheels murdered the little black devils.

Fresh prints of a large tiger were found close to our tent on the 3rd April. We started at once in search of their maker. First searched a 'nullah,' or small dry water course, full of 'corinda' bushes, near at hand. Here Little caught a view of the tiger as he was making off, but out of shot. We now followed up his footsteps for at least two miles, when the ground became too rocky to follow them any longer, so we beat a few patches of cover about, but without success, and returned to again try and follow up the foot-marks. While so doing, we suddenly came most unexpectedly upon this tiger, but he moved off round some rocks before we could get a shot. We now ran hard across an open space about five hundred yards to a ravine, for which he seemed to be making, and got there just in time; for we soon saw him coming quietly towards where we were lying hid, with our spare guns by our side. We allowed him to come within fifteen yards, up to a mark we had agreed should be the signal for firing a volley, when he was at once knocked over; another bullet or two settled him. This tiger measured twelve feet two inches, and was stout in proportion: the largest we had as yet killed.

On the following day, the men we had sent out early to look about for fresh prints of large game, reported that they had found the prints of two tigers, in the bed of a river in which there were some dense covers of 'corinda' bushes, willow-trees, and reeds. We tried for a long time to find out where these beats could be hiding, having beaten the cover, but without seeing them. At length a man thought they must be hid in some very thick 'corinda' bushes on the opposite bank of the river, so I waded across the stream, which was about four feet deep, and by carefully looking under the bush, while standing in the water with my rifle all ready, thought I could at last make out what appeared to be part of a tiger, but the bush was so thick and dark that nothing could be distinctly seen, though I was but a very few yards off. On returning to my friend on the opposite bank, we agreed to fire a volley on guess into the bush, so sent all the men well to the rear, to be out of danger. After our first shots all was still, so I fancied that no tiger could, after all, be in the bush; but presently there was a slight movement, we poured in more shots, and directly afterwards a tiger appeared, which we shot dead by a ball in his skull; the beast fell half in the water down the sloping bank.

Thinking that all the sport was now over, the men, in a body, went round to the dead tiger by a ford higher up the river, and were standing close round it, pointing out to each other the shot-holes, admiring the skin, and talking over its death and the mischief it had done to their cattle, as they usually do, when all of a sudden a most appalling roar was heard to proceed from the very midst of them, as it were. The effect this caused was ludicrous in the extreme, for with one accord they precipitated themselves into the stream with a great splash, and regained the opposite bank in the utmost terror, each struggling to be first across the river. On hearing this roar, all our guns being unloaded at the time, and knowing there must be another tiger close by, we also sought safety in flight. Little got under a bush, while I quickly gained the top of a large thorn-tree nearest me, well scratched in the process, for at the time my costume was extremely scanty, as, on coming out of the water, feeling much chilled, I had taken off my clothes to dry in the sun, and was standing in merely my brown shirt. The next moment we were horrified to hear that a man had been killed. The panic now being over, we reloaded and hastened to the spot, where we were delighted to find that the man was not dead, but had been merely knocked down by a tiger and severely clawed.

It seems that on hearing we had killed the first tiger, this man, who had been posted to look out by himself up a tree, hastened to join the rest of the beaters, while they were rejoicing and talking over the dead tiger. All this time there was another tiger still in the very bush in which we had killed the first one; but as long as the men remained in a body, though only about two yards from them, it kept quite still. On seeing this man approach the bush alone, the beast rushed with loud roars out upon him and knocked him down, actually running off with his turban, which fortunately was a very large one, and no doubt saved his head from the tiger's blows. We had the wounded man at once conveyed to the village, and then tried to follow up the tiger's prints, but could not manage it far, so again began beating some covers about, out of one of which we started the tiger, about fifty yards in our front. He went off at a racing pace; we fired, but apparently missed, and the brute dashed away straight for the hills.

It was too late to look for it there to-day, so we returned to our tent and gave directions for the wounded man to be attended to. After all he seemed more frightened than hurt, excepting some ugly scratches, but he was now quiet enough, having been dosed with quantities of opium, according to native custom in these cases. He soon recovered, and twice afterwards came out hunting with us in following seasons. This accident would never have occurred had he kept close with the other men, for the tigers never attack a large body of people well together, but are sure to select any solitary individual. There was much diving afterwards in the river to recover shields, swords, shoes, spears, and many other articles lost in the panic. The beast we had killed was a fine young tigress. This was the second accident that had happened to our men, but in neither case could we prevent it, nor did we consider ourselves at all to blame, for in both cases they had neglected to observe our most positive caution to keep well together. This we always before starting to hunt took good care to earnestly impress upon the men, and used frequently to send back those who neglected to obey this order, whereby they forfeited part of their day's pay. Before 'following up,' all old men, or very young ones, were carefully excluded, for they were mostly the offenders, lagging behind either to smoke, or from mere carelessness. Had anybody ever been killed at this sport, we should of course have had to pay his relatives a pension.

Independently of all this, it is a most horrible sickening feeling that comes over you on hearing the cry raised that a man has been killed. Natives are very apt to shout out and repeat this cry on seeing any one covered with blood, when perhaps no very serious damage has been done after all. On one occasion, I remember, when hunting alone, this cry was most needlessly raised; the men began to howl and loudly bewail the supposed loss of one of their number; we had almost cleared a thick patch of willow bushes, through which cover we had beaten without seeing any tiger, though fresh prints were very plentiful, and being so near the end, the Bheels were slightly scattered, when at our very feet up jumped a splendid tiger, and with terrific roars dashed back through the very midst of us into the dense cover we had just been beating. I fired two shots, and afterwards knew he was hit by the quantity of blood we discovered on the bushes. There was at the time some little panic; many nimbly climbed small trees near at hand, while the greater number threw themselves flat on the ground—a capital plan of avoiding danger in these cases. After calling out and answering each other's names, suddenly all began to cry out that one was missing or killed; we supposed he had been carried off by the tiger, for on counting heads over and over again, the proper number of men could not be made up. After a horrible suspense, we were joined by the missing man; the lazy rascal had loitered behind to smoke his pipe. His reappearance caused a general feeling of intense joy, mingled with anger at the fright he had caused us all, so that we were in doubts whether to almost kiss or kick him.

Next morning, 5th April, we were agreeably surprised with the news that there were fresh prints in the very same cover that we killed yesterday's tigress in. We took up a different position, and again beat the cover; soon a fine tigress came bounding by us at a long gallop. She was instantly rolled over and killed. On inspecting her we were astonished to find that this beast was the same that had escaped us on the previous day. This we knew to be the case by finding two fresh gun-shot wounds on her body—one in the shoulder, and another in the belly; from both these places she had licked off the hair around each wound as large as the palm of a man's hand. We were glad to find these marks, as it proved that we had shot true enough the day before.

Excerpted from Tiger Shooting in India: Being an Account of Hunting Experiences on Foot in Rajputana During the Hot Seasons from 1850 to 1854 *(London: Smith, Elder & Co. 1857).*

EDWARD BRADDON

Sport in Lower Bengal

When he joined a mercantile firm in the great city of Calcutta, Edward Braddon found his life taken over by 'sordid details relating to Indian produce, sale and barter.' But the white man's world in eastern India had other activities that fascinated him. There was pig-sticking, regarded as superior even to fox hunting in England; snipe, teal and duck abounded in jheels *or tanks. Braddon was only one of the many who tried his luck with gun and spear but he was among the early raconteurs of sport for a wider audience. He published his tales in the* Blackwood's Magazine, *recording his experiences in the Santhal rebellion of 1855 with as much verve as he did tiger-hunts in the Tarai. As the century wore on, the lines of division between whites and Indians became much sharper. But in Bengal, with its large community of planters, traders, army and civil officers, the hunt had all along been a mark of social status. And there was much sport to be had even at the door-step of the great metropolis.*

Many were the songs sung by pig-stickers of Bengal in honour of the creature they hunted,—songs with a refrain that was generally to the effect that 'the boar, the boar, the mighty boar' was blessed with all the virile virtues. Possibly in moments of enthusiasm and wassail this animal may have been overpraised. He has not a pleasant temper, his habits are open to unfavourable criticism, he may fail in his family relations—but he has plenty of pluck. He will fight anything that comes in his way; not even a tiger daunts him, and, what is more, the tiger sometimes succumbs to the terrible tushes of the boar. I have seen a boar bearing away from such heroic battle the marks—deep and frequent marks—of a tiger's claws, and that boar swam the Ganges in flood,—a sufficient feat for an unwounded animal, and one that should set at rest the question whether pigs can swim.

A dangerous brute is the Bengal boar. Throughout the whole of my sporting career only two of my beaters were killed, and one of these was cut to death by a boar; a leopard killed the other: not one was either killed or mauled by tigers.

But my first experiences in this line were, I regret to say, less connected with the mighty boar than with the sow, which, though it cannot rip up a horse's flanks or belly as can the boar, can gallop a little, and, instead of ripping, can bite. This chase of the female swine I saw what time I was out with the Calcutta Tent Club in their beats on either side of the Hooghly, between Calcutta and Diamond Harbour. A memorable club was this in its palmy days, and of some importance when I saw it in its decadence. It is celebrated by a large engraving from a picture by Mr William Taylor, B. C. S. (brother of the better known artist Frederick Taylor), which once was a familiar object on the walls of Indian sportsmen. In that presentment of the Club were shown several lights of the Indian turf and sporting world—Baron Hochpied de Larpent, Jim Patton, the two Brackens, and others; including that distinguished member (the central figure of the group, unless the prostrate boar be so considered) Billy Pitt, the huntsman of the Club.

I first attended a meet of the C. T. C. as a guest of William Bracken, a sportsman known principally in connection with tiger-shooting. In that pre-railway time, when the Mofussil beyond Barrackpore was only to be reached by slow and wearying travel by palanquin, or slower journeying by river in the old-time boat-house of India—the *budgerow*—the great majority of Calcutta men passed their lives without penetrating into the interior farther than a buggy would carry them; but William Bracken made an annual expedition into the tiger country along the Ganges between Bhagulpore and Maldah, and there spent a month in the pursuit of big game. That was *the* month of the twelve for him, and for the other eleven he made out his time by an occasional jaunt with the Tent Club and week-end gatherings at his country house at Budge-Budge, near which quail and snipe, and perhaps an alligator, were to be shot; and where also in their season the mango fish, dear to the epicure, was to be eaten in its prime. There was a billiard-table, too, almost as a matter of course, for every Indian house of any account possessed one; and there was occupation for the lazy or meditative ones in watching from the wide verandah of the upper floor the argosies that passed to and fro along the river—those argosies that then were to be seen carrying their white sheets from stem to stern and from deck to topgallant yards, and gliding majestically over the waters with silent strength—not the latter-day titanic craft of many funnels and dismal smoke and racket, that puff their way along with volcanic strength that is destitute of grace.

It was on the Budge-Budge gram-fields, when we were shooting quail, that I had my first gun accident. As I was loading the right barrel while the left was loaded, the latter contrived to discharge itself; some of the shot knocked off the peak of my wide-brimmed *sola topee*, and that was the sum total of the damage done on that occasion. But not long after that, while I was shooting snipe in a Howrah jheel, and when the snipe were more plentiful than I had ever seen them, the same mishap occurred; and that time it was the end of my thumb, not my hat, that was carried away. I made a desperate effort to continue shooting when the flow of blood was stayed and the mutilated digit bandaged in a pocket-handkerchief, but with only partial success. Breech-loaders, I need hardly say, were unknown in those days; and even when they had come to be common, I perversely stuck to my muzzle-loaders for two or three years. I lost a good many snipe by this ultra-conservatism; but my old chum Jacky Hills profited thereby, in that when we shot Oudh jheels together, the pedestrian advantage I enjoyed through length of limb and lighter body was neutralised by my having to stop to load.

William Bracken had experienced and survived a much more perilous incident than mine above described—an incident which was commemorated in his library by a tiger's skull, in the jaws of which a shoe was held. That shoe, with Bracken's foot in it, had been held clutched by the cruel teeth of that skull while the tiger lived. Bracken had wounded the tiger, a fighting one that charged the elephant upon which Bracken rode; the elephant fell and threw its rider; the tiger seized upon the foot that wore that shoe; the latchet of the shoe was partly severed by the tiger's fangs, and Bracken was able to pull his foot out of its dangerous position, leaving the shoe behind for the tiger's delectation, until a bullet from another howdah made an end of the tiger and the episode. Bracken's foot was sorely wounded—the effects never wholly disappeared; but this accident stayed not those annual expeditions to the Ganges Churs.

In my first Doorgah Poojah I graduated finally in pig-sticking. The Doorgah Poojah is a Hindoo holiday very strictly observed in Lower Bengal, and I observed it with the keenness of a Brahmin. This festival gave me almost a fortnight's freedom from mercantile affairs, and enabled me to go with Jack Johnston after the pigs of Berhampore and Kishnaghur. It was a sufficient privilege to be his companion. It was bliss unalloyed to share his sport.

Any man might well be proud of having served his novitiate under so perfect a master. The king of spears he was called: with him it was in very truth a case of a *cuspide corona*; and where he rode few were the first spears that went to others. What a happy fellow he was, and how much he did to make his companions happy! He was a man of fortune when associates of his own age were owing their way on pittances of Rs 400 or Rs 500 a-month. He had a stud of fourteen or fifteen horses, all but one of which were the best Arabs that money could buy; and every horse in his stable, except one or two racers, did he drive in his buggy or lend to his friends as if it had been a Rs 300 hack instead of a creature worth Rs 2000 or Rs 3000. And while still a young man he was free to come home to England, where he has so far succeeded with his English stud that he has won his Derby.

What a revelation of sport was that fortnight spent with the jovial and hospitable indigo-planters of the Kishnaghur borderland! There was no question of hunting sows thereaway; boars were there sufficient for the purpose, if not in quantity to satiate; and every day brought to the six or eight horsemen engaged two or three chances of blooding their spears, if not always the first chance.

And that expedition led the way to another and more ambitious one of some months later when in very much the same country Henry Torrens, the Resident at Moorshedabad, held his great gathering of pig-stickers. I think that meet must have been then, and must still remain, unique in its way. Nearly a hundred elephants marched in line through the long kassia grass, where the pigs had their lairs; six or seven horsemen rode on either flank in pairs or threes, ready, when the pigs should break from their cover, to separate the boars from the sounders and ride them down and on some half-a-dozen howdah elephants were sportsmen of a less enterprising kind, or enterprising sportsmen without available mounts, who helped to drive the pigs by a fusilade directed against hog-deer, black partridge, florican, leek, and hare. Twelve days did this incomparable chase endure, and ninety-and-nine were the boars whose skulls and tushes recorded the hunters' prowess. I seem to remember that, on the last day of those happy dozen, we were all eaten up by anxiety to bag the round hundred, and how, when that day was spent and night bore down upon our happy hunting-ground, we were unanimous in preferring our tally of 99, because, as we argued, people to whom we narrated the history of the great Torrens' hunt might be incredulous if we said the boars killed were a hundred, neither more nor less. Possibly we were moved to argue thus by that old-time story of the Indian colonel who, being asked why he had not given a full thousand as his day's bag of snipe instead of 999, gravely observed,—'Sir, do you think I would perjure my immortal soul for a single snipe?' However this may be, we rejoiced heartily on that last night of Torrens's hunt: we drank toasts and made speeches, of which none were worth remembering save that of our witty host, and most were worth forgetting straightaway; and we sang songs, principally in honour of the noble boar, with rattling choruses in any tune and in any time, oblivious of the fact that a dirge to that animal would have been more appropriate and quite as tuneful; and, finally, we carried our host in his palanquin to the *ghat*, some two furlongs off, where his boat was moored—an agonising exploit for our unpractised shoulders—and shouted ourselves hoarse until the founder of the feast was carried down the stream out of earshot.

Apropos of the above snipe yarn, I wonder if the Indian colonel continues to the present day to play the Munchausen. Somehow, it always happened in my time that the colonel monopolised this *rôle* in public estimation, and one might have justifiably believed that the army was traduced, and the civilians let off too easily in this connection. It was of a colonel (the brilliant if erratic Teddy Oakes) that people told the tale of how, being at sea in a violent storm when hope was abandoned, and the passengers were bidden to pray, his nearest approach to orison was, 'Oh, Pilot, 'tis a fearful night!' It was a colonel who, according to fable, declared that on a voyage round the Cape his ship was spoken (thousands of miles from land) by a man in a tub who would not come on board the ship, but took in a supply of biscuit and water and was left in mid-ocean. It was a colonel from whom the wily snake escaped by entering a bamboo tail first, after that colonel had twice pulled it from that refuge by the tail. And according to popular belief, a colonel told that story about the quail which nearly resulted in his prompt discomfiture. For the colonel had described a flight of quail that clouded the sky, and then, having settled, covered the parade-ground in close-packed swarm; and he told how he got out an 18-pounder cannon and loaded it almost to the muzzle with powder and No. 10 shot, and trained the gun to volley its contents into the thick of the birds, and then he asked of those who sat at mess, 'How many do you think I shot?' and a subaltern of more wit than veneration answered him 'a million.' Whereupon that colonel changed his tactics to meet the situation, and said, 'No, by G—, Sir, not one!' These things were old, old chestnuts a generation since: perhaps, like many another of their kind, they have had a neogenesis.

In those Calcutta days there was often a day or half-day when I could get away into a snipe jheel, and on many occasions I was able to make bags of twenty and twenty-five couple. The Howrah jheel was the nearest, as it was the best, of those within easy reach. It was not as well known as others, and it was as much in the country as if it had been fifty miles away. The E. I. Railway, which for many years had its terminal station at Howrah, was not then, or had only recently been, constructed. The Hooghly had not then been bridged, and one made one's way from Calcutta to my jheel by dinghy across the river, and then about a mile on foot. Many a pleasant picnic of one have I had on the banks of that swamp, where the shade of luxuriant tropical vegetation made the mid-day halt restful to eye and limbs, and where a refreshing draught of cocoa-nut milk was to be had in season straight from the trees that spread their broad leaves above. The jheel was just enough for one gun; it could be compassed in an hour and a half or so; and birds that were roused in one part of it would generally, if the gunner permitted, settle in another. In later years I fancy the snarers have spoiled this and other jheels round the Bengal metropolis by netting snipe for the Calcutta market; but the Howrah jheel was a really good one in my day, and dear to me for other reasons than because it was the tomb of—part of me. Why sport should be spoiled by this netting of snipe I am at a loss to conceive: to the Calcutta Khansamah any bird of attenuated bill and legs—snippets, water-rail, &c.—passes for a snipe, just as any bird not bigger than the house-sparrow answers to his conception of ortolan. I have seen the impostor snipe—a very mocking bird—on dinner-tables outside Calcutta, and the fraudulent ortolan enters into the *menu* of most of India's provinces. For the genuine ortolan—that delicious mouthful—is, as far as my experience goes, very strictly localised. I have seen, shot, and eaten them in only one district (Kishnaghur); but I have had ground larks, sand-martins, and many other small fowl offered to me in the name of the ortolan in twenty districts and in three provinces. The sport provided by this winged delicacy is, I need hardly say, poor; it is in fact demoralising, for there can be no question of aiming at this bird or that: the shooter has to fire his charge of dust shot into the brown of the swarm that whirls over the dusty plain like unto a cloud of dust. But if one cannot get ortolan save by shooting them, then I should feel inclined to shoot.

After a year or so of town life, which was brightened by occasional spells of sport, and dimmed more frequently by wearying consideration of freights and customs dues, and grey shirtings and madapolams, and other items of commerce, I broke the bonds that bound me to a desk in Clive Street, and made for the Mofussil. I could not shake the dust of Calcutta off my feet in a literal sense, because I quitted that capital during its mud season, but I did so figuratively, and from that hour ceased to be a townsman.

I made for Kishnaghur, the happy centre of a series of snipe jheels, that came almost up to the compounds of some of the bungalows: I travelled in a *bauleah*—a smaller edition of the *budgerow*—but I did not make direct for my destination. I had an important engagement with myself (now my only master) to keep by the way. I had to visit a snipe jheel (then famous above all among the happy initiated) where a record of fifty couple in a day had been made, and I had to break that record if I could. At that season—the month of October—this jheel was accessible by water, with some amount of running aground in the navigation of the nullah that ran past it and into the Hooghly, two or three miles distant. It was touch and go with us as the *bauleah* was hauled up this shallow stream; indeed there were innumerable touches, but, happily, there were as many goes, and ultimately my ark was brought to anchor at the very verge of the shooting-ground just as night fell, and my floating home was lighted up for a short evening; and then dinner came, and mosquitoes and countless winged creatures that dashed themselves against the candle-shades in battalions, and upon the burning wicks in platoons, and generally make night terrible for any one but the most ardent entomologist in a new field of research. I was not a scientist that way, if in any way, nor were these flying torments novelties to me, so I sought early slumber in the darkness.

Next morning I was up betimes, to make as long a day as possible for my record-breaking expedition. The Kanchrapara jheel was about a mile in length, and of a breadth that admitted of three or four guns shooting in line. I had it all to myself, and had to get over fifty couple of snipe out of it. It was an epoch-making occasion, and, refreshed by a long night's sleep, I felt equal to the task and in a mood to shoot my best. My first half-hour among the birds encouraged confidence and hope. Snipe were plentiful; at one time I had five couple down upon the ground, all killed before my coolies commenced picking up: when half my allotted time was done I had bagged just thirty couple, and I had then some untried portion of the jheel ahead, and all the birds that had gone back upon my course, to reckon with. At the close of the day, when I had shot to the end of the jheel and back to my boat, I had on the snipe-sticks 51½ couple: I had broken one record and set up for myself another that, in several years of steady shooting, I have never again accomplished. Other Indian shots (including colonels) may have got bigger bags—notably in the best days of that splendid snipe country that lies along the Oudh and Rohilcund Railway line, some twenty miles from Allyghur; but that is my record, and I am satisfied with it.

I devoted the following day to the jheel and the ambitious attempt to outdo my first effort. Ambition on that occasion, as is not uncommon with it, got me into trouble. I attempted to beat up some dangerous ground in the middle of the swamp—treacherous yellow bog that shook under foot, and, its crest being broken, absorbed one slowly but with disagreeable certainty. I went up to the middle in this; saved myself from sinking deeper by spreading my arms out on the unbroken surface; and was eventually dragged out, covered with mud and ignominy, by my coolie attendants. That was a grand day's shooting also, and I brought to bag forty-nine couple, including two or three birds wounded on the previous day. I thought then, and I have always since thought, that there is no shooting to compare with that of a good snipe jheel.

Many a day thereafter did I have in that Kanchrapara jheel with Montresor, then of Kishnaghur, Elphinstone Jackson, Dacoity Commissioner, stationed at Bandel near Hooghly, and C. S. Belli, Hooghly's Collector, but never with such results as to the head of game shot. Thoroughly enjoyable were those outings, however, even though our snipe gave an aggregate tale of thirty or forty couple only, and involved a variety of travel not always luxurious. Setting forth by boat up the river, we used to land at a point distant some two and a half or three miles from the jheel, and this distance had to be covered either by riding the mare known, for some recondite reason, as Shanks's, or such country *tats* as could thereaway be collected. Little choice was there among these ponies: one might prove on closer acquaintance a more vicious kicker or more persistent stumble than another but all were equally unkempt, ragged, and deficient in every quality that makes a park-horse: all were alike ill-caparisoned, with saddles of uncomfortably restricted seats; stirrup-leathers that whether they were leather or rope, defied all effort to lengthen or shorten them; narrow reins that cut into one's fingers as though they had been bands of metal, and headstalls wherein strings predominated. What matter? *Vogue la galèr.* Those sorry steeds saved our legs for the half-day tramp in the jheel—saved those limbs by some five or six miles of tramping, while they inflicted upon back and arms and other portions of our anatomy tortures and wounds that were unforgotten for days. What of that? We had our shooting from noon till nigh unto sunset, with an interval for lunch, and then we put ourselves upon those equine racks again and rode homeward rejoicing. And I think the one of us who got the greatest amount of enjoyment out of the whole proceeding was Belli, who, as likely as not, died not kill a bird, although he blithely fired away at dozens.

Do any of my readers know whence the snipe come and what instinct directs their migrations? They come into Great Britain and Ireland in the early winter, say in November: that is the season also, in which they arrive in Northern India—i.e., in November and the early winter; but they arrive in Australia and Tasmania in the same month, although that month commences the Australasian summer. This seeming incongruity of migratory purpose one might explain by the assumption that the time of their arrival in the places named does not depend upon the climate of those localities, but upon the necessity that drives them forth from their summer home (i.e., their habitat between May and November): that they are obliged to leave that Oxygean home about November, and betake themselves to any spot whatever where their feeding-grounds are to be found. But the question is further complicated by the erratic conduct of Indian snipe. These birds arrive in Northern India, as I have said, in November, when the weather is fairly cold, or at least cool; but they come into Bengal and Southern India in September, when it is blazing hot. On two 1st of Septembers I have shot them (some two or three couple) in the Kishnaghur jheels. Do these snipe of the south start from that unknown habitat of theirs two months before the rest of their kind? and if so, why? I am told that the birds in Southern India are of a different variety from those of the north, although I could never distinguish one from the other; but that affords no explanation of the snipe's vagaries, but rather the reverse—for if all Indian snipe were of one variety, the question in hand might be disposed of (as far as India is concerned) by the bold assertion that these migrants first settled in the south and moved up northwards later on.

Their migrations are very bewildering. Thirty years ago there was snipe-shooting in Tasmania almost equal to that of India. Officers attached to the Tasmanian garrison of that time have told me of their bags of twenty-five and thirty couple. To-day snipe are found in but few places, and only in small numbers. One of my Tasmanian friends was good enough to reserve a small snipe-shooting for me one year: there were only three snipe in this preserve, and the curious thing is that every year three, or perhaps four, snipe come to that same patch of marsh. The decadence of Tasmanian snipe-shooting cannot be attributed to any physical or climatic change in the country—the birds have not been driven away from that land by drainage, as has happened with our Lincolnshire fens. There is nothing but the snipes' caprice to explain it.

This capriciousness on one occasion sold me terribly. I was shooting over a chain of jheels in Oudh, and about sunset came to the last of the series, a small one close to my camp. Although small, it often held a fair number of birds, and might generally be reckoned upon as good for at least six couple; but that evening it was alive with snipe. Even in the failing light of a land that knows no twilight I might have bagged eight or ten couple if I had shot there then, but I resisted the temptation, and fired not at all: to-morrow I promised myself a splendid bag out of that swarm. When I went there next day, brimful of hope, there was not a snipe to be seen, nor were there any number to speak of anywhere that I went after them. And this did not happen in the season when snipe gather together for their departure into space. Even the poor apology that migratory reasons compelled them to disappoint me was not forthcoming. Curiously enough, it happened shortly after I penned the above paragraph that I took up vol. xvii. of 'Longman's Magazine,' in the March number of which is a paper by C. T. Buckland, B. C. S., at one time of Hooghly. He speaks of the Kanchrapara jheel as a magnificent shooting-ground, known then to few besides the men of Hooghly, and tells how a friend of his could bag his fifty couple there. He also mentions a spot in Chittagong where he could always get a couple of snipe on the 1st September.

For some three years I made the most of such shikar as was to be had in Kishnaghur and the neighbouring districts. I had charge of zemindarees and indigo factories, and my work was mostly in the saddle, where also was a good deal of my recreation. I had a tolerable stud of five or six horses, a cast-iron constitution, and a passionate love for field-sports, compared with which my attachment to business was decidedly platonic. Wherever pig-sticking was to be had, in Kishnaghur, Berhampur, or Burdwan, I rode with the hunt. Wherever there was a snipe jheel, I paid it frequent attention. Quail and jungle-fowl were not neglected, but I got no forwarder with big game other than boars.

Indeed the only chance I had of making my *début* as a slayer of the larger feline creatures resulted in a crushing failure. A panther was marked down for me in a small thicket, and I went forth to do for it. When I reached the ground the panther was still there, and a keen-eyed native pointed it out to me. 'Hitherward was its head,' said this man, 'thitherward its tail. Doesn't the sahib see it? There, there!' and he pointed to a spot about three yards off. But I did not see the panther—either its head or tail or anything that was its; I saw only a mass of light and shade under a dense overgrowth of greenery, dead leaves, and grass, that were yellowish where the pencils of light broke in upon the gloom and, otherwise, mysterious shadow that told nothing to my unaccustomed eye. All that I looked upon in that greenwood tangle was equally panther;

I could pick out no particular patch as being any more pantherish than the rest; of head or tail I made out nothing where all was equally one or the other,—and still that native of keenest vision besought me to see the panther's head and tail and right forefoot, and many other details of its anatomy. Then there came a roar out of the thicket, and a rush which was like the volcanic upheaval of the ground at my feet, and, as it seemed, several tons of upheaved matter hit me on the chest and other parts, and I was catapulted on to the broad of my back a yard or two from where I had stood. That upheaval was the panther. The brute hadn't had the patience to wait until I saw him, or the modesty to take himself off peaceably in some other direction: he had resented my staring his way, even though I saw him not, and so had emerged out of his lair like an animal rocket, and knocked me down in his flight. As he failed to claw me, I came off scathless; but not so my attendant, who foolishly embraced the panther in view to arresting his flight: he got himself rather badly mauled, and did not come a whole man out of hospital for some weeks. That was my disastrous commencement with panthers.

It was about this time that an unfortunate beater of mine lost his life by a foolhardy act like that above described. We were beating pigs out of the long grass on the left bank of the Bhagiruti, and a boar getting up at this man's feet, or from under his feet, he jumped upon it. Why he did so it is impossible to say: it furthered no object of anybody's, for we were awaiting the pig at the edge of the higher jungle, and quite ready for it, and, in fact, we did get it. But as soon as we had speared this boar, we were made acquainted with the sad accident that had befallen the beater. The boar had ripped him across both thighs and both arms with those clean deep cuts that the boar inflicts when its tushes have not been blunted by age; and although the poor man lived to reach a hospital, he died there in spite of every attention, and the necessary amputation of one mutilated limb.

Excerpted from Thirty Years of Shikar *(London: William Blackwood and Sons, 1895).*

JAMES FORSYTH

Game Animals and Birds of the Plains

For a man who died at the age of 32, Forsyth's was a rich and varied life. He opted to serve in India to get a chance to do more than 'shoot seals in the Shetland isles' and to get a shot at the denizens of the Indian jungle. He was active in the suppression of the Rebellion of 1857–8, and soon after served as a forester in central India. But Forsyth stands out not so much for his military prowess or administrative abilities as for his first-rate skills as a naturalist. His observations of people and trees, the fauna and bird life of the central Indian plains are still a must for wildlife enthusiast and specialist alike. Here, he tells of the now-extinct Indian cheetah, and of tribal techniques of hunting the blackbuck. Forsyth's is one of the few windows into a world that is all but lost to posterity.

Few things are more enjoyable than marching along during the cold season in a rich open country like the Narbada valley with a well appointed camp, and plenty of leisure to linger over the numerous objects of interest or amusement presented by such a tract. Very little of this sort of thing fell in the way of the forest officers of those days, however. Our work lay in the depths of distant forests, or at most in the half-reclaimed frontier belt lying between the hills and the plains, where timber transactions generally took place, and the chief depôts for forest produce had been established. When by chance our direct route from forest to forest led across such an open region, our movements were as rapid as man and beast could make them; and at the earliest possible moment we hurried again from the face of civilization, like ghosts at cock-crow, to bury ourselves again in the depths of the wilderness. In after years, when employed in revenue work in a populous district, I saw the reverse of the picture. Marching by fair roads and easy stages, with a duplicate set of canvas houses (for such our large Indian tents really are), one of which goes on over-night and is pitched ready for your arrival in the morning, in the deep shade of some mango grove, near a populous village which supplies all your wants; starting after the morning cup of hot coffee to ride slowly along through green fields and grassy plains; and looking on the forest-covered hills on the blue horizon only as an agreeable vanishing point in the landscape, or as unpleasantly complicating the questions of liquor excise and police administration! It is amazing what a difference the point of view makes. The man who has dwelt for years among the forests, and their simple wild inhabitants, will regard nearly every question that arises in a wholly different light from him whose experience has lain only among the corn fields of the plains, and their tame and settled tillers. And each of them will probably arrive at a conclusion as little comprehending the whole bearings of the question as the other.

Snipe and wildfowl begin to arrive in these central regions of India, voyaging from the frozen wilds of Central Asia, early in October; and, before the end of November, every piece of water and swampy hollow affords its contingent to the gun. The common teal, and the whistling teal, are the most numerous as well as the first to make their appearance. The lovely blue-winged teal is scarcely less common; and of larger ducks, the red-headed pochard, the wigeon, the pintail, and the gadwall, are found throughout the winter on nearly every tank of tolerable size. On the main rivers, and on the larger reservoirs such as those of Bhandara and Lachora in Nimar, which, though owing their existence to the hand of man (the giants of past days, who knew the requirements of India better than their successors), yet approach the dignity of lakes, many other species of wild fowl will be found, including that king of ducks the mallard, the common grey goose, and the black-backed goose. The latter species is extremely common; the others, which are much superior for the table, are comparatively infrequent. Numerous wading birds, storks, herons, and cranes, haunt every pool and marsh. Few of these offer much temptation to the sportsman, except the Demoiselle crane, generally known as the Coolen, which is much sought after, and is therefore difficult to approach. Few extensive wheat or *gram* fields in the Narbada valley will be found at this season without a flock of these delicious birds stalking across it in the morning and evening grazing on the young shoots.

If encamped in the neighbourhood of a river or swamp, the traveller will probably be aroused at daybreak by the quavering and sonorous call of the giant Sarus crane, a bird revered by the Hindus as a type of conjugal affection. They are nearly always seen in pairs, and, should one of them be shot by the ruthless gunner, the companion bird will return again and again to the spot, to hover and lament over its slain friend in a manner that generally prevails on the hardest hearted to grant immunity to the race for ever after. A contrast to this happy union of lovers is found by the Hindu in the Braminy ducks, which also associate in pairs, but, by a cruel fate, are compelled to pass their nights on the opposite banks of a stream, wailing forth their unavailing love in the melancholy 'Chukwa, chukwi,' which few travellers by the rivers of India have failed to hear in the dusk of the evening. Their unfitness for the table, probably more than the Hindu adage against their slaughter, protects them from the gun.

Of other winged game, the grey quail—best of Indian game birds, in my opinion—will be found in good numbers in most grain fields. I have never seen them here in such swarms as in some parts of upper India, where eighty or a hundred brace may be bagged in a day; but the sport is none the worse for that. Twenty brace is a first-rate bag in Central India; and generally the sportsman has to be contented with much less. The common grey partridge, which closely resembles in appearance the English bird, abounds in many places. It hugs the vicinity of villages, and feeds foully. I have seen a covey of them run out of the carcass of a dead camel, and speed across the plain like so many hares. These nasty habits, and its skulking nature, much belie its appearance as a bird of game. Far different is the gallant painted partridge, which here takes the place of the black partridge of upper India. I have seen the latter in Bandelkand; but I am positive that it nowhere occurs in the Central Provinces. The appearance of the two species is so alike, and their habits are so identical, that assertions to the contrary have no doubt arisen from mistake. No game bird could afford more perfect shooting than the painted partridge. Of handsome plumage, and excellent on the table, his habits in the field admirably adapt him for the purposes of the gun. He frequents the outskirts of cultivation, in spots where bushes and grass-cover fringe the edge of a stream, for he seems to be very impatient of thirst. The proximity of some sort of jungle seems to be as necessary as the neighbourhood of crops. Morning and evening small coveys or pairs of them will be found out feeding in the stubble of the cut autumn crops, that latest reaped being the most likely find. On being disturbed they seldom run farther than to the edge of the nearest cover, from which, on being flushed, they rise like rockets, with a great *whirr*, straight up for twenty or thirty yards, and then sail away over the top of the cover to a distance of a few hundred yards; this time plumping into the middle of the cover, from which it is not so easy to raise them again. This beautiful bird is most common in the extreme west of the Central Provinces, and in good spots a bag of ten to fifteen brace to each gun may be made in Nimar and the Tapti valley.

The most common way of shooting quail and partridges is by beating them out with a line of men; but it is a poor sport compared to shooting them over dogs. I have used both pointers and spaniels in this sport. The former secure the best of shooting in the early morning and late in the evening, while the birds are out of cover and the scent good, and four hours' shooting may thus be had in the day. But a team of lusty spaniels is, I think, on the whole preferable, as they are useful also for many sorts of cover shooting where pointers could not be worked. They also keep their health better, and degenerate less in breeding than any other imported dog, which is probably due to their desent from a race originated in a warm climate. They make the best of all companions, and are not so liable to 'come to grief' in many ways as larger dogs. Fresh imported blood is however required, at least once in every two generations, to keep all English sporting dogs up to their best in India. The spaniels should either be large Clumbers, or of the heavy Sussex breed, as a small dog like a cocker cannot penetrate the jungle cover. The noble Clumber, otherwise faultless, has the fault for this particular purpose of giving no tongue on game: I commenced the breed, which I maintained for twelve years in India, with a strain of pure Clumbers in the never-to-be-forgotten 'Quail'—a dog that for looks and quality surpassed anything of the breed I can now discover in England. All his descendants were more or less crossed with Sussex or cocker blood; but none of them ever gave tongue till the fourth generation, when symptoms of it began to appear. On the whole, then, I think I would prefer the heavy Sussex breed.

On one occasion the whole of my spaniels were very nearly being 'wiped out' by one of a class of accidents that must be looked for in India. I was shooting quail in a grainfield near Jubbulpur, with 'Quail,' 'Snipe,' 'Nell,' and 'Jess,' when on a sudden they all began to jump violently about, snapping at what seemed to me to be a large rat. But coming nearer I made out that it was a huge cobra, erect on his coil, and striking right and left at the dogs. I lost no time in pelting them off with clods of earth, and then cut the brute's head off with a charge of shot; when I found that the snake had been in the act of swallowing a rat, of which the hind legs and tail were protruding from his jaws, so that his repeated lunges at the dogs had fortunately been harmless. All these spaniels were famous ratters, and had no doubt been attracted by the cobra's mouthful, for they generally had, like all dogs of any experience in India, a wholesome dread of the snake tribe. I never lost any of these dogs by an accident, though exposed to all the dangers of panthers, hyenas, wolves, snakes, and crocodiles; and all of them lived to a good age, in excellent health. As with men, English dogs keep healthy enough if properly treated in accordance with the climate.

Of larger game, the principal animal met with in the settled parts is the black antelope, which has probably followed the clearings made by the immigrant races. The aversion of this animal to thick uncleared jungle has made it, in the Hindu sacred literature, a type of the Aryan pale, of the land fitted for the occupation of the fair-skinned races; and the appropriate seat of the devotee is still upon its black and white skin. It is too well known to require any minute description. Suffice it to say, that not even in Africa—the land of antelopes—is there any species which surpasses the 'black buck' in loveliness or grace. In Central India, although this antelope attains the full size of body, the horns of the buck (the female is hornless and of a fawn colour) rarely exceed a length of 22 inches. I have shot one with horns 24½ inches, and seen a pair that measured 26 inches. The longest horns are probably attained in Gujerat, and about Bhurtpur in Northern India. In all the corn districts of Central India it is found in considerable herds, and does much damage to the young crops. I have seen herds in the Sagar country, immediately after the mutiny of 1857, when they were little molested, which must have numbered a thousand or more individuals. A tolerable shot could at that time kill almost any number he chose. In most cultivated districts, tracts of the poorer land are kept under grass for cattle-grazing, &c., and these preserves are generally the favourite mid-day resorts and the breeding-grounds of the antelopes. Thence in the evening they troop out in squadrons on to the cultivated lands in the vicinity; and all the night long continue grazing on the tender wheat shoots, returning in the grey of the morning to their safe retreat. Many will, however, remain in the fields the whole day, sleeping and grazing at intervals, unless driven off by the cultivators. In such places the voices of the watchers in the fields will be heard in the still night shouting continuously at the antelopes; but they seldom succeed in effecting more than to move them about from field to field, doing more damage probably than if they were left alone, for a buck killed in the morning will always be found filled nearly to bursting with the green food. Although many of them are shot by the village shikaris at night, and more snared and netted by the professional hunters called Pardis (who use a trained bullock in stalking round the herds to screen their movements), the resources of the natives are altogether insufficient, in a country favourable to them, to keep down the numbers of these prolific and wary creatures; and it is a perfect godsend to them when the European sportsman hits on their neighbourhood as a hunting-ground.

There are many ways of circumventing them. Living quite in the open, they rely principally on the sense of sight for protection, although at times warned also by their power of smell. One way is to drive up to them in one of the bullock carts commonly used in agriculture. The native shikari often gets near them by creeping up behind a screen of leaves which he works before him. Where they have not been much harassed the European sportsman, in sad-coloured garments, can usually stalk in on them when passing between the grass plains and the crops. In the very early morning, if a station be taken up in their usual route, they are nearly sure to come within shot, the grunting of the bucks warning the sportsman of their approach some time before they emerge from the darkness. One of the most successful and interesting plans is to ride a steady shooting horse nearly up the herd. When within say four hundred yards, slip off and walk on the off side of the horse in such a direction as will lead past the herd within shot, if possible on the down-wind side. If they have been so shot at in this way as to be shy of the horse, take a groom and pass them further off; and when a convenient bush or hillock intervenes drop behind, and let the man lead the horse on, passing well clear of the herd. They will probably be so intent on watching them out of the way, that you will generally be able to creep in on them without much difficulty. Shots at antelope in populous districts are seldom got much under 150 yards now-a-days, which is however near enough for modern rifles to make sure work. One great advantage of employing a horse in stalking is that it will often enable you to follow and spear a wounded buck which might otherwise escape. If you have a brace of good greyhounds in the distance ready to slip, the chances will be still better. A wounded buck often gives a beautiful run with greyhounds, which have never been known to catch an untouched and perfect antelope on fair hard ground, though under conditions unduly favourable to the dogs they have sometimes done so. A shooting horse, like several which I have possessed, who is quite steady under fire, does not need to be tied, and will come to call, is a perfect treasure for many sorts of sport in India. As in all good qualities, the Arab is the most likely to develop such a character; but most horses are capable of being taught something of the business. Should neither horse nor hounds be at hand, a wounded buck should not be followed up too quickly. If left to himself he will probably lie down in the first cover he comes to; and by watching the line he takes you may often follow up and secure him.

In upper India they are frequently shot by approaching them on a riding camel. The more bells and gay trappings he has on him the better, as the antelope on this plan fall victims to their curiosity and amazement. I brought down to Central India with me a trained camel, with which I had thus bewildered many an antelope into rifle distance; but after getting some dangerous tumbles owing to the yawning cracks that form in the black soil in these provinces after the rains, I had to abandon the camel as a shooting vehicle. As a sport antelope-shooting palls upon the taste. There is too much of it, and it lacks variety. So I should think also would be the case with much of the African sport we read of. To the beginner in Indian sport, however, there is no pursuit more fascinating. The game being nearly always within sight, the excitement is maintained throughout the day's sport. Simple as it seems, it takes a good man and a good rifle to make much of a bag when the antelope have been much disturbed. The old hand is apt to smile at the enthusiasm of the 'griff' when he dilates on the glories of antelope-stalking; but the time was when he too passed through the stage at which the acquisition of a particular long spiral pair of horns was more to him than the wealth of all the Indies, and when nothing impressed him so profoundly with the vanity of all human affairs as the miss of 'a few inches' under or over, which so frequently terminated the weary stalk. Perhaps I may be allowed to quote a description of the pursuit of a master buck, written many years ago, when I myself was in the throes of the 'buck fever.'

'I had frequently seen in my rambles over the antelope plains a more than ordinarily magnificent coal-black buck. I had watched him for hours through my 'Dollond,' but my most laborious attempts to reach him by stalking had as yet proved futile. His horns were perfection, of great size, well set on, twisted and knotted like the gnarled branch of an old oak tree. As the sun glanced on his sable coat, it shone like that of a racehorse fit to run for the Two Thousand Guinea stakes—in fact, he was the *beau ideal* of a perfect black buck. Of course, the more difficult the task appeared, the more determined was I that these superb horns should be mine, and that in future I would disregard every buck except the one. He was constantly attended by two does, to whom he confidently entrusted the duty of watching over his personal safety—and faithful sentinels they were. They seemed to relieve each other with the precision of sentries, and clever indeed would be the stalker who could approach within many hundred paces ere the warning hiss of the watchful doe aroused the grand signior from his siesta. It was then grand to see the majestic air of the buck, as, after stretching his graceful limbs, he slowly paced towards the object of his suspicion, still too far distant to cause him any alarm. Now he stops, and, tossing his nostrils in the air, snuffs the breeze that might convey to his delicate sense the human taint. Now he lazily crops a blade or two of grass, or scientifically whisks a fly from his glossy haunch with the tip of his horn; anon he saunters up to one of his partners, and seems to take counsel regarding the state of affairs. Again, as some movement of the distant figure catches his eye, his sudden wheel and prolonged gaze show that, despite his careless mien, not for a moment has he lost sight of his well-known foe. But soon the does begin to take real alarm; and after fidgeting round their lord, as if to apprise him of the full extent of the danger, trot off together towards some other haunt. Now they halt a moment, and look round appealingly to the buck, and again with feigned consternation start off at a gallop, every now and then taking imaginary ten-barred gates in their stride. At last the buck, after remaining behind a decent time to maintain his character for superior courage, follows them at a pace that mocks the efforts of every animal on the face of the earth but one—the hunting leopard.

'Such was the invariable result of my best efforts for upwards of a week. I would not risk a long shot, as it might drive him for ever from that part of the country. His favourite haunt was a wide grassy plain, intersected here and there by dry watercourses, up which I had many a weary crawl, *ventre à terre*. I soon found out his usual feeding and drinking places; and observed that to reach the latter he almost daily crossed a deepish dry nullah about the same place. This struck me as affording the means of circumventing him, so I took up my position in the nullah; but as luck would have it *my* buck took his water in some other direction for the next two days. Many other herds of antelope constantly passed within easy shot of where I was ensconced; but not until I was almost giving up hope on the third day, and was taking a last sweep of the plain with my binocular, did the well-known form of the master buck greet my vision, as he slowly wound his way with his two inseparable companions towards the pool to which he had watched so many of his species passing and re-passing in safety.

'The wind was favourable, and the buck came steadily on till he arrived within a long rifle shot of where I was posted. Here he suddenly threw up his head, and, after standing at gaze for a few moments, turned sharp to the left and started off at a canter for a pass in the nullah, about a quarter of a mile from where I was; I knew he could neither have seen nor smelt me, and was at a loss to account for his sudden panic till, on turning round in disgust, there was the cause behind me, in the shape of a small parcel of does, which had evidently been returning from the water, but, having discovered my unprotected rear, were now pulled up in a body, and staring at me with an air which had telegraphed the state of affairs to the old buck in an unmistakable manner. I felt very much inclined to sacrifice one of the inquisitive does to my just wrath, but preferred the chance of a running shot at the buck; so off I started at a crouching run (somewhat trying to the small of the back) up the bed of the nullah, in the hopes that the buck might have pulled up ere he crossed, and would still afford me a shot. Nor was I mistaken, for, on turning a bend of the tortuous nullah, there he stood broadside on, in all his magnificence, not eighty yards from my rifle; but, alas! who could shoot after a run, almost on all fours, of some 500 yards or so? When I attempted to bring the fine sight to bear on his shoulder, my hand trembled like an aspen leaf, and the sight described figures of eight all over his body. There was no help for it, however: he was moving away, and I might never have such another chance. So, almost in despair, I fired. I was not surprised to see the ball raise the dust a hundred yards or so on his further side, and with a tremendous bound of, I fear to say how many yards, straight in the air, away went the buck like an arrow from the bow. In for a penny, in for a pound! once fired at, I might as well have the other shot; so stepping from my cramped position, I held my breath as I tried to cover his fleeting figure with the second barrel. He had gained at least 150 yards ere I touched the trigger, but the ball sped true, and over rolled the buck in a cloud of dust. Short was my triumph, however, for ere I had well taken the rifle from my shoulder he had regained his feet, and was off with hardly disminished speed. It is very rare that an antelope thus suddenly rolled over does not succeed in regaining his legs. Their vital power is immense, and nothing but a brain shot or broken spine will tumble them over for good on the spot. When shot in the heart they generally run some fifty yards and then fall dead, and I much prefer to see an antelope go off thus, with the peculiar gait well known to experienced shots as the forerunner of a speedy dissolution, than to see even the prettiest somersault follow the striking of the ball.

'In the present instance I watched the antelope almost to the verge of the horizon. Now and then he slackened his pace for a few seconds, and looked round at his wounded flanks, and then, as if remembering that he had not yet put sufficient distance between him and the fatal spot, he would again start forward with renewed energy. The two does, as is generally the case when the buck is wounded, had gone off in a different direction; and were now standing on the plain, a few hundred paces from where I stood, gazing wistfully from me to their wounded lord. Such are the scenes that touch the heart of even the hardest deer-stalker, and for a moment I almost wished my right hand had been cut off ere I pulled trigger on this the loveliest of God's creatures.

'When he dwindled before the naked eye till he seemed as a black speck on the far horizon, I still continued to watch him through my glass, in the hope that he might lie down when he thought himself concealed, in which case I might steal in and end his troubles by another shot. Suddenly I saw him swerve from his course, and start off in another direction at full speed. Almost at the same instant a puff of smoke issued from a small bush on the plain—the buck staggered and fell, and many seconds afterwards, the faint report of a gunshot reached my ears.'

The person who came to my aid in so timely a fashion was a native sportsman, whom I then saw for the first time. He was more like the professional hunter of the American backwoods than any other native of India I have ever met. His short trousers and hunting-shirt of Mhowa green displayed sinewy limbs and throat of a clear red brown, little darker than the colour of a sun-burnt European. An upright carriage and light springy step marked him out as a roamer of the forests from youth upwards; and the English double-barrelled gun, and workman-like appointments of yellow sambar leather, looked like the genuine sportsman I soon found him to be. Many a glorious day did I afterwards pass with him in the pursuit of nobler game than black bucks.

The chikara, or Indian gazelle, is another antelope very common in Central India. It is called often the 'ravine deer' by sportsmen; and, as regards the first part of the name, is so far well denoted. Its favourite haunts are the banks of the shallow ravines that often intersect the plain country in the neighbourhood of rivers, and seam the slopes of the higher eminences rising out of the great central tableland. These are generally thinly clothed with low thorny bushes, on the young shoots and pods of which it browses like the domestic goat. Of course it is wrong to call it a 'deer,' which term properly belongs only to the solid-horned *Cervida*. Considerably smaller than the black antelope, the gazelle also differs much from it in habits. It prefers low jungle to the open plain; and trusts more to its watchfulness and activity than to speed, which however it also possesses in a high degree. It is very rare to catch a gazelle, or still more a herd of them, off their guard; and it is surprising how, on the least alarm, the little creatures manage to disappear as if by magic. They have probably just hopped into the bottom of a ravine, sped along it like lightning for about a hundred yards, and are regarding you, intent and motionless, from behind the straggling bushes on the next rising ground. Should you follow them up they will probably repeat the same manoeuver, but this time putting three or four ravines between you and them instead of one. They also resort to the cultivation to feed, though not so regularly as the black antelope; and their numbers are not sufficient to do any notable damage. In the morning they may often be found picking their way back to the network of ravines where they stay during the day. Should you disturb them at this time, they will most likely seek their cover at top speed; and what that amounts to will amaze you if you let slip a greyhound at them. Chikara have not yet learned the range of the modern 'Express' rifle; and consequently they still often let one get almost within the killing distance of the old weapon, and are easily knocked over with the 'Express.' The depth of their slender bodies is so small, that a bullet must be planted in a space little wider than a hand's-breadth to make sure of stopping them. Shots are generally got at a distance of from 100 to 150 yards; and the difficulty of such fine shooting at uncertain distances, together with their peculiar 'dodginess' in keeping out of sight, makes the stalking of them a more difficult, and I think more interesting, sport than the pursuit of the larger antelope. Their art has little variety in it however; and there is something to the experienced eye in the features of the ground which will almost infallibly tell whereabouts one is likely to have stopped after his first disappearance. Unless they have been seen to go clean away, they should always be followed up on the chance of being found again.

The last of the antelopes met with in the open country is the Nilgae, the male of which, called a 'blue bull,' will stand about 13½ hands high at the shoulder. The female is a good deal smaller, and of a fawn colour. Their habitat is on the lower hills that border and intersect the plains, and also on the plains themselves wherever grass and bushes afford sufficient cover. The old sites of deserted villages and cultivation, unfortunately so common, which are usually covered with long grass and a low bushy growth of Palas (*Butea frondosa*) and Jujube (*Zizyphus jujuba*) trees, are seldom without a herd of nilgae. They are never found very far from cultivation, which they visit regularly every night. When little fired at, the blue bull is very easily approached and shot. It is very poor eating, and affords no trophy worth taking away, so that it is not much sought after by the sportsman. The beginner, however, who is steadying his nerves, or the inventor who wants a substantial target for a new projectile, will find them very accessible and convenient. The blue bull is an awkward, lumbering, stupid brute; and it is highly ludicrous to observe the air of self-satisfaction with which a blockhead of a bull, who has allowed you to walk up within fifty yards of him, will blunder off to the other side of a nala, then turn round and stand still within easy range of your rifle, and look as if he thought himself a very clever fellow indeed for so thoroughly outwitting you. He is a favourite quarry with the unenterprising Mahomedan gentleman. The antelope his style of dress and powers of locomotion do not allow him to approach; the rugged ground and thorny underwood prohibit his succeeding with the forest deer; the tiger he likes not the look of, and the pig he may not touch; so he gets him into a bullock cart, and is driven within a few paces of an unsuspecting blue bull, whose carcase, when shot and duly cut in the throat after the rules of his faith, makes for him the beef which his soul loveth. Awkward and inactive as he looks, however, the blue bull when fairly pushed to his speed will give a good horse as much as he can do to overhaul him. It is in vain to attempt it in or near the jungle; but if you can succeed in getting at him when he has a mile or two to go across the open plain, a real good run may be had with the spear. I have never heard of a blue bull attempting to charge when brought to bay, in which respect therefore the sport of riding them is inferior to pig-sticking.

Such are the principal animals which form the objects of the sportsman's pursuit in the open country. As, however, in a state of nature, there never are herbivorous creatures without their attendant carnivora to form a check and counterbalance to them, so we find various natural enemies attendant on the herds of antelope and nilgae, whose acquaintance the sportsman will occasionally make. The nilgae, is a favourite prey of the tiger and the panther. But it is in the low hills where he retires during the day, rather than in the plains where he feeds at night, that he meets these relentless foes; and the chief carnivorous creatures of the open country are the hunting leopard, the wolf, and the jackal.

I have several times come across and shot the hunting leopard when after antelope; but they cannot be called common in this part of India. They live mostly in the low isolated rocky eminences called

Torias, that rise here and there like islets in the middle of the plains, and on the central plateau, and which are frequently surrounded by grassy plains where they hunt their prey. They are of a retiring and inoffensive disposition, never coming near dwellings, or attacking domesticated animals, like the leopard and panther; and I never heard of their showing any sport when pursued. Their manner of catching the antelope, by a union of cat-like-stealth of approach and unparalleled velocity of attack, has often been described. A few are kept tame by the wealthier natives, but more I think for show than real use in hunting.

The common jackal, always ready for food of any description, seldom fails to make a meal of any wounded animal, and I have seen a small gang of them pursue a wounded antelope I had just fired at. The fawns of the antelope and gazelle frequently become their victims.

The wolf is extremely common in the northern parts of the province; frequenting the same sort of ground as the antelope and chikara. I have very seldom met with them in forest tracts; and I think that in India they are clearly a plain-loving species. They unite in parties of five or six to hunt; the latter being the largest number I have ever seen together. More generally they are found singly or in couples. I have several times observed them in the act of hunting the antelope; their method being to steal in on all sides of a detached party of does and fawns, and trust to a united rush to capture one or more of them before they attain their speed. Fast as the wolf is (as you will learn if you try to ride him down), I do not believe he is capable of running down an antelope in a fair hunt, though doubtless old or injured animals are thus killed by him. When game is not to be had, the wolf seldom fails to get a meal in the neighbourhood of villages, in the shape of a dog or a goat. They are deadly foes to the former; and will stand outside a village or the travellers' camp at night, and howl until some inexperienced cur sallies forth to reply, when the lot of that cur will probably be to return no more. Unfortunately the wolf of Central India does not always confine himself to such substitutes for legitimate game; and the loss of human life from these hideous brutes has recently been ascertained to be so great that a heavy reward is now offered for their destruction. Though not generally venturing beyond children of ten or twelve years old, yet when confirmed in the habit of man-eating, they do not hesitate to attack, at an advantage, full-grown women and even adult men.

A good many instances occurred, during the construction of the railway through the low jungles north of Jubbulpur, of labourers on the works being so attacked, and sometimes killed and eaten. The attack was commonly made by a pair of wolves, one of which seized the victim by the neck from behind, preventing outcry, while the other, coming swiftly up, tore out the entrails in front. These confirmed man-eaters are described as having been exceedingly wary, and fully able to discriminate between a helpless victim and an armed man.

My own experience of wolves does not record an instance of their attacking an adult human being; but I have known many places where children were regularly carried off by them. Superstition frequently prevents the natives from protecting themselves or retaliating on the brutes. In 1861 I was marching through a small village on the borders of the Damoh district, and accidentally heard that for months past a pair of wolves had carried off a child every few days, from the centre of the village and in broad daylight. No attempt whatever had been made to kill them, though their haunts were perfectly well known, and lay not a quarter of a mile from the village. A shapeless stone representing the goddess Devi, under a neighbouring tree, had instead been dabbed with vermillion, and liberally propitiated with cocoa nuts and rice! Their plan of attack was uniform and simple. The village stood on the slope of a hill, at the foot of which ran the bed of a stream thickly fringed with grass and bushes. The main street of the village, where children were always at play, ran down the slope of the hill; and while one of the wolves, which was smaller than the other, would ensconce itself among some low bushes between the village and the bottom of the hill, the other would go round to the top, and, watching an opportunity, race down through the street, picking up a child by the way, and making off with it to the thick cover in the nala. At first the people used to pursue, and sometimes made the marauder drop his prey; but, as they said, finding that in that case the companion wolf usually succeeded in carrying off another of the children in the confusion, while the first was usually so injured as to be beyond recovery, they ended, like phlegmatic Hindus as they were, by just letting them take as many of their offspring as they wanted: An infant of a few years old had thus been carried off the morning of my arrival. It is scarcely credible that I could not at first obtain sufficient beaters to drive the cover where these two atrocious brutes were gorging on their unholy meal.

At last a few of the outcaste helots who act as village drudges in those parts were induced to take sticks and accompany my horse-keeper with a hog-spear and my Sikh orderly with his sword, through the belt of grass, while I posted myself behind a tree with a double rifle at the other end. In about five minutes the pair walked leisurely out into an open space within twenty paces of me. They were evidently mother and son; the latter about three-quarters grown, with a reddish-yellow well-furred coat, and plump appearance; the mother a lean and grizzled hag, with hideous pendent dugs, and slaver dropping from her disgusting jaws. I gave her the benefit of the first barrel, and dropped her with a shot through both her shoulders. The whelp started off, but the second barrel arrested him also with a bullet in the neck; and I watched with satisfaction the struggles of the mother till my man came up with the hog-spear, which I defiled by finishing her. In the cover they had come through my men said that their lairs in the grass were numerous, and filled with fragments of bones; so that there was little doubt that the brutes thus so happily disposed of had long been perfectly at home in the neighbourhood of these miserable superstitious villagers.

Dogs that are in the way of hunting jackals will readily pursue a wolf, so long as he runs away. But the wolf generally tries the effect of his bared teeth on his pursuers before running very far, and only the most resolute hounds can be brought to face them. I have several times had my dogs chased back close up to my horse by a wolf they had encountered when out coursing foxes and jackals; and only once saw the dogs get the better of one without assistance from the gun. On that occasion I had out a couple of young greyhounds, crossed between the deerhound and the Rampore breed; and along with them was a very large and powerful English bull-mastiff, rejoicing in the name of 'Tinker,' whose exceedingly plebeian looks in no way belied his name. He was an old hand at fighting before ever he left the purlieus of his native Manchester; and in India had been victor in many a bloody tussle with jackal, jungle cat, and pariah dog. His massive head and well-armed jaws combined in a high degree the qualities of a battering ram and heavy artillery; and his courage was in full proportion to his means of offence. On the present occasion the three dogs espied the enemy sitting coolly on his haunches on the top of a rising ground; and the young dogs, taking him no doubt for a jackal, went at him full speed, Tinker as usual lumbering along in the rear. Soon, however, the hounds returned in a panic, with their tails well down, and closely pursued by the wolf, a large dark grey fellow, snapping and snarling at their heels. The greyhounds fled past Tinker, who steadily advanced, dropping into the crouching sort of run he always adopted in his attack. No doubt Master Wolf thought he too would turn from his gleaming rows of teeth and erected hair, as all his canine assailants had done before. But he never was more mistaken, for the game old dog, as soon as a pace or two only remained betwixt him and the enemy, suddenly sprang to his full height and with a bound buried his bullet head in his advancing chest. I saw the two roll over and over together; and then the gallant Tinker rose on the top of the wolf, his vice-like jaws firmly fastened on his throat. At this point of a combat he usually overpowered his antagonist utterly, by using his immense weight and power of limb to force him prostrate on the earth, the while riving at the throat with a force that often scooped a hollow in the earth under the scene of action. His efforts were now directed to effect this favourite manoeuver; but the wolf was too strong for him, and repeatedly foiled the attempt. But the young hounds, who were not at all without pluck, soon returned to his assistance, and seizing the wolf by different hind-legs, made such a spread eagle of him that Tinker had no difficulty in holding him down while I dismounted and battered in his skull with the hammer-head of my hunting-whip. None of the three dogs had been bitten, Tinker having got his jaws in chancery from the very first. I am sure that the three, or even Tinker alone, would have killed him in time without my assistance; for Tinker never let go a grip he had once secured, and though not so large, was not much inferior to him in strength.

 The catalogue of amusements offered to the sportsman in the open plain would be incomplete without a mention of the 'mighty boar.' He is to be found almost everywhere—in the low jungle on the edge of cultivation, and sometimes in the sugar-cane and other tall crops; and with a liberal expenditure of self and horse may be ridden and speared in a good many places. Generally, however, the country is highly unfavourable to riding, the black soil of the plains being split up into yawning cracks many feet in depth, or covered with rolling trap boulders, both sorts of country being almost equally productive of dangerous croppers. The neighbourhood of Nagpur affords the best ground; and there there is a regular 'tent club,' which gives a good account of numerous hogs in the course of the year. The sport has been so voluminously described that I believe nothing remains to be said about it. The hogs that reside in the open plains are not much inferior in size to those of other parts of India; but those met with in the hills are generally much smaller, and far more active. A brown-coloured variety has sometimes been noticed among them. The common village pig of the country shows every sign of having been derived from the wild race originally.

Excerpted from The Highlands of Central India: Notes on the Forests and Wild Tribes, Natural History and Sports *(London: Chapman and Hall, 1879).*

FRANK B. SIMSON

Rhinos, Javan and Indian

Frank Simson's Letters on Sport in Eastern Bengal *were based on diary notes kept during his postings in various districts over a period spanning nearly four decades. Things had 'changed greatly' over these years: Kassimbazar, the best place to find tigers had none left and there were no wild boar around in Noakhali. But the most spectacular decline was that of the two species of rhinos he had shot. His is one of the few good accounts of the Javan rhino which he encountered in the Sunderbans: it disappeared altogether soon after. The salt and tidal lands retained much of their cover, but even a limited amount of pressure proved too much for the Javan rhino. The larger Indian one-horned rhino had contracted in terms of its range but was still seen in the Brahmaputra valley's* chaurs *or wet grasslands in Assam. Simson was an acute observer of not only the big and small game but also of the varied terrain of Bengal. The Javan rhino now survives in only two places in South-east Asia—a small reserve in southern Vietnam and in Udjong Kulon national park, Java, Indonesia. Simson was among the few who shot and wrote about it in India.*

LETTER NO. 47

I wish I had had more experience in shooting the rhinoceros, for it is to be got in Eastern Bengal by any one who has time and opportunity though only in the distant outskirts of the province or almost on the sea shore. There are two kinds of rhinoceros in Bengal, *R. indicus* and *R. sondaicus*. The first is the larger and is found where Mymensing joins Assam, on the east of the Brahmapootra, and in

Assam, and exists all along the base of the Himalayan slopes to the north of Rungpore and Purneah but westward of Purneah it does not belong to the area of Eastern Bengal.

When I was appointed to that inferior sporting station Backergunge I thought to myself that I might at any rate shoot a rhinoceros; and I will tell you how the deed was accomplished. I knew the animal abounded in the Soonderbunds and was to be found all the way from the mouth of the Hooghly to the mouth of the Megna; this is the chief habitat of the lesser rhinoceros. Occasionally one has been killed to the south of Tippera and Chittagong. The one now in the London Zoological Gardens was captured when very young between Chittagong and Arracan, as it was crossing a muddy river. The whole establishment of Government khedda elephants happened to be on the march when news was brought that a little rhinoceros was in difficulties in some mud, and with the aid of these elephants the animal was taken alive and did well in captivity.

I found that no rhinoceros has ever been known near headquarters at Backergunge, but that the Soonderbund folk and the English planters who had taken up Soonderbund lands reported that they were plentiful between their settlements and the shores of the Bay of Bengal, and I persuaded the judge to accompany me in a trip to the Bay. Budderuddeen was not with me in those days, nor any shikarry of any value to compare to him; so I got what information I could through the police, who knew shikarries who used to procure venison from the more southern tracts of mangrove and soondry trees. At last I procured the services of two men who for good consideration agreed to take us up to some rhinoceroses, but strongly advised us to confine our attentions to spotted deer and jungle-fowl, as the rhinoceros was a fierce beast, not to be stopped by a bullet.

The country for the sport was most peculiar, for the most part devoid of any water that was not salt; the greater part of the soil was covered by the spring-tides: it was intersected by deep tidal nullahs with muddy banks. The good ground was at least as far off as boats with relays of rowers could reach in two nights' and a day's row, not far from a village which had been established by a colony of Arracanese Mughs, known as 'Isla Foolzurree.'

We got a comfortable large boat for ourselves and a commodious and fast-pulling boat for the attendants, to serve also as a cook-boat. We loaded both boats with large jars of good water for drinking and cooking purposes, and started down stream about 4.30 P.M., rowing all night and taking advantage of tides under the direction of the shikarries who knew the streams. We soon left regular cultivation and well-populated regions and sailed and rowed all day down streams thickly fringed with trees. Here and there we came to marshy places covered with very high grass, and flags, and marsh-plants. In these tigers, buffaloes, and rhinoceros were to be found occasionally; crocodiles were very numerous, and lovely kingfishers of a kind I had never seen before were common. On the second day we got to the rhinoceros land.

All the way down the judge and I had discussed as to where a rhinoceros should be hit. The judge had killed plenty near the Brahmapootra churs, but off elephants, and he argued for aiming at the shoulder and the neighbourhood of the heart. I declared that I would not fire if I could help it save just behind the ear. We both thought that the hide of a rhinoceros was proof against anything but a heavy bullet projected from a strong charge of powder. I have since come to the conclusion that, during life at least, the hide of a rhinoceros is not much more impenetrable than that of an old urna buffalo. My battery carried bullets fourteen to the pound. The judge had a heavy single rifle that carried a thing like a small cannon-ball, and required a coolie to carry till the time for use arrived. We had no luck the first afternoon, but saw spotted deer and plenty of tracks of both rhinoceros and tigers. I shot a large crow with ball after we got to the boats, and this rather raised our credit with the shikarries; they were to scout for game early next morning and we were to await their reports.

Next morning, after breakfast, one shikarry returned and said rhinoceros had been found, and that the other shikarry was on the watch. We had to go some distance in small canoes, and then we landed and had to walk over a most extraordinary hard kind of ground. We marched under thick green foliage and among trunks of fair-sized trees; but from the very wide-spreading roots of these trees there grew crops of stumps bigger than ordinary carrots, most difficult to travel among, and very trying to the ankles. Here and there were dense thickets of grassy or shrubby underwood, and large patches of that huge crackling fern called 'dinkybon;' the soil was generally covered with an inch or two of mud, and sometimes there were patches of short turf. Generally you could see under the trees for some distance; and if beaters could have been got, deer might have been driven past gunners posted in ambuscade.

After a very long trudge in this trying walking ground we came upon the second shikarry, and a council of war was held. We were each to take only one attendant to carry a spare gun, and to go as silently as possible. I took my most sporting chuprassy and two smooth-bores, one by Sam Smith and one by Joe Manton. The judge, I think, had a double rifle and his man carried the heavy single rifle before mentioned. After a little the shikarry made signs to stop, and after a little reconnoitring he came back and pointed to a patch of dinkybon, and whispered that there was a rhinoceros in it, and that he would now make his salaam and climb up a tree, and leave the rest to us. On this all the other natives declared that they must go up trees, and they all salaamed and went up trees accordingly.

The judge and I agreed to go, one on one side and one on the other of the patch, and to come together at the first call, for both of us knew that two weapons might be required; I went to the left and the judge to the right. Peering about, presently I saw that something was moving in the dry fern-bush, which was about ten feet high, so I stood by the trunk of a tree about twenty-five yards from the fern, and put the Joe Manton gun against the tree on full-cock, and held the Sam Smith ready. Soon I saw a nose poked out, then the eye, and then the ear of a rhinoceros; as soon as this came out I let fly, and you can scarcely conceive the row which followed—something between the roar of an elephant and the neigh of a horse, but far stronger. The smoke hung, and as it passed I saw a rhinoceros standing, looking directly towards me. He stood a minute. I knew it was no use to fire at this lowered head; there is no vulnerable place there, and the ball would have glanced. I called to the judge and said, 'Here he is in front, looking at me!' but I did not move from the trunk of the tree, and I do not know if the rhinoceros distinctly made me out; they are said to have bad sight. The animal turned round, and as he did so I shot him high in the shoulder and he bolted. I followed, but was instantly stopped, for there lay a rhinoceros stone dead. There had been two in the bush, and my ball behind the ear had killed the first. After the judge and I had made sure that the beast was really dead, we went after the other. The blood had gushed out on the trees and was frothy; we argued that the ball had penetrated the lungs and would be mortal. After going a little way further we considered that we should certainly lose ourselves, so we stopped and retraced our steps to the dead rhinoceros, and, after much shouting, got our attendants to come to us.

The next thing was to cut off the head and send it to the boats. This was no easy task. I had an ordinary sheathed hunting-knife, and the judge had a large pocket-knife; but these were far too small for the purpose, and it took us nearly an hour cutting, hacking, and twisting before the head was got off. It was then tied with jungle creepers to a thick stick cut for the purpose, and sent off to the boats on the shoulders of two men.

We then spent some hours trying to get more rhinoceros. We saw, I think, six; but they were on the move, and either smelt or heard us, and we never got near enough for deadly action. The judge, who was an excellent shot, fired and wounded some animals; but no good was done. The walking was at times most difficult: the trees and foliage were characteristic of Soonderbund jungle, such as are to be found nowhere else in India proper; tigers and deer evidently abounded. It was impossible to stalk animals from a distance—first, because the walking was difficult; secondly, because of the fern which crackled as man or animal passed through it. I was transferred to Noakholly soon afterwards, and never revisited these parts: but the judge went again; he could get no gentleman to accompany him, and had a few natives with him. He put up a tiger and severely wounded it; in searching for this animal, a chuprassy came upon it and was killed. After this the judge went no more. These places are said to be most unhealthy for Europeans; fevers of the worst type are soon contracted, and the Soonderbunds are considered deadly if any stay is made in them.

But there is no doubt that between the northern parts of Backergunge and the shores of the Bay of Bengal there spreads a country full of large game, and where a sportsman, if he could retain his health and learn the localities and plan out the proper methods for shooting, might obtain grand sport. There are no elephants to be got; there is no drinkable water; there are no villages where supplies can be got; there are no roads nor bridges, and ever recurring deep muddy tidal salt-streams. An enterprising sportsman, with a good steam-launch, might perhaps be able to work the country; but during my time I never heard that any Europeans had been able systematically or effectively to hunt up any portion of it. Every now and then some such short expedition as I made is undertaken, and occasionally some English gentleman connected with the ownership of Soonderbund grants kills a few tigers and rhinoceroses and deer; but generally fever puts a stop to his ardour for sport. The forests are, however, gradually disappearing before cultivation, in spite of the insalubrity of the soil, and villages are springing up, and steamers go through the deep channels, and native craft through the smaller streams.

I never heard anything more of the rhinoceros I wounded or those fired into by the judge; but immediately after I left the district the horn of a rhinoceros was offered for sale in the bazaar of Burrisawel, the head station of the district, and the judge had an opinion that this was the horn belonging to the second animal of the two to which we had been taken by the shikarries.

This rhinoceros seems to be a harmless animal, feeding on branches of trees and the rank succulent herbage of muddy swamps. It never appears to visit cultivated places or to damage crops; unless roused it has no murderous propensity; it kills no other animal, and its size and thick hide protect it from the tiger. The unhealthy climate of the Soonderbunds has prolonged its existence to this date; but ere long, as these salt and tidal lands are brought under rice- and jute-cultivation, the animal will disappear. In the struggle for existence it will not be able to fight the battle by adapting itself to change even to the extent that tigers and hogs seem to do; these shift their quarters, and constantly appear in new localities.

LETTER NO. 48

The last letter referred to *Rhinoceros sondaicus*; but the *Rhinoceros indicus* generally referred to as the rhinoceros of Bengal by sportsmen, is a different and a larger animal, inhabiting extensive swamps and marshes, where the grass is the tallest and densest to be found, and where the jungle called in Mymensing 'taradhām' occupies large spaces of soft muddy soil, which I should think would only suit heavy animals in the dry season. In these thick grass jungles no man on foot can make his way: he must either follow paths made by elephants, buffaloes, and rhinoceros, or cut his way through them, or go on elephants; and it is on elephants alone that the sport of rhinoceros shooting has been carried on in the regions of the

Brahmapootra churs and in the valleys of Assam, and in the Terai at the base of the Himalayas about Rungpore and Purneah.

I went after a rhinoceros at times; but neither in the neighbourhood of the Brahmapootra nor in the north of Purneah did I ever see one. If you can accompany any sportsman who is acquainted with good rhinoceros-ground and have leisure and desire to kill these animals, you will have better luck. I neither knew the country nor Europeans nor natives in it, and contented myself with hog-hunting and tiger-shooting; I cannot therefore relate tales of rhinoceros-slaughter in which I took part.

But numbers of other sportsmen and friends killed them yearly, and their skulls were to be seen in the verandahs of many houses in Mymensing and Dinagepore and Rungpore. From looking at these it was at once seen that the brain took up only a small space in the huge bony head, and that bullets in the head need not necessarily prove mortal. Well-directed shots, however, seemed easily to have pierced to the brain; and from all I could learn the chief thing in rhinoceros-shooting was to manage to get as close to the animal as possible, and the only difficulty in doing this was the great fear elephants had of approaching a rhinoceros at all. Sometimes these animals are pugnacious, and when they charge the line of elephants the latter animal, as a rule, takes to flight and cannot be brought to face the rhinoceros at all. I have heard stories and seen pictures of elephants having been knocked over by rhinoceroses; still I never heard of any serious accident that had occurred. An elephant that will face a fighting rhinoceros is a most valuable animal. Much depends on the mahout; but elephants are curiously uncertain in temper and courage. My best elephant had a great reputation for good behaviour with rhinoceroses, and I killed scores of tigers from her back; nevertheless, even with Sowdaugor, the pluckiest mahout I ever had, she ran away—positively declined to be brought close up to a plucky rat: the courageous little animal stuck out its fur, squeaked, and jumped towards the ponderous elephant, which tucked up its trunk, screamed, backed away, and could not be brought forward, though I tried for quite a quarter of an hour to force the elephant to close quarters.

Excerpted from Letters on Sport in Eastern Bengal *(London: R. H. Porter, 1886).*

J. H. BALDWIN

Hunting the Houbara

The houbara or McQueen's bustard is a ground-living bird that was shot by zamindars, princes and officers because it was a 'gourmet's delight'. But the bird had excellent camouflage and stalking it was a challenge for the most experienced shikari. Baldwin, a veteran hunter, had ranged across Bengal, central India and the North West Provinces (now Uttar Pradesh). Only our obsession with big game of the four-footed kind blinds us to how much of shikar consisted of tracking and shooting birds. Partridge and grouse, peafowl and houbara were especially welcome to the British, inveterate meat-eaters who wanted a change from chicken and mutton!

The birds of the open grasslands included the great Indian bustard, Asia's heaviest land bird, now a rare species. Though never a delicacy like the houbara, it was a large ground-dweller weighing up to 14 kg, making it a large quarry. The florican was another such bird of the grasslands. While wildlife and forests are often seen as synonymous, an amazing variety of birds lived in scrub, desert and grassland.

The chase of the bird requires such caution and patience as would sorely try the temper of even a practised deer-stalker.—WHITE'S '*Selborne.*'

THE INDIAN BUSTARD

Description

Male. Height—From 3½ to 4 feet.

Bill.—2 inches in length, of a yellowish colour, nearly straight and pointed.

Iris.—Pale yellow.

Legs.—About 20 inches, and yellow. Three toes, with claws shaped like those of common fowl; no hind toe.

Neck.—16 or 18 inches.

Tail.—12 inches.

Wings.—Broad and rounded, when closed do not reach to within 6 inches of tail.

Upper part of head, where there is almost a plume, glossy black. Neck, which in old cock birds is covered with long feathers, white, duller behind than in front. A line, 2 or 3 inches broad, of dark brown feathers, extends across the upper part of the breast, from shoulder to shoulder. Remainder of breast down to end of abdomen, white, mottled with light fawn colour. Back and upper parts of wings reddish, beautifully variegated with very thin lines of dark brown. Tail the same, but with slightly darker tinge. The feathers not unlike those of a turkey. Primaries and lower feathers of wing black, with a tip of white on the end of each feather; altogether wanting the long moustache of the European bustard (*Otis tarda*).

Female.—Considerably smaller in size than the male, plumage not so bright. The feathers on the neck not so long, nor of so pure white, but speckled throughout.

There are many varieties of bustard, and among the largest and most noble in appearance is the Indian. This bird is, in my opinion, the king of game birds, and the value of its feathers, its excellence as a bird for the table, and last, though not least, the very great difficulty in shooting it, render it a prize to be much coveted; to slay an old cock bustard is or should be the ambition of every young sportsman, and an achievement only brought about by perseverance and a knowledge of the habits of the bird: many a sportsman spends the whole of his service in India without bagging a single one.

I first met with the bustard on the road between Mirzapore and the Kuttra Pass. I again shot it when marching from Allahabad to Saugor. And during my three years' service at Jhansie, I constantly came across this splendid bird; as also in Lullutpore and in Gwalior. Between Morai and Jhansie, in the neighbourhood of Dubra, there is a plain behind the Dawk Bungalow, a favourite spot for bustard. It is common in the Saugor country, and, I have been told, in the neighbourhood of Hansi; occasionally in the Meerut district. Once I heard of it at Mozuffernugger. I have never met with it in Bengal Proper, Assam, nor anywhere on our Eastern frontier. The Indian Bustard is not usually found on the dead level plains, as is often supposed to be the case, but frequents low, rocky, undulating hills, or high grounds interspersed with patches of bushes and scrub jungle.

Generally a pair may be seen stalking about together, sometimes a solitary one, occasionally four or five; and I once counted, within two miles of the city of Jhansie, no less than fourteen bustard, nearly all females, feeding in a field of young corn. The birds are not often met with in the hot weather; they are sometimes seen in the cold season, but the best time of all for finding them is early in the 'rains.' After the first downpour they may be seen on ground such as I have described.

The bustard is the most wide-awake and cunning of all feathered game, and seldom permits the sportsman to approach within close range. The eyesight of the bird is wonderfully keen; and they will hardly ever feed or settle near grass or cover high enough to screen an enemy, or to allow a stalk to be made. In attempting to approach the birds for a shot, the same tactics as I have already described for approaching black buck should be adopted: circle gradually round, keep your gun hidden behind you on the off-side from the game, walk leisurely, and never look towards the quarry, but so manoeuver as to appear to be passing by and not taking the slightest notice of them.

Many scores of times I have tried, by getting a bush in a line with the bird, to creep up stealthily within distance, but have seldom succeeded. When about a hundred yards off, the sharp-eyed bird had almost invariably discovered me and taken to flight. Perseverance, however, will be rewarded in time, and on one occasion I grovelled up a shallow ravine on my stomach to within forty yards of five cock bustard; one fell dead to the right barrel, loaded with B B shot, and a second dropped to a similar dose from the left, but again recovering himself, took to flight. I watched him till out of sight, but still keeping my eyes in the direction in which the bird had flown, I fancied, though not at all certain, that I saw him throw up his wings and settle in a large patch of high grass. Having picked up and duly admired the slain bird, I made straight for the high patch I have mentioned, and making my syce lead my horse and walk in a line with me, we crossed backwards and forwards three times, and beat the whole of the grass without finding what I hoped for. Just as I was wiping the perspiration off my face, and remarking to my attendant that I must have been mistaken, the bird rose within ten yards of us, and this time fell dead enough. I could see by the dry blood marks on the feathers that he had been severely wounded by my first shot.

The flight of the bustard much resembles that of the common vulture, especially when at a distance, but the former bird may always be distinguished from the latter in the following manner. The vulture makes several powerful strokes, and then sails along with wings extended; again he strikes out as before, and once more glides along motionless. The bustard, however, slowly flaps his wings in a measured manner as he flies; but never once, from the moment he takes wing till the time he settles, however long a flight he may take, does he ever glide along with the wings extended as the vulture does; and the knowledge of this fact has often enabled me to decide between the one bird and the other at a very great distance.

The bustard generally flies tolerably straight when flushed, till he reaches the ground where he intends alighting, when, probably to see that all is safe, he sometimes circles round before he drops. It is a good plan, if the bird flies right away out of sight, to mark the line of flight taken, and probably, by following the direction carefully, you will again find him. The bird does not often rise any great height from the ground, is very awkward in getting under way, and carries its head and neck fully extended; the flight, though heavy, is powerful.

The food of the bird is principally locusts—when removing the skins of specimens I have found the stomach full of these insects—grasshoppers, various kinds of beetles, worms, centipedes, and such like. It is said they will even eat lizards and small snakes, but I am doubtful about this, and have never once found anything of the kind when stuffing the birds. I have also seen them in the early morning busily engaged in a field of corn, eating the young green shoots. Jerdon tells us: 'In default of insect food it will eat fruit of various kinds, especially the fruit of the Byr and Coronda.' Doubtless this is correct, though I have never discovered anything of the kind in the stomach of the bustard. I once flushed one, when shooting quail, out of a high piece of standing corn; he had the temerity to fly directly over my head and got two barrels of quail shot, which made him drop his legs and swing about; but again recovering himself, he made off. I thought that he had made good his escape, but about half-an-hour after, when we had finished our quail shooting, I was delighted to see my syce approaching with the bustard in his hand.

We had left our horses about a quarter of a mile from our shooting-ground, bidding the syces wait our return. Our grooms had heard the report of my gun, and from where they were seated, on higher ground, had noticed that the bustard staggered and was badly hit; and, the bird flying in their direction, they were able to watch it for a longer time, and to keep it in sight a further distance than I could. They saw the bustard collapse suddenly, and fall dead on the open plain.

The finest old cock I ever shot, or rather assisted in shooting, was close to Jhansie; a friend and I flushed him, and marked him down again; we followed him up, and, by taking advantage of the ground, succeeded in getting within long range, about 60 or 70 yards, when he rose, and our four barrels were just sufficient to bring him down. This was a magnificent old cock; I stuffed him, and my friend, who took part in slaying him, sent him home to England. The bird was of such great length that it took a box the size of a coffin to contain him.

I have found it an excellent plan, where the birds are very wild and unapproachable, to drive them. The sportsman, who will day after day find the same birds on the same spots—very likely in company with sundry old black bucks for mutual protection—should mark well once or twice the exact direction the birds take when flushed, which will generally be over the same line of country. Next time he visits the place, to which the birds will be almost sure to have returned, let him ensconce himself behind a bush or rock under the line of flight he marked on the previous day, but without allowing the game to witness what he is about, and then direct an attendant to move round and quietly put the bustard up; very probably he will from his ambush have the pleasure of seeing them make straight for the ambuscade laid for them. Of course this stratagem does not always succeed, but I have several times practised it with success.

One of Scindiah's officers at Jhansie had a buffalo trained for stalking bustard, and by this means killed several. The way he managed was as follows:—Guiding the buffalo by his tail in the direction of the game, he concealed himself behind the beast, and thus was able to get within shot of the birds, who, accustomed to see cattle grazing about, suspected no danger. This appears to be rather a poaching dodge, however, and I have never tried it.

The birds may often be seen stalking about, with natives cultivating their fields close by; and yet, directly a European sportsman comes in sight, the bustard immediately take notice of him, and utter their hoarse cry of alarm, sounding something like the word 'hook.' A fine-sighted, accurate-shooting small-bore rifle is a good weapon for bustard;[1] though I generally trusted to B B shot, with rather a large charge of powder.

The first bustard I ever killed, a fine old cock, was with a rifle bullet, but a desperate 'fluke,' as the story will show. Many years ago, before I had been in the country a year, I obtained a month's leave to accompany two others on a shooting trip. I was then stationed at Benares; we travelled rapidly to Mirzapore, and commenced marching in the direction of the Kuttra Pass, on the Saugor road, till we reached a place called Lalgunge, where we put up at a small bungalow. My companions were both old sportsmen, though I at the time was a regular Griffin. I had brought a double Enfield rifle with me—about the worst sporting rifle that I am acquainted with—and was most anxious to shoot one of the numerous black bucks which we saw constantly on either side of the road. Within a week, after a goodly expenditure of ball cartridges, I had bagged two bucks. One of my companions, Captain P—y, had mentioned more than once that bustard were occasionally to be met with in the neighbourhood, but never having seen one, I had only a meagre idea what the birds were like.

At length we moved our quarters to another bungalow farther on, a march of eight or ten miles; and, instead of keeping to the road, I made a detour in search of game. Within an hour of starting I severely wounded a fine black buck in the hind-quarters, but the creature made off. I followed him alone for a very long distance, until, quite tired out with running, I sat down on the brow of a hill to recover breath. While mopping my face with my handkerchief, I was startled by hearing a deep, harsh, guttural 'hook' behind me. I looked round and observed for the first time four large birds stalking away, and looking back over their shoulders at me. Not knowing what they were, I was watching them unconcernedly, when it suddenly struck me that they were the very birds Captain P—y had so often spoken about, and had been describing to me

---

[1] I can strongly recommend those made by Holland of Bond Street, as particularly suited for bustard-shooting.

only the previous evening, viz. bustard. My gun was with my syce far behind, but I determined on taking a shot with my rifle. There was no time to spare, already the birds were preparing to take wing; so glancing along the sights, I took aim at the nearest, and fired just as one of the others took to flight. The bullet fell short, but struck the bustard as it ricochetted, and to my delight the bird dropped motionless; on running up I found that I had killed a magnificent old cock. The bullet had passed through the throat, just below the ear—a great piece of luck. Presently my syce and horse came up; we slung the bird over the saddle, and made for our halting ground. My companions were both astonished at my good fortune—and perhaps a little envious as well at the success of the Griffin—and wondered how I had managed to hit him so neatly; but, as was natural for a youngster, I did not furnish them with the whole facts of the case, but merely stated that I had knocked the bird over with an Enfield bullet.

The skin of the bustard is very thin, and stuffing one, or rather taking the skin off, is a laborious task, and requires much care, especially if, as is often the case, the body is covered with rolls of fat. At certain times of the year it is a good bird for the table, especially young hen, boiled like a turkey, or baked, as some prefer. At the same time if the bustard of Europe (*Otis tarda*) is not superior when cooked to the Indian species, I think that its great excellence has been exaggerated, for I venture to say that the flesh of the Indian bustard at the best of times cannot compare in delicacy with that of the floriken, nor, as I am informed, with that of the oobara. The feathers, I need hardly mention, are highly prized by the salmon-fisher for certain flies.

Natives often catch these birds; they watch and mark them down just before dark, and then draw a net over them. I have often seen them brought into stations for sale at a rupee each.

One day, when on horseback, I flushed three bustard; as they were making off, a large tawny eagle, with wings half closed, made a swoop at them; the bustard rose higher in the air, and I distinctly heard them uttering cries of distress, The eagle having regained a position above the pursued, came down again with a tremendous rushing noise like a bolt, but missed his aim a second time, and after a third unsuccessful attempt desisted from molesting them further. Having a partiality for anything in connection with hawking, I became excited at so grand a sight, and clapped spurs to my nag to keep pursuers and pursued in sight. Possibly I approached too near, and thus was the cause of the eagle giving up the chase.

I have often read about the pouch said to be in the bustard's throat for the purpose of holding water. Certain it is that the Indian bustard has no such appendage, for I have again and again most carefully examined the throats of both cocks and hens, and could discover nothing of the kind.

It has been said by certain writers that the bustard is 'swift of foot,' and more extraordinary still, 'has been coursed successfully by greyhounds.' I am only able to speak as to the Indian bird, though I imagine there is not any great difference in habits, if there be in plumage, between it and the African Pauw, or the Great Bustard of Europe. I myself have never seen the bustard run swiftly, or move in any other manner than a stately walk. Sometimes, when suddenly alarmed, they move away from the object that has frightened them at a rather quick pace just before taking wing, but that is all that can be said.

As to the second assertion, 'the coursing with greyhounds,' as regards our Indian bustard, I will answer for its being a fallacy, and a single day's trial would speedily convince the most sceptical on this point.

The bustard builds no regular nest, but about the end of July or early in August the female selects some suitable spot on the ground generally a hollow in an open field, amid the sprouting shoots of some young crop. There she deposits two eggs of a mottled brown colour, and rather larger in size than those of a turkey.

I have been informed on reliable authority that about the breeding season the cock bustard often utters a peculiar long-drawn droning note, and that this singular call is generally heard after nightfall.

Bustards may often be seen feeding in open fields; they are especially partial to patches of linseed, called by the natives *tillee*. It is not the crop itself which forms the chief attraction to the birds, but swarms of locusts and grasshoppers, which invariably abound in these fields of linseed, and, as already mentioned, form a favourite food of the bustard.

Before concluding my rather lengthy remarks on the bustard, I must mention that I have several times observed that the bird—all the *Otidæ* I believe are given to this practice—instead of making off on the approach of man or other disturber, sometimes crouches down and attempts to conceal itself, and I believe that this manoeuver is practised when the wary bird has come to the conclusion that there is no imminent danger, and squats down simply to avoid being noticed. Once, when crossing a vast plain on the back of a camel in the Mirzapore district, I observed while yet a long way off, on a low range of hills to my right, the white neck of an old cock bustard, near the track we were pursuing. The bird, after eyeing us for some time, as we neared the spot where he was standing, at length squatted down behind a bush with the design of allowing us to pass by without noticing him; and I dare say the same trick had often been practised before, but this was the last time, for I had observed through a telescope the manoeuver of the wily Otis. Directing the native on the saddle behind me to take the guiding string of the camel and to pursue his course slowly for the purpose of attracting the keen eye of the old bustard, doubtless still peering at our movements from his place of concealment, I slid down on the off side of the camel into a soft clump of grass. We had gone through the ordinary performance of making the camel 'baith' or kneel down to enable me to dismount, our halting, the usual, delightful customary gurgling of the brute, as he doubles in his legs for the position required, and our whole behaviour, would to a certainty have awakened the suspicions of the crafty gentleman behind the bush, and induced him to seek safety in flight; but I was in hopes that by dropping down quietly in the manner I have described, with the camel moving steadily on to take the eye of our friend, I might, by taking advantage of the ground, steal within gunshot range. Most unfortunately I had only brought with me cartridges loaded with No. 5 shot, far too small for such game, but that could not be helped. There was a low, rocky ridge, behind where the bird had crouched. On the crest of this ridge I noted a certain high bush as a landmark, as nearly as possible opposite to where the bustard lay hid; and then stooping down to avoid being seen, I hastily turned back and retraced my steps in the direction we had just previously come. After walking two or three hundred yards, an opportune nullah branched off in the direction I wished to take; along the bottom of this I took my course, and at length fairly turned the enemy's flank by getting behind the ridge I have described, which screened me from view. Glad to be able to straighten my back again, I now rapidly made for the spot I had marked for taking my shot, and placing my mushroom-shaped hat on the ground, slowly peeped over, and there within thirty yards I had the satisfaction of beholding the bustard, 'fairly outdone, cunning as he was. He had his back to me, and had risen to his feet again, and was standing erect, with his eye still fixed on the camel, now some distance off; and doubtless was congratulating himself that he had been passed by unperceived. Poor deluded fellow! Chuckling to myself I strode towards him to gain a few yards for the small shot, and as he in the customary way of his species made two or three awkward flounders forward to get under way, and raised his heavy wings, I seized the favourable opportunity, and a single barrel dropped him dead.

THE OOBARA

The Punjab is, I believe, the only part of the Bengal Presidency where the oobara is found, and it certainly does not frequent our North-Western Provinces; but as it is a very fine game bird, and may perhaps be found in certain localities in Bengal that I am not aware of, I have included it in its proper place among the bustard tribe. I myself have never had the good fortune to come across a single one; but the following interesting notes, supplied by a brother sportsman from the Bombay Presidency, will amply make up for my deficiency as regards this particular bird. My friend's notes are as follows:

'So far as my experience goes, I have only met with the oobara in any numbers in Sind, although I have occasionally come across a few in the plains of Rajpootana, in the vicinity of Nuseerabad; but as a general rule I believe it may be assumed, that they are to be found in certain parts of Northern India where large tracts of sandy plains exist.

'In appearance the bird (which belongs to the otidæ tribe) is very like a small bustard, and about midway in size between that and a floriken. The plumage is of a mottled brown, white under the wings, and white-breasted. Both male and female have a ruff of white feathers round the neck, but in the male this is far larger, and in an old bird it has almost the appearance of a beard when in full plumage.

'The oobara, like birds of the same tribe, has only three toes. It feeds principally on wild berries, and in Sind on the leaves of the jamba or wild mustard, but will, I believe, also eat insects, and especially locusts. It is also justly considered a great delicacy for the table. The feathers of the oobara are much prized, and are extensively used in the manufacture of salmon flies; and there is also a small piece under the shoulder of the wing, which is much valued for ladies' hats.

'The oobara is a very shy bird, feeding only in the morning and evening. In the heat of the day it lies down generally under the shade of a bush. It seldom allows itself to be approached within range by a sportsman on foot, but can be stalked on a camel specially trained for the purpose, and then only in the heat of the day. When flushed it takes a long flight, and generally does not settle under half a mile, except in the very hottest part of the day.

'Oobara are generally to be found in pairs, but I have frequently seen as many as eight or ten together. They are very fond of ground covered with small sandy hillocks, in the shade of which they scratch about and repose during the heat of the day.

'The oobara I believe to be to a certain extent a migratory bird, leaving India about the end of February, and coming in with the first approach of the cold season. The earliest date I can remember to have flushed one was on August 28th; but this was unusully early, and the bird was a solitary one and quite out of condition.

'I had long laboured under a mistaken impression that oobara-shooting must be very tame, simply from the fact that they had to be potted off a camel's back; but when I once took to it, and found what care and patience were necessary before a bird could be brought to bag, I altered my opinion, and after this enjoyed several days' stalking, and came to the conclusion that the sport itself was not only good, but exciting.

'The way oobara are shot in Sind is off the back of a camel driven by an experienced shikary. The camels are specially trained for the purpose, and taught to circle round very slowly, obeying the slightest touch of the rein, and to stand perfectly still when stopped, so as to afford a steady shot. On coming on to a plain known to be frequented by oobara the shikary at once reduces the speed of the camel to a very slow jog-trot, and carefully examines the different tracks of the birds, which are to be seen in every direction. A good shikary will at once tell a fresh foot-print from one a day old, and how long since the bird has gone on. As soon as he has hit upon a perfectly fresh track he turns his camel short off, and proceeds to wheel round it in a large circle, until he has succeeded in determining that the track has not extended beyond the circumference of the beat. Being now assured that the bird is within the circle, he then proceeds very gradually to diminish the circumference each round, carefully examining each tuft of grass or small bush as he passes. The bird, the moment it becomes aware that it has been seen, crouches down, and, like the ostrich, fondly imagines that when its head is rammed into a bush its whole body is out of sight. The shikary eventually brings you to within twenty or thirty yards of a bird hidden in a bush; or, as I have often seen it, squatting down on the bare plain.

'The colour of the bird, which seems peculiarly assimilated to the nature of the country it frequents, renders it most difficult to seen and it often requires great practice and an experienced eye to discover it when squatting, even when within twenty yards. I have often been driven round a bird half a dozen times, with the shikary vainly attempting in a low whisper to point out where it was, before I could distinguish it myself. When the bird has been brought within range the camel is at once stopped (slightly to the right hand, of course, so as to allow a good shot), and if a hot day the bird will simply allow itself to be shot. I have, however, always found it far better to allow it to rise, feeling that there is far more satisfaction in bringing it down on the wing than in knocking it over on the ground. The moment it is dropped the camel is made to kneel down, and the shikary runs up to the oobara, and gives it the *coup de grace* by cutting its throat, after the manner of the followers of the faithful, and, slinging the bird behind the camel-saddle in less time than I have taken to relate it, jumps up and starts off after another. Sometimes, however, towards the afternoon, when the birds are beginning to get lively, they will not lie, and in this case the best way of getting a shot is to circumvent them in a very large circle, and while the birds' attention is taken up with the manoeuver, to drop off the camel without stopping, and then run in sharply on foot. In this manner I have shot many a bird that would never have allowed me to approach within range on a camel's back.

'The largest bag I have made is eight oobara in one day in Lower Sind, but I have frequently known from ten to fifteen bagged by one gun, and I am told even more than this may be shot in a day in Upper Sind, where they are far more plentiful. It depends, however, entirely on the training of the camel and the experience of the shikary, for although it may seem simple enough, it is by no means as easy as it looks to stalk one scientifically.'

The Bengal Floriken

Description

Male. Height.—About 22 or 23 inches.

    Beak.—1¼ inches, nearly straight, and rather stout.

    Iris.—Gold colour.

    Legs.—Yellow, 10 inches in length. Toes, three only, small in proportion to size of bird.

    Tail.—6 inches in length.

    Wings.—Short, not extending when closed to within 3 inches of end of tail.

    Long, loose feathers, hanging in the front of the neck, from the throat to the beginning of the breast, and also a kind of plume from the back of the head to nearly the lower part of neck. Black on the head and neck, and for two inches down the back; black down the breast and abdomen, where there is a reddish tinge. Lower parts of body to the tail, black. Back pale rufous, mottled with black. Wings white on outer parts, with here and there light brown feathers, and gradually merging into beautifully streaked brown, red, and black variegations; outer edge of three first primary feathers, black. Tail-coverts dull brown, variegated with streaks of a light fawn colour. Tail black, tipped with white.

    Female.—Rather larger than her consort; a peculiarity of the florikens, though not of the bustards. Though perhaps less strikingly, can hardly be said to be less beautifully marked than the male. She has no plume on the back of the head, though the feathers are longish in that particular part; no long feathers down the front of the neck.

    Head and neck, light fawn or yellow, variegated with streaks of brown across the feathers. The back beautifully mottled with numerous brown spots and splashes. Wings of a lighter tint. The primary quills darker, but no actual black or white on wings.

This floriken, one of the finest of our Indian game birds, is generally found in the open grass patches in the Terai, especially where the high jungle has been recently burnt, and the young grass shoots have again begun to sprout. In Assam it is very plentiful, especially in the sandy churs (or islands) covered with grass in the Berhampooter. Also in certain parts of the Purneah and Dinagepore districts; in the Nepaul Terai, Bareilly, and Philibheet country, the bird is common, especially near the banks of the Sardah. I have never once met with it in the Jhansie district, though the lesser floriken or leek was tolerably common, nor anywhere in the neighbourhood of Saugor or Jubbulpore.

I never shot the bird till in 1865, when quartered at Tezpore, in Assam, where, as I have already stated, it was common. On two occasions I killed five splendid birds in a day, and several times two or three. The old cocks, from their conspicuous black and white plumage, may be seen a long way off when stalking about, as they often do, near the edge of the grass jungle. When much hunted, they naturally become wild, but in out-of-the-way places I have had little difficulty in shooting them. It does not take much to bring one down; No. 3 shot is about the best size, and if a floriken rises within range, it can hardly be missed, so slow is the flight—a steady measured flap, not unlike the flight of an owl—hence its name 'ooloo moor' in some parts of the country.

It is a good plan always, if possible, to keep following up the same bird; whether it is that it becomes tired or not, I cannot say, but I have found that often at first, on the sportsman approaching, a floriken will rise very wild, and take a long flight; the same thing will likely occur a second or perhaps a third time, and then, after two or three of these flights, the bird will settle in some high standing grass, and not get up at all, allowing you almost to tread on it without moving.

A dog is most useful for this sport I have constantly marked a floriken down, but have not been able to flush or find him again through want of a good setter, or bustling spaniel, to discover his whereabouts. Moreover, a winged bird if it falls in thick cover, is sometimes most difficult to find without a retriever. Sometimes if on putting up a floriken the sportsman conceals himself immediately, the bird, after making a long flight, will gradually edge round and settle again near where it originally took wing.

It is a remarkable fact that in both descriptions of floriken, the hen bird is larger than the cock. This is not uncommon with certain kinds of hawks and falcons, but it is the only instance that I know of among game birds, except the quail and painted snipe. The floriken often makes a kind of cluck as it rises, and on settling I have frequently seen it run for some distance in a stooping position.

It is not unusual to come across two or three old cocks together by themselves. Their chief food is grasshoppers, locusts, centipedes, beetles, white ants, worms, and such like. I have observed them sometimes jumping up in the grass, and this habit often betrays them. When beating with a line of elephants in the Terai, I have seen twelve and fifteen of these birds in the course of a day, but usually they are not fired at, for fear of disturbing large game. The floriken is one of the very best birds for the table, having an excellent flavour, the meat being particularly rich and delicate. In Assam turkeys are scarce, but a floriken, at the head of a table, is an excellent substitute; and I have often been asked by ladies who had a dinner party coming off, to procure them a floriken. I have never once found the nest or seen the eggs of this bird. In Assam I believe they went away to breed, for at certain times of the year not a bird was to be seen for months together.

I always imagined that the Bengal floriken, like the bustard, was polygamous, but doubtless am mistaken when so good an authority as Hodgson tells us: 'The floriken is neither monogamous, nor polygamous, but the sexes live apart at no great distance.'

I have seen a peregrine falcon, belonging to a native gentleman, slipped at a floriken; but the latter, though twice flushed by beaters, persisted in throwing itself into the long grass so soon as ever the falcon attempted to gain a position above the quarry. At last the floriken took refuge in a very dense patch of grass, and baffled all our efforts to detect his hiding-place.

The Lesser Floriken or Leek

Description

Male. Height—About 12 inches.

Very similar in general resemblance to the large floriken, though much smaller, except in one remarkable point: the cock bird is decorated with a plume of beautiful thin curling black feathers, about three inches in length, commencing from behind the ear, and curling gracefully forward over the head; the tips of these feathers are flat and circular, and altogether this plume gives the bird a most elegant appearance.

The head, neck, breast, upper part of legs, under the tail, and abdomen, black; wings white, speckled here and there with light brown feathers, white also just above the round of the wings; the back, a beautiful mottled brown of a very gamey description. The eye full and large; a white spot on side of the head close to the ear.

Legs.—Rather long, and the toes same as bustard and larger floriken, viz. three, wanting the hind toe.

The Female is slightly larger than the male.

Of a pale yellowish brown, beautifully speckled on the wings. The down at the roots of the feathers, as in the larger species, is of a lovely pinkish hue.

Until quite recently it was a matter of constant dispute among sportsmen and naturalists whether there were not two distinct species of lesser floriken; but this matter has now been set at rest to the satisfaction of all; and Jerdon, who to my personal knowledge took great pains to clear up all doubt in the matter, in the 'Birds of India,' speaks most decisively and to the point, giving his opinion most distinctly, and putting the matter, I venture to say, beyond further controversy, that there is only one species, and I have not the slightest doubt but that he is right. What gave rise to this difficulty was, that certain brown floriken were shot in various parts of the country, which, on being dissected, proved to be male birds, and then some asserted that the brown and black were distinct species of floriken; but the fact was, these brown cock birds were simply specimens of *Sypheotides auritus*, in a plumage which adorns the bird at certain seasons of the year. It is in the breeding season, as with all birds, I believe, that the lesser floriken appears in full dress; and when in perfect feather there can hardly be a more lovely bird, with his coal-black neck and breast, and strikingly beautiful upper plumage; but at other times, as already explained, he wears a garb more resembling the female.

Young cocks not unfrequently, like young black game, have a patchy coat of black and brown, reaching the complete plumage, I believe, in the second year.

I have not met with the smaller nearly so frequently as with the larger floriken, though in many parts of the country it is very common; in the neighbourhood of Nuseerabad, for instance. The first one I ever saw was close to the race-course at Lucknow, but afterwards I shot several at Jhansie, where they appeared in limited numbers soon after the first fall of rain, at the same time as the bustard. After two or three months, they disappeared again till the same time, about the first week in August, the next year. In habits this bird much resembles the larger kind, as also in the description of food it subsists upon. It also, I have been told, has the habit of jumping up in the grass, and thus drawing attention to its locality. Jerdon and other writers have stated, possibly correctly, that this peculiar habit of rising in the air a few feet above the grass, flapping its wings, and then dropping down again, is for the purpose of attracting the females. I am by no means sure, however, that this habit may not arise from the birds jumping up at flying insects just out of their reach.

I have never personally observed the lesser floriken going through this performance, though I more than once noted the larger kind doing so in Assam, and always about July and August, which is said to be the breeding season. But on the other hand it is the time of year when flying insects are most numerous. I may add that floriken are particularly addicted to feeding on the cantharides.

It is often found in vast plains covered with a short light-coloured grass; and at Jhansie I flushed and shot it on rising ground covered with grass and short scrub bushes, just where one would expect to find bustard. I once saw four together, but never more, and generally speaking one rose here and another there, at some little distance apart. If within distance, there can hardly be a simpler or more easy shot. It often, when alarmed suddenly, utters a croak. The flight is just like that of the larger kind, and the bird has the same habit of circling round, and also stooping and running with its tail held erect after settling. Like *S. bengalensis*, the flavour of this bird is justly considered most excellent. The flesh of the body is dark; that of the legs a lighter colour. It is a curious fact, but I believe strictly the case, that although the two descriptions of floriken are found over large tracts of country, yet nowhere are they to be met with in one and the same spot. The same remark applies also to the black and painted partridge, and the two varieties of jungle fowl. Jerdon tells us:

'The floriken lays three or four eggs of a thick stunted oval form, very obtuse at the large end, and of a dark olive colour.' I never have seen a nest, nor could I ever procure an egg, although I constantly have offered shikaries a reward to bring me one.

Excerpted from Large and Small Game of Bengal and the North West Provinces of India *(London: Henry S. King, 1876)*.

P. D. STRACEY

Shooting Elephants

Unlike in Africa, elephants in India were caught more often than shot at. The Raj imposed controls on the shooting of elephants so that there would be enough to catch for the army, for transport and for hauling timber. But crop-marauders or 'rogues' were hunted down and tea or coffee planters could do as they liked on their estates. P. D. Stracey was a forester with a long acquaintance with the country's largest animal, both in captivity and in the wild. In his book, Elephant Gold, *he recounts his narrow escapes when he hunted elephants. Stracey not only stayed on in India after 1947 but also became one of the few foresters to take a keen interest in game management and wildlife protection.*

Elephants can be very destructive creatures and in parts of Assam they are really troublesome. But this is only to be expected in a country where the forests are so closely interlocked with inhabited areas, forming a hinterland around the fertile valleys which have been cleared for cultivation and tea—a natural buffer, as it were, between the mountains and the plains. Here the elephants can live unmolested except for the periodic catching operations, though the rapid clearance of the jungles outside the reserves has exposed them increasingly to contact with man. It is this encroachment on their habitat which is mainly responsible for the existence of crop raiders and rogue elephants.

Just about the time I had finished with elephant catching, Milroy introduced the Elephant Control Scheme and he urged me to take to elephant destruction under it. I was reluctant to do so until I could obtain a good rifle, as I had had such disappointing experiences with the rather inadequate weapons in our kheddas. At that time, we had only an old 8-bore black powder rifle and although I dug out a couple of .577 low-pressure cordite rifles from the Deputy Commissioner's stores, they were not really satisfactory. Eventually I was posted to Goalpara, the crack division of Assam and in those days still a hunter's paradise. Just about this time I managed to get a heavy, high-velocity rifle and, although Milroy was now dead, I felt I had to accede to his wishes and take up elephant control.

There were several crop-raiding elephants in the area, living in the stretch of *sal* forest which had villages all along the edge and tongues of cultivation projecting into it. Only troublesome elephants could be shot, though according to the licence any mature male elephants wandering by themselves in the vicinity of crops were potential victims. I was never very happy about this rule, as it permitted one to follow up the tracks of any solitary or semi-solitary animal near villages and kill him, even though he might not have been really troublesome. It was compulsory to shoot a makhna for each tusker. The scheme was devised by Milroy to supplement the killing of troublesome elephants with the prospect of a reward in the shape of the tusks. On the whole it was a sound scheme, except that in Assam the proportion of tusked to tuskless males is not as much as fifty per cent, so that an elephant shooter is generally hard put to find as many makhnas as tuskers. But Milroy was an astute man and knew that if he did not insist on makhnas being killed, sportsmen would only go after the tuskers. About forty to fifty elephants are killed every year in Assam under proclamation and the control scheme, as well as a few in stockade operations, and this seems to look after the problem of the troublesome male elephant fairly well.

At the time when I first went to Goalpara, I wish someone had suggested that I should dissect elephant skulls, for it would have saved me several unhappy experiences. If one does not know the vital shots, it is impossible to kill elephants cleanly. Looking back, I am a little surprised that the old Doctor Babu did not suggest practical experiments to supplement the quite inadequate drawing of the outline of an elephant's head, with three dots on it to indicate the brain shots, which appeared in the standard reference book on elephants and their diseases by Evans. I did remove the heart from a poor prostrate cow elephant which I had to destroy, but somehow it never struck me to do the same with the brain. I had climbed into a tree in order to shoot downwards at the place where I thought the heart would be, but I was surprised later to find it hidden almost between the forelegs. I was also astonished at its large size—almost as big as a football and so hard with muscle that it was like gutta-percha.

The punch of a heavy bullet fired at close quarters into the temple of an elephant will stun, even if it does not kill immediately. For its perfect execution the elephant must be facing at an angle of forty-five degrees away from the shooter and it is worth holding one's fire to get this position. One sportsman who had shot a number of elephants used to send a man round to attract the elephant's attention and if it was in a paddy field at night, to flash a torch on the animal to make him face in that direction. It is remarkable how long an elephant will lie stunned on occasions and this has misled many a hunter. I learnt by experience that after firing the first shot I must attempt to get up to the fallen elephant immediately in order to be ready for a second shot should the animal not be dead.

A useful shot when an elephant is lying prone is the back-of-the-brain shot, in the spot where the two lobes of the skull fall away to the hollow of the neck. This should be taken with the rifle held parallel to the ground, the bullet being directed downwards into the head. I worked out this shot when studying the sectioned skulls of elephants. The skull is only half an inch thick at the back of the brain, just before the point where the spinal cord begins, and I think one could reach this through the thickness of the top of the neck with a powerful pistol. The shot that I favour least of all is the so-called earhole shot; apart from the difficulty of locating the earhole passage correctly, this has to traverse about eight inches of hard bone before reaching the brain.

I got my first elephant in Goalpara just after I was married. One winter night I was awakened and told that a troublesome tusker that I had been looking out for was feeding in the fields some two miles from the Raimona bungalow. I left my wife still asleep and went out into the pale light of a waning moon. When I reached the field it was about three o'clock in the morning and the elephant was not to be seen. I sat by a fire with the men who were watching the fields and at first light I took up the tracks of the crop raider. They were clearly visible in the dewy grass and led into the jungle. Within a short distance in the forest we saw him leisurely moving away. I had sent for my elephant Mohan Prasad but kept him some distance behind me while I tracked on foot. The forest was open and as I was trying to outflank the wild elephant to get a head shot he heard me and turned, extending his ears in a challenging manner. At that time I had no clear idea as to the position of an elephant's brain but I fired at the frontal bump as he faced me some forty yards away. To my chagrin he merely rocked back on his feet, staggered and then dashed across my right front, going like an express train through the jungle before disappearing into some high grass. It had been impossible to take a second shot.

When I had recovered my breath I called up Mohan Prasad and mounting him, cautiously followed the track of the wild tusker. Standing up on the saddle I could see right over the top of the twelve-foot-high grass. There he was under a tree, facing us and obviously waiting for us. I had read all there was on the subject of following up wounded elephants and I was not going to fall into any trap. My mahout and I remained silent, watching him, and after a minute or two the big tusker turned and moved away towards the river which lay between me and the main forest block. We made a rapid detour and when we reached the river I jumped down and with a couple of villagers who had followed me, went down into the dry sandy bed and began to look for his tracks. While we were scouting about I happened to look up and there was the elephant with his head sticking out of the jungle watching us from the farther bank. We were walking right into him. I threw up my rifle and dropped to my knee, but before I could fire he turned and went off. Then followed a long chase. He moved steadily, but halted now and then to lie in wait for us and though I was eager to finish him, my inexperience was tempered by caution which would not allow me to go up to him in the thick stuff where he invariably paused, which was just as well.

We spent nearly two hours on his tracks until I was longing for breakfast. We stopped in a little open space to dry out dew-drenched clothes and to hold a council of war. It was quite obvious that I had not seriously wounded the elephant and I was very dejected. I had missed the small passage to the brain which is concealed under the frontal bump at the upper end of the elephant's trunk. This is a very misleading shot and if the elephant is not standing perfectly still or if one is not absolutely accurate it is easy to miss the actual opening. Its diameter is no bigger than that of a cigarette tin and it is surrounded by comparatively hard bone. We decided to follow the elephant a little longer and started off again. That we were dealing with a cunning animal was obvious and I did not relax vigilance.

We had hardly gone a couple of hundred yards, the trackers in the lead followed by myself and a man carrying my rifle, when with a trumpeting scream the elephant shot out of some cover straight into our midst. The trackers scattered, the leading man leaving his cloth hanging on a small bush. I sprang to one side, striking my right leg hard against a fallen log and stumbling over it. The tusker stood there squealing with rage and uncertain whom to attack. I turned for my rifle and mercifully my gunbearer was just behind me. I fired into the elephant's temple at about thirty feet and knocked him down like a ninepin with the heavy .475 rifle. I immediately replaced the expended cartridge and shifted my position as he lay stunned, only the tip of his trunk moving a little. I stood over him and was determined not to let him get away this time.

After a while which seemed an eternity to me, the tusker started climbing to his feet unsteadily. I let him get up. It was obvious that all the fight was out of him and he started to turn away like a feeble old man. I raised my rifle and fired into his head once again, at where I thought the brain was. I reloaded and fired again but still he would not go down. It was like a boxer manoeuvering for the kill and repeatedly punching his reeling opponent. He was staggering forward like a drunken man, I within a few yards of him dodging about in the jungle to place my bullets, he intent only on getting away. The rifle was hot in my hands and he was momentarily resting his tusks in the fork of a small tree, when all of a sudden one of the bullets found his brain and he threw up his head with a jerk before crashing to the ground stone-dead. I felt shaken and nauseated then, though strangely enough in those last few moments I had been as cool as ice, although tensed up at the same time. I remembered seeing through the corner of my eye my trackers searching around in the jungle for the cartridge cases as I reloaded and fired, oblivious to the death-drama that was being enacted before them.

The elephant was a menacing-looking animal, with that peculiarly enlarged base of the trunk where the tusks begin which somehow seems particularly formidable in elephants. His tusks were heavy but not symmetrical. I was proud that I had got him after all, though it worried me that I had had to expend as many as seven bullets on him. His head was brought in on a bullock cart and several months later, when the skull was clean, I used a crosscut saw to cut up the skull in sections, very much like cutting wedges out of a cake. There at last was the brain cavity. Nine to ten inches across and four to five inches deep and set well back in the skull, it resembled nothing so much as a giant clover-shaped bun, the ear holes of the skull draining, as it were, the shallow bottom portion. The correct shot, obviously, was the half-front position between the eye and the ear and just above the cheekbone, into the natural hollow of the elephant's temple. At last I knew where the brain of the elephant was and if in future I made a mistake, the fault would be mine. Years later, when I came to be head of the Forest Department, I ordered that sectioned elephant skulls should be maintained and displayed in places where shooters could learn how to kill an elephant cleanly.

During the five and a half years which I spent in Goalpara I was shooting hard to keep myself busy during my spare time, but I managed to combine work with pleasure. Tracking troublesome elephants took me into out of the way corners of the forests, which proved useful in my supervision of work in the division. I shot only crop-raiding elephants and cattle-lifting tigers, with the exception of one elephant and one tiger which I suspect may have been harmless.

Many a night I went out into the paddy fields at the call of distressed villagers to scare away elephants. Most times a shot over the raider's head would be enough to scare him away, but sometimes only a twelve-bore loaded with small shot would be successful. But if they were really determined crop-raiders, there was nothing to do except kill them.

I shot elephants by daylight in the forest, as well as in the fields at night, but the surest way was by tracking the marauder in the daytime. In this way I shot the Gangia tusker, which gave me my record pair of tusks.

He was a persistent and cunning raider and, on hearing that he had been coming out for several successive nights, I camped in the village specially to get him. One side of a long strip of cultivation was open to the forest and I had a small hut built on the edge of it. It was a night of the full moon in November and he came out early while I was having my dinner—a great mass with two gleaming white tusks, gliding like a ship out of the jungle towards the paddy in the light of the rising moon. He moved like a thief on tiptoe, slowly and very stealthily, and I crouched in the shadow of the trees literally holding my breath. I let him get well into the paddy and start feeding before daring to move. Then I started looking for a line of approach, but there was none. I knew it was useless trying to get up to him directly, so I made my way cautiously to the top end of the cultivation. I was now between him and the jungle. Where the ricefields petered out, I at last found a line of approach through an abandoned house site. My rifle had a large torch light clipped to the barrel and I held it close to my side to avoid the reflection of the moonlight on its bright surface. I crept stealthily towards the feeding elephant but I could no longer see him and I stopped in the shadow of a small mango tree to take a breather.

I was wet with dew and could feel the leeches on my legs, but I concentrated entirely on the elephant. I could locate him by his rhythmic feeding noises—'swish, swish, swish' as he beat the pulled-up paddy on his knee and then a pause as he put it into his mouth. I listened intently and suddenly the sounds appeared to be coming nearer—he was feeding towards me. It was almost too good to be true. Then I caught sight of him as he cleared the bushes to my left. I thought I noticed a certain peculiarity in the way he fed. He never turned his back on the village and always faced in the same direction. Then I lost sight of him and I waited patiently. Suddenly, from the swishing sounds which were my only guide, I sensed that he was moving away from me. Realizing that I would never get a chance again, I controlled my excited shivering and braced myself for the show-down. On the tips of my toes, placing my feet carefully and avoiding fallen leaves and shrubs, I went forward through the patch of abandoned homestead to get as close as possible to the elephant. My luck was holding, as the cover was just right.

I halted behind the last large bush which was between me and him. He must have been a hundred feet away and once I saw his trunk come up above the cover as he tested the air. Then gradually I began moving to the right and simultaneously raised my rifle. I was almost clear of the bush when suddenly he stopped feeding. I knew at once that he had detected me. 'Now or never!' I thought and stepped to the side for the shot. At the same moment there was a snort and the elephant turned away. I pressed the button of my torch as I pulled the rifle to my shoulder, only to be blinded by the reflection of the light thrown back by a smaller bush which I had not noticed before. Immediately I reversed my movement, running to the left so as to sight the elephant. He was in full stride for the jungle by now and I aimed for his left temple, guided by his eye in the beam of the torchlight. I pulled the trigger once and when he did not go down, a second time. But he hardly faltered in his stride and it seemed as though I had missed him completely. I reloaded my rifle, running after him and shouting to make him turn and charge as I had been taught to do, but with no effect. I had just time for one more long shot somewhere in the direction of the side of his neck and then there was only the splashing as he crossed the stream at the edge of the black enveloping jungle.

I stood there, exhausted and crestfallen to the verge of tears, my clothes damp with sweat and my legs soaked to the knees, while the villagers gathered round. I felt convinced that I had lost the biggest tusker of my life. I explained to them what had happened, going over in pantomime the drama of the last hour or more. They tried to console me but I could sense their disappointment. Mine was worse, because they were my villagers and I had let them down. However, there was nothing to be done and after I had sent off a messenger for my elephant Mohan Prasad to be at the village at the crack of dawn, I went back to my hut. Needless to say I spent a restless and nearly sleepless night, as I went over again and again the events which had ended in my losing the giant tusker. I could only conclude that the elephant had spotted the reflection of the moonlight on the bright metal of the torch on my rifle, though why he had let me get so close to him was a mystery. There was also that strange way of feeding in the paddy field—never turning round and always facing in one direction. I was to know the answers the next day.

Early next morning my faithful Mohan Prasad was there and we took up the tracking from the spot where the raider had splashed across the stream. It was rather difficult at first, following across the grazed land and light grass round the village where there were many old tracks, but once we got into the thicker stuff it was easier. We had not gone very far when the realization dawned on me and my mahout almost simultaneously that there was something unusual in the tracks left by the wounded elephant. They were far too broad to be normal and soon we realized that he was dragging his left leg badly. Immediately my hopes rose sharply. We knew now that he was badly hit and we hoped to catch up with him soon. Sure enough there he was, hobbling along within a mile of the village. He had hardly covered any ground at all during the night.

He was magnificent standing there in the forest facing us with his ears spread forward, full of fight in spite of his crippled shoulder. It was quite obvious that he had a broken bone somewhere and I could not have shot so badly as all that the night before. I could only guess that while attempting to swing the heavy rifle against my natural swing as he ran to the left, I had gone behind and my shot had caught the shoulder, high up in line with the eye at which I had been aiming. I was eager to finish him off, both to save him further agony and because I wanted to have done with it. It was one thing stalking him at night in the bright moonlight, but quite another having him before me in that crippled condition and at my mercy. He went down fighting, trumpeting furiously after the first shot which did not find his brain, but I made sure of things when he struggled to his feet.

When I examined him I found that his right eye was blind. This then was the explanation of my being able to get up to him so close in the bright moonlight, for I had been on his blind side the whole while. It also explained why he always fed facing in one direction, his blind eye towards the jungle from where he could expect no danger and his sound eye towards the village. I had indeed been lucky, both in my approach and in the shot which had anchored him so close to the village. But night shooting never appealed to me after that—there are too many uncertainties.

The elephant was an enormous animal, as befitting the possessor of such a large pair of tusks. He measured 10 feet 6½ inches from sole to shoulder as he lay and was in fine health, though old. I reckoned he must be nearly a hundred, judging partly by the tips of the tusks which were beginning to show signs of decay. They measured 7 feet 1½ inches and weighed 77 lb. each; the thicker of the two was 18¾ inches in circumference, an inch and a quarter off the world record for the Asian elephant. The longest recorded are a pair of tusks from the Goalpara division of Assam, not very far to the east of where I shot my big tusker; they measure nine feet two inches. The heaviest are from an elephant shot by the Raja of Talcher in Orissa, weighing 120 lb. and 114 lb. respectively. As far as I know, this is a world record for the Asian elephant. The next biggest are also from Assam; a pair from an elephant shot by the late Lord Lytton measured eight feet nine inches and eight feet two inches; while each tusk of another pair which I collected in Goalpara, just north of where I got the Gangia tusker, were eight feet long. I wonder if it is more than a coincidence that the three heaviest pairs recorded from Assam in recent years are all from the Goalpara district, south of the Bhutan Hills, and all within a radius of fifty miles. This region produces the best elephants in Assam and Milroy believed that this is due to the rather dry feeding and unequalled range available for the herds, which move up to the hills during the rains and down to the plains in the cold weather.

Although the Gangia tusker was really blind in one eye, it is extraordinary how short-sighted perfectly normal elephants seem to be at certain distances. I once took my koonki right up to a large wild makhna which was feeding in some high grass on a river bank. We were walking in the dry, sandy river-bed and I was astonished how close he let us come; then suddenly he got our wind and was off in a flash. On another occasion I had a load of young ladies on my elephant when in the distance we saw a makhna, standing near the edge of a swamp and stuffing water hyacinth into its mouth. I approached with some trepidation, but he took absolutely no notice until we were within twenty or thirty yards, when he suddenly went off and was soon lost to sight in the high grass. On another occasion in Kaziranga a party of six or seven persons on two elephants approached a wild tusker feeding in a swamp to within forty or fifty yards across open ground and it was not until the mahouts clapped their hands that the tusker made off. The fact that it was a sanctuary might have accounted for the elephant's lack of fear, but there are other instances which would indicate that the species has very poor sight. On one occasion a friend and his wife were having a picnic lunch on the Diyung river in the North Cachar Hills when an elephant made its appearance on a sand spit about a hundred yards away. Perry, who was the Deputy Commissioner of the district and a person who has had considerable experience with wild elephants, says that he tried approaching him and succeeded in getting to within a remarkably short distance. The tusker took no notice of him until he was close up, although he was right out in the open and there was nothing to obstruct the animal's vision.

What is more, the several remarkably narrow escapes which elephant hunters have had when charged, coupled with my own observations on the behaviour of wild elephants in certain circumstances, makes me inclined to think that they have difficulty in focusing on an object which is not at a certain optimum distance away. Apart from my own narrow escape which I attribute to this weakness, at least two sportsmen say they owe their lives to some unaccountable failure on the elephant's part when by all the tokens they should have been run down in the charge. A tea-planter named Baldwin was once faced by a charging elephant, which he had provoked into turning on him in order to secure the head shot, when his rifle jammed and he was left helpless and convinced that his last moment had come. He could not account for the elephant missing him, as he stood still in its path. Walsh, another tea-planter, described to me how he was following an elephant in thick bush when it suddenly whipped round and charged him; he tripped as he took a step backwards in the act of raising his rifle and sat down at the base of a tree and perhaps this may have made the elephant lose sight of him and charge on and past him. Charlie de la Nougerède had wounded an elephant in thick bamboo jungle in the Garo Hills when it turned on him and brushed past him, knocking him down and walking over his shoulder and arm. Would there be such a high proportion of near misses with any other animal? And could it be accounted for by the setting of the elephant's eyes almost at the outer edges of the wide forehead, rendering it incapable of focusing them on objects outside a certain optimum distance?

Long before I took up elephant control work, of course, I was sometimes forced to shoot elephants because they were too sick and weak to stand a chance; and occasionally it was necessary to shoot them in a stockade because they were uncontrollable. This did not often happen, since it was generally possible to release such animals. Large bulls, in fact, seldom get caught in khedda operations; they usually manage to escape out of the ring of drivers, who will do their best to exclude them from the stockade. Apart from the damage they can do both to the koonkis and to the other captives, particularly the young ones, there is the trouble of disposing of them if it is necessary to shoot them. It is no joke trying to get rid of a four-ton elephant's carcase within the comparatively small area of a stockade. Where there are nearby villages of tribal peoples, some of whom eat elephant meat readily, there is no problem as these people will come down from their hills and cut up and remove a dead elephant in no time. But elsewhere labour must be employed to bury the remains at some distance. The resultant fouling of the stockade may be so great as to spoil the chances of further catching for the rest of the season.

In connection with mela shikar I mentioned the biggest elephant I ever saw captured by this method, a solitary more than nine feet high at the shoulder which took three koonkis to bring him in and three phands to hold him. Normally such an animal would have been left severely alone, but apparently the Phookans had wanted a large elephant for a koonki—although I did not know this at the time—and their phandis had tackled him successfully. In due course I auctioned him, but there were no bidders for what was obviously a handful and I could not sell him. A neighbouring planter, a keen conservationist who had played a large part in the creation of the local sanctuary, came to the depot and when he saw the animal he expressed the desire to purchase him as a shikar elephant. We settled the price, which was double what we were normally getting for such large makhnas, and the deal was closed. When the Phookans learnt that I had parted with the elephant they were very upset, but it was their fault not to have been frank with me from the beginning. Probably they had expected to get him at the customary cheap rate of three hundred rupees.

That evening he was pegged out and tied by his hind-legs to a tree, with the neck rope or phand stretched to another tree in front of him. He struggled violently and I had a suspicion even then that he was stretched out too much, but I was inexperienced and this was my first independent khedda. Next morning, when we untied him to take him down to the river, we found him tottering and weak in the hind-quarters. He had obviously hurt himself severely during his struggles by tugging against the ropes. We got the biggest koonkis in the depot to prop him up and led him gently down to the river. The Doctor Babu and I followed with fear in our hearts, for he was staggering like a drunken man and could hardly move his back legs. When he reached the river edge he lurched forward, with his trunk stretched out and collapsed on the water's edge. We tried to lift him with the biggest tuskers, but failed. We untied all his ropes in the hope that he would try to get up on realizing that he was free. Men brought him titbits of sugar-cane and plaintain leaves, but though we practically put them into his mouth he would not take them.

By this time the whole depot had heard the news and the riverbank was crowded with people, all elephant men and all deeply concerned for the plight of the big makhna. I made them stand away and they remained silently watching. It was possible to go up to him and, provided one kept away from his questing, restless trunk, it was quite safe to touch him. He was a mountain of an elephant and we had the opportunity of examining him closely. The day passed slowly. The Doctor Babu looked up his books and came out with a long name—I think it meant some form of paralysis. He said he could do nothing. I had my own suspicion that it was complete rupture of all the ligaments and muscles in the loins, but I secretly hoped that he would get up during the night and walk away to freedom.

The next morning there he was still lying in the same place and our fears became worse. I began to look up Evans' *Care and Treatment of Elephants*, for the vital shots with which to kill an elephant. The only weapon I had in the camp was an old eight-bore B.P. Rifle, a weapon handed down by a succession of khedda officers and seldom used. I did not even know whether the large paper-cased, black-powder cartridges were still fresh. When I broached the question to the Doctor Babu that we might have to destroy the elephant, he would have none of it. He was a Brahmin and hated taking life. I said to him, 'Doctor Babu, we cannot let him lie there starving and it is my duty to destroy him.'

However, we let him remain there a second night and next morning I made my preparations with a heavy heart. Killing an elephant in the upright position is difficult enough, but to kill one cleanly as he is lying on the ground is a task which at that time utterly defeated me. I could only think of the obvious heart shot, as at that time I did not know the shot into the back of the brain. This is the spot which the mahout pokes at with his goad, while he sits on the neck of the elephant a little farther back. A bullet placed here, parallel to the legs and trunk of the elephant as it lies, has to penetrate some ten inches of flesh and muscle of the neck and then there is a thickness of only a quarter of an inch of bone enclosing the brain.

I decided to try the heart shot and I called for a packing case. The silent ring of spectators seemed to be offering up a prayer for the dying makhna. I thought that the heart would be easy to reach, but all that took place in response to the shot was a spasmodic and violent twitching of the legs and body and a gush of blood out of the nostrils, which stained the water red. I fired again and then again. I was beginning to panic. I went round to the river edge and lay down in the water for a frontal brain shot. I must have fired five shots. Finally he lay still, with the blood oozing out of his trunk and his eye glazing. I staggered to my tent and was violently sick. Orwell has written a most moving account of killing an elephant in Burma under similar circumstances.

For the next twenty-four hours I felt ill. I had a great blue bruise on my right shoulder, the result of the kick of the heavy rifle. The same rifle came to be known later on as the 'jaw-breaking' rifle, because it nearly broke the jaw of one of my assistants and cut his cheek and upper lip badly. He had been afraid of it. I had also been afraid of it, but only for the elephnat's sake.

This task of destroying a sick or old elephant is a most painful business. Somehow if the first bullet does not go home—and this often happens—the situation gets out of control and one generally has to fire again and again. Milroy and the Director of the veterinary department at the time once had a dreadful ordeal in destroying an old cow elephant which they felt was better put down. Something went wrong and they could not kill her, try as they might. Their ammunition ran out and they had to send in thirty-five miles from Kulsi to Gauhati for more. I think it took over twenty shots to finish the job and by that time they were both in a state of near collapse.

I had an equally unnerving experience with Phookan's famous 'sakhni', the name given to female with tusks. This was a rarity that not even the oldest elephant men could remember having seen or heard of before and she was without doubt the most extraordinary animal I ever saw captured. She sent our biggest koonkis flying and we had to try and control her by putting in Joyram Goonda, a powerful makhna which happened to be in full musth at the time. A young phandi had volunteered to take on the double job of manoeuvering him and subduing the wild sakhni. She put up a good fight, but eventually Joyram cowed her, his temper aroused by his musth condition. Even so she required two elephants to bring her out of the stockade. There was such excitement when it was discovered that she was a female with tusks that Phookan made up his mind to keep her for himself and he refused to sell her at any price. The elephant men said she was barren and she was certainly very ugly, with rather stout tusks projecting about a foot from her lip, and she was as savage as she looked.

From the start Phookan had bad luck with her. The next winter he was catching on the north bank and took this elephant with his other koonkis across the Brahmaputra. She gave a lot of trouble when swimming the river and had to be roped to another koonki. It was while she was on the north bank that she killed her first victim, her own grass cutter who apparently had approached her too silently, crouching low in the grass while trying to make out her footprints from among those of other elephants. Three months later she shook off her mahout when she panicked over something that ran across her path and promptly trampled him to death. When Phookan was bringing her back to the south bank in the early part of the winter for my last khedda, she killed a third man. This time it was an umbrella which slipped off her back and which startled her so much that she threw her rider, whom she promptly attacked and crushed to death with her forefeet. After that the men refused to have anything to do with her and the wretched animal was tied to a tree and abandoned.

There was something definitely wrong with the elephant and Phookan decided to destroy her. This was a most unusual thing for a koonki owner to do, but I suppose he was convinced that the sakhni was an unlucky animal, better dead than alive. He came to me because he had no suitable rifle. We discussed it for a while and when I saw that he had made up his mind I agreed to go out and perform the unpleasant task. I motored out to the Brahmaputra early next morning and had to wait for a ferry to take me across to the Majuli island. It was a cold rainy day in December and I found the dispirited elephant men huddled round a small fire. I asked them to show me where the sakhni was tied and one of the men went with me rather reluctantly. I was shocked at her condition. She was up to her belly in mud and filth and had not been untied for a week. A few plantain stems and leaves were lying about.

The doomed elephant was in a furious temper. She lunged at me when I approached her and would definitely have attacked me if she had not been tied to the tree. It was the first time I had seen her after her capture two years ago and she looked more vicious than ever. For a moment I wondered if there had been some fault or neglect in her training—but then Phookan was very thorough in such matters.

But I had come for another purpose and with the usual tight feeling in my chest I took up my position at an angle and fired at her head. It was difficult, because she kept moving her head sideways trying to face me. It was not surprising that she failed to drop to the shot. I cursed myself and began to get jittery. Next time I tried the heart but here again it was difficult to judge the angle, bogged down as she was in her own filth. She went over sideways and then got up and beat her head in rage and pain against the tree to which she was tied, trying desperately to get at me. Thank God my third shot, which I again directed at her head, succeeded. I got back late that night with a violent attack of shivering, the result of the tiring trip and the unnerving experience, plus my old friend, malaria. I never want to repeat such an elephant execution again. Phookan gave me one of the sakhni's tusks as a souvenir and I had it for several years, until it was stolen from me.

I did not kill very many elephants during my shooting life of about fifteen years. After the first excitement had worn off, I found myself growing careless. So careless, in fact, that I nearly met my end at the hands—or rather tusks—of the tallest elephant I ever encountered and the last one I ever followed.

My mahouts and I had suddenly come upon his tracks high up towards the Bhutan border and the only reason why we decided to follow them was their enormous size. I was always looking out for that eleven-foot elephant and this looked like it. The forest floor was dry and tracking was slow; he had obviously been going fast, not stopping to feed, and we spent the whole afternoon without catching up with him. At the time I was camping at the end of the forest tramway a couple of miles from the Bhutan border; the next day was Sunday and bazaar day and we had to get back to base, I to supervise the payment of labour and the men to do their weekly purchases of foodstuffs. However, we decided to give it another try early the next morning and if we did not catch up with him soon, to abandon the tracking.

Dawn was just breaking as we crossed the sandy river-bed where we had left his tracks the previous evening and we spotted a tiger having his early morning drink before retiring for the day to the forest. My men said it was a good omen and that we would come up with the elephant. The tracks led us up to the abandoned Sanfan bungalow and the elephant had walked all around it. Then he had stepped over a large fallen tree lying at such a height that only a very big elephant could have negotiated it. My mahouts whistled softly and whispered that he was a prokanda (enormous) elephant. I was tremendously keyed up as I visualized a record elephant. Just on the edge of the jungle was the spot where he had laid down, probably in the early hours of the morning, and the earth was moist with the dribble from his mouth. I jumped down to examine the impression of his tusk in the damp earth and my spare mahout, whom I used to bring out as a tracker, did the same. He pointed to the elongated depression, about three inches wide and said in a whisper that the animal was a tusker. I shook my head and whispered back that I thought he was a makhna—the impression was too narrow to have been left by the tusk of such an enormous elephant. There is a difference between the tusks, which are tremendously enlarged incisors, and the tushes, which project directly downwards, unlike tusks which normally turn upwards. This difference in the two types of male Asian elephants is an obscure phenomenon and the pattern is not the same everywhere. In Assam makhnas are very common, while in south India they do not constitute a very high proportion of the males of a herd. In Ceylon strangely enough, almost all the males are tuskless.

In this case there was only one way to settle the argument and we went on. Shortly afterwards I should have noticed that the elephant's tracks had begun to zigzag sharply, a sure sign that he knew he was being followed, but my mind was too full of the prospect of a record elephant and a makhna too. We passed through some marshy ground with high grass and as was my practice, I climbed back on Mohan Prasad and stood up on his pad to peer over the top. Immediately after we ran into some very thick jungle matted with thorny creepers and after trying to negotiate it on the elephant, I decided to get down and pressed my mahout's shoulder to tell him my intentions. By this time we had become a perfect team—Mohan Prasad, Ontai and I—and they would go with me anywhere and do anything I wanted them to do. Handing my rifle to the mahout after pulling the safety catch back from the firing position, I jumped down as lightly as I could while the spare mahout slid down the tail of the elephant. Taking back my rifle, I motioned to Ontai to stay where he was and went forward with my tracker. Almost simultaneously we came into a slight clearing with an indistinct sort of path running through it and the man went forward to try and pick up the trail. There was an eerie feeling about the place, which was one of the loneliest and wildest spots I had yet come across in my division, what we foresters call a 'non-productive blank' given over to rank growth and climbers and seldom visited by anyone—the perfect hide-out for solitary elephants.

I was following a little behind with my eyes to the ground trying to pick out the tracks of the big elephant, when suddenly I noticed a slight movement in the jungle ahead and before I could catch my breath out stepped on enormous elephant, his ears cocked and his trunk questing us. He was surely the biggest elephant I had ever seen, but his tusks were miserably thin and not even long for his size. No wonder the impression they had left in the earth where he had slept was so misleading. In my surprise and because my mind was bemused with the thought of this record elephant, which I hoped and half expected was a makhna so that I could shoot him, I did something which no elephant shooter should do: I turned to where my mahout Ontai was on Mohan Prasad and said, in Assamese, 'Hey! It is a tusker after all!' Hardly were the words out of my mouth and I had turned to face the elephant when he was in full stride, coming at me absolutely silently but with deadly purpose. My voice had been just the indication he had wanted of our location. I had thrown away all the advantage by speaking. 'Now for it', I thought as I raised my rifle to my shoulder, automatically stepping a little to the side to obtain protection from a tree which was standing near me. I sighted on his frontal bump as he came pounding along and when he was about fifty feet away I squeezed the trigger. There was no result. I checked for a fraction of a second and pulled the second trigger, still with no result. Then my mind went blank. I had half lowered my rifle and I am sure my mouth must have been agape. The tusker was almost on me. His head was lowered and he was skidding to a stop. In another second he would have grabbed me. I could look right into his eyes (which I remembered afterwards were light coloured, the sign of a dangerous elephant) when I came to life. Dropping my rifle at his very feet I dived into and under some tangled creepers which draped the tree.

Instantaneously I picked myself up and dashed madly forward, up the very track the elephant had made as it charged me. Fear lent wings to my heels. I was wearing light-soled rubber and canvas boots. I had been a sprinter in my youth, with exceedingly good reflexes which never let me down at starts, and I had at one time cherished ambitions to run for India in the Olympics. It was that long-dormant instinct, that natural talent for a quick start and burst of explosive energy when required, which saved me now.

But as I ran I expected every moment to be grabbed from the rear. It was a forlorn hope I told myself, with a frightening clarity of vision, it was the end! When I reached the wall of jungle from which the tusker had emerged I took an automatic header into it, screwing up my eyes and lowering my head against the cruel thorns. Scrambling to my feet I dashed on, tripping, falling, tearing my way desperately through the cane and thorny climbers. All the while I was unconsciously making a circle towards the high bank which I knew bounded the dead ground into which we had followed the elephant. When I was almost out of breath I heard a shot ring out, and then another. I stopped, only just able to stand for exhaustion, and looked up at a tree under which I found myself thinking 'Can I climb it?' I listened, trying to still the tearing gasps which were forced out of my mouth. Everything was quiet. I ventured a cautious hail to my men. Back came an answering coo-ee. 'What has happened, has the elephant gone?' I shouted and could have laughed with relief at the answering shout—'Yes, it has gone.' 'Shall we come to you?' called Ontai but I shouted back 'No, stay where you are, I am coming', relief and wonder in my heart.

Judging my position from the direction from which the shouts had come I slowly, for I was in a very exhausted and tremulous state, made my way through the dense jungle, climbing under and pushing down the creepers and thorns with my hands and feet. At last I came to the little opening and there was Mohan Prasad, placidly pulling down creepers and stuffing them into his mouth while Ontai and the other men were perched on his back as usual. I could hardly believe my eyes. They greeted me with grins which turned into sympathetic clickings of the tongue and murmurs as they saw my state. They handed me the rifle and I looked impatiently at it, for all the while I had been fleeing from the elephant I had at the back of my mind the question of what had happened to my rifle, that trusty weapon which up to that moment had never let me down. One glance and I saw the reason why the weapon had failed to function—the safety catch was still on. A momentary aberration had nearly cost me my life, for when I had dismounted from the elephant after slipping the catch to the safety position I had failed to change it to the ready.

We stood in the little clearing going over the incident and they told me what had happened. When I had dived under the creepers, more or less disappearing from sight, the tusker had searched for me practically at his feet, tearing down the jungle and kicking it about, all the while squealing with rage and completely oblivious of the presence of Mohan Prasad at the edge of the clearing. The tracker had run up the path and hidden himself and the elephant had not caught sight of him. It was concentrating entirely on me. Mohan Prasad had shied away instinctively from the sudden charge of the wild elephant and Ontai had found himself momentarily up against a tangle of creepers, but he had managed to hold the elephant with his goad. 'As soon as I saw the elephant come out of the jungle I put a couple of cartridges from your belt into your shotgun', he told me. Wedged as he was against the jungle Ontai had yet managed to turn his body around sufficiently to enable him to discharge first one and then the other barrel into the side of the enraged elephant as he kicked and tore at the jungle looking for me. The effect of the cartridges at such close range and from so unexpected a quarter had the effect of propelling him into the jungle head-first as if ejected from a mighty cannon. Ontai said he could hear him tearing away through the jungle and mercifully in the opposite direction to where I was.

Then we started laughing. It was the natural reaction from tension and I laughed until the tears came to my eyes, while the men laughed with me, occasionally stopping to elaborate on the miraculous escape I had had—we *all* had, for that matter, for if the tusker had spotted our elephant he would certainly have attacked it. My hat was gone somewhere in the jungle, my shirt-back and shoulders were in ribbons and my scalp was full of thorns, while my face and arms were lacerated by the terrible tearing I had undergone in the jungle. Suddenly the narrowness of our escape came upon me and I looked at Ontai. It was he who had saved the situation with his presence of mind and yet there he sat, with a grin on his face as if it was all in the day's work. I was suddenly so moved that tears came into my eyes.

We hurried to get out of that awful jungle and to make our way back to camp. We were in time to catch the forest train with its load of labour and forest staff going back to the base settlement for bazaar day. The story soon spread of our encounter with the 'goonda'—that is what he was really, for he had attacked us in an unprovoked manner, unless you consider tracking him into his lair constitutes unnecessary provocation. I was the subject of much anxious questioning and of respectfully amused wonder as I sat in the guard's van having the thorns picked out of my scalp. And later when I reached the settlement and the doctor got on to my head with a pair of tweezers and attended to my lacerations, my staff and the villagers had to be told the story again and again. As for me, that was my swan-song in the matter of elephant shooting. I was sensible enough to realize that I had really lost interest in what had once seemed an exciting sport. It had been useful in keeping me from brooding over the long separation from my wife and our child whom I had never seen—and whom I never did see, for they were both torpedoed on the way out from England. When an elephant shooter loses interest in a sport that is the dangerous stage, at least according to the books. And so I packed up and stuck to tigers after that.

Excerpted from Elephant Gold *(London, 1963).*

F. T. POLLOCK AND W. S. THOM

Sport in Upper Burma: Sumatran Rhino

Though now extinct in India, the Sumatran rhino was still around in the late nineteenth century. Pollock and Thom's standard work on Burma recounts how the animal, already persecuted for its horns, was also eaten by hunters. The excerpt is one of the few that describes the habits or the habitat of a species that has managed to survive in South-east Asia, but has long been extinct in the subcontinent. Unlike the great Indian rhino, the Sumatran has two horns and often lived in lowland jungles that were not easily accessible for sport. Pollock was a senior police officer and Thom had been in the staff corps: no doubt, their postings gave them the time and opportunity to track this already rare quarry.

The Sumatran rhino now survives mainly in Malaysia and Indonesia: it has long vanished from India, where it was found in parts of north Bengal and the north-east. When Pollock hunted it, the species was already uncommon: it now ranks among the handful of mammals to have vanished from India.

RHINOCEROS SUMATRENSIS—THE ASIATIC TWO-HORNED RHINOCEROS [W. S. THOM]

To those desirous of becoming better acquainted with this rhino's haunts, habits, and appearance, Blandford's description of the animal in his *Fauna of British India* may not be amiss here. He describes this animal as the smallest of living rhinoceros and the most hairy, the greater part of the body being thinly clad with hair, and the ears and tail more thickly covered. The two horns are some distance apart at the base; both are slender above, except in the case of females, which have mere stumps a few inches, and the anterior horn of the male in fine specimens elongate and curve backwards. The skin is usually rough and granular; the folds, though much less marked than in the one-horned species, are still existent, but only that behind the shoulders is continued across the back. Colour, varying from earthy-brown or black. Dimensions, somewhat variable. The type of Sclater's *R. lasiotis* was 4 feet 4 inches high at the shoulder, and 8 feet long from snout to root of tail; its weight about 2000 lbs. An old female from Malacca was only 3 feet 8 inches high; the average height of adults is probably 4 feet to 4 feet 6 inches. The largest known specimen of the anterior horn measures 32 inches over the curve. Skull, 20 inches; basal length, 11.25 inches zygomatic breadth.

Varieties.—Specimens from Chittagong and Malacca were living at the same time in the Zoological Society's Gardens, London, in 1872, and the former was distinguished by Sclater as *R. lasiotis*, by its larger size, paler and browner colour, smoother skin, longer, finer, and more rufescent hair, shorter and more tufted tail, the ears having a fringe of long hair and being naked inside; but above all by the much greater breadth of the head.

Unquestionably the difference was considerable; but by far the most remarkable, the shape of the head is shown by Blyth to be variable in both *R. unicornis* and *R. sondaicus*, for he figured and described a broad and narrow type of each, as well as *R. sumatrensis*.

The other distinctions scarcely appear to me of specific value, and I am inclined to regard the two forms as varieties only.

Distribution.—Rare in Assam, though one specimen has been recorded on the Sankosh river, in the Bhutan Duars (*P. Z. S.* 1875, p. 566). Another was shot 20 miles south of Comillah in Tipperah, in February 1876. From Assam the species ranges to Siam, the Malay Peninsula, Sumatra, and Borneo.

Habits.—Very similar to those of the other species. This rhinoceros inhabits forests, and is found at a considerable elevation, having been observed 4000 feet above the sea in Tenasserim, by Tickell. It is a shy and timid animal, but easily tamed when adult.

Details obtained by Mr Bartlett concerning a young animal born in London, induced him to regard the period of gestation as probably a little over seven months. This differs greatly from Hodgson's account of the period in *R. unicornis*, but no details are furnished in the case of the last-named species, whilst the evidence is stated in that of *R. sumatrensis*. Still, for so large and apparently so long-lived an animal, seven months of uterine life appears short.

Anderson, in his *Fauna of Mergui and its Archipelago*, mentions having heard of a two-horned rhino seen swimming in the sea near High Island in the Archipelago. Probably all rhinoceros are good swimmers. (They certainly are.—F.T.P.)

The story of the Chittagong rhinoceros that was unable to swim must be, I think, a mistake. The account given by Mason, and repeated by Blyth, of this or any other rhinoceros attacking fire should be received with great caution. To my personal knowledge Mr Blyth's principal informant had a weakness for relating *shikar* stories which were frequently good, but not always authentic.[1]

I was fortunate enough, during my five years' sojourn in the Ruby Mines district, to come across four of these animals, three of which I succeeded in getting. I had also many opportunities, when spending my short leave out in camp, of studying the habits of these ungainly mammalia. They are not easy to find, and are always very timid and shy, but when found they are easily stalked and killed, provided you are armed with a heavy rifle. They are capable, when alarmed, of dashing away through the densest jungle at a great pace, and often travel for many miles over the roughest country before they come to a halt. Mud-wallows, swampy ground, and dark, damp, cool jungles amongst the hills, up to an elevation of 4000 feet, are the spots they usually frequent. (I found them at the base of the Arrakan range, near Cape Negrais, at a very slight elevation only.—F.T.P.)

Three or four animals may sometimes be found in one locality within a mile or so of each other. But as a rule they are solitary; I have on one occasion come across two females wallowing in the same mud-hole.

These mud-holes are usually found at the source of some small stream, where the soil is swampy, or of a clayey nature. A spring or a marshy piece of ground by some stream is often utilized in the same manner and one rhinoceros may have two or three wallows, or mud-holes, which he visits in turn: principally during the months of May, June, July and August.

[1] I do not agree with this statement. Mr Blyth to my certain knowledge was a most painstaking and able naturalist, and did not accept all he heard, but most Karens and Burmese assert that this rhinoceros rushes at a fire and scatters it.—F.T.P.

The sportsman will be notified of the near vicinity of a wallow by the caked mud which has been rubbed off the rhino's body by the bushes and tree-trunks as the animal passes.

Should the rhino be in his mud-bath, the sportsman will sometimes be made aware of the fact by hearing peculiar, low, rumbling, humming sounds, the noise being very similar to that made by a species of large hornbill when soaring through the air, or like the sound made by a vulture's wing when stooping to the earth.

These rhinos sometimes wander great distances to feed, but are most frequently found within a mile or so of their wallows. They feed principally on bamboo-leaves, shoots, young cane, thorny shrubs, and a bush called 'Kyau-sa.' It was in the Sagadaung jurisdiction of the Shan State of Momeik, Ruby Mines district, that I first made my acquaintance with these animals. I had previously, in other parts of the district, when out looking for gaur and elephant, come across old tracks, but had never had the luck to find any fresh ones, or to light on a fresh mud-hole.

One evening, in the beginning of the rainy season, Moung Hpe turned up in the 'zayat,' or rest-house, whilst I was lolling in my long arm-chair, under the soothing influence of a good dinner, cigarette, and a cup of coffee, and said—'Thakin, hnepen taung-daw thwa gya-zo, mane-ga wakok the-ma hnit-yauk kyan-kyi-ya ah thit twe ge de' (Sir, let us go into the hills tomorrow; two bamboo-cutters saw fresh rhinoceros tracks yesterday). This was excellent news, and Moung Hpe was immediately made the recipient of a bottle of Younger's Monk Brand beer, an old shooting-jacket, and five rupees.

Poor Moung Hpe, as fine a tracker as ever stepped! I wonder what he is doing now? Ah, those glorious days, gone like a fleeting dream!

As the locality in which the rhino's tracks had been seen was distant in the hills above Sagadaung only some eight or ten miles, I made up my mind to pay a flying visit to the spot, leaving early next morning, and camp out only one night in the jungle, as I had a good deal of work on hand.

At 6 a.m. next day I was well on my way into the hill, having first arranged with my camp-followers and servants as to the spot to be chosen for our camp. Moung Hpe and an old retired Shan *shikarie*, whose knowledge of the hills was extensive, accompanied me. He was a curious old fellow, this Shan, and never better pleased than when smoking away at a long Shan pipe and drinking raw spirits.

I was armed on this occasion with only an old single 450 sporting Martini-Henry carbine, which belonged at one time to poor Tucker of the police, as daring a sportsman and as good a shot as Burma has ever seen, and a double 12-bore shot-gun by Joseph Lang, which burned a maximum charge of 4½ drams of powder, and carried a spherical ball with great precision and penetration up to 30 or 40 yards.

After reaching the outlying spurs of the Shwe-u-taung range we struck the Tunkachoung stream, along whose banks we walked. As luck would have it, we had not gone a quarter of a mile before Moung Hpe, jumping down into the sandy bed of the stream, exclaimed—'Thakin, thakin, kyi-ba-thee ma, kyau-kyi-ya, ah-thit gane ma net saw zaw thaw-ge-de' (Sir, sir, look here, fresh rhino tracks; the animal passed early this morning). On examining the tracks I noticed that the water in them was still a little clouded at the bottom, and that they must be very fresh tracks. I knew that Moung Hpe very rarely made a mistake as to the freshness of a track, and I was convinced, from the decided way he had spoken about the tracks, that he was quite certain in his own mind that they had been recently made.

We decided to wait the arrival of our camp-followers before taking on the tracks, as there was a suitable piece of ground for pitching camp on the banks of the stream, and I wanted to warn them about making any noise which might disturb the rhino.

After waiting about an hour and a half our followers turned up, when we informed them of the fresh tracks, and after cautioning every one not to cut down bamboos or make any preparations for chopping firewood, etc., till they had heard me fire, we started off on the trail.

The ground in the neighbourhood was very hilly, and at times we had to push our way through cane-jungle and over swampy ground.

After covering some two miles of country the track showed that the animal had been feeding round in circles,[2] and at times tracking was made to my eyes almost impracticable, as the animal had crossed and recrossed its own tracks dozens of times. So mixed in fact had

---

[2] All rhinoceros feed in circles.—F. T. P.

they become, that sportsmen unaccustomed to tracking would be under the impression that two or three animals instead of one had been in the vicinity. We succeeded eventually in hitting the right trail out of this maze, and after travelling for over three hours, came on unmistakable signs of a mud-wallow. Moung Hpe now fell back and requested me to lead, warning me at the same time to be careful, as a wallow was not far off, and, indeed, numerous traces of it were apparent. The surrounding jungle was composed of cane, bamboo, and tree forest.

Following a small game track I pushed my way cautiously along the side of the hill through the foliage, which fortunately was not very dense here, Moung Hpe and the old Shan hunter following in my traces.

I felt as if my heart was going to burst from suppressed excitement, at the thought of seeing for the first time a real wild rhinoceros. Of course I had seen the larger species of rhino in the Zoo, but had never seen one in its wild state. Moreover, the animal I was tracking had two horns, and, if I remember rightly, no specimen of the *Rhinoceros sumatrensis* was in the Zoological Gardens when I paid my visit.

Rounding a clump of overhanging canes on a sloping ridge, I caught sight of the edge of the basin or wallow on the brow of the hill, a few feet above the level of my head, and at a distance of about 25 yards. There were numerous fresh signs of the rhino's presence in the splashes of yellow clayey mud all round the edge of the basin and on the surrounding trees, but no sound emanated from the wallow, which appeared to be deserted.

Turning round and holding up my hand as a signal to Moung Hpe and the Shan to stay where they were, I cocked both barrels of the smooth-bore, bent low, and moved cautiously forward about ten paces. My coat unfortunately caught on a hooked thorny cane-creeper, and knocked down a small decayed branch, which fell to the ground with a faint rustle.

The scene was changed in a moment. An ugly, small-eyed, piggish, horny-looking-beast reared itself up out of the wallow in a sitting posture, only exposing its head and shoulders, and blinked at me stupidly for a few seconds in an undecided manner, as if debating in its own mind what manner of animal I was. I did not give it time to consider long, but jumped from my stooping attitude and aimed hurriedly at the huge head, firing both barrels in quick succession. A tremendous commotion in the wallow immediately ensued, followed by sounds like the drawing of corks from very large bottles, the sound being caused by the rhinoceros pulling its feet out of the clayey soil as it rushed out of the wallow and bolted, in full flight, down the hill-side, through cane-brake and bamboo, carrying everything before it. At first I was greatly disappointed, being under the impression that I had made a clean miss.

Moung Hpe, who was also greatly disgusted at my shooting, came up with a very long face, taking it for granted that I had missed, and said, 'Thakin, thakin, thee-tokan-now ka-bedaw ma ya bu kyan hue dine ma twe hnine bu ma, hman bu tin de' (Sir, sir, you will never have such luck again; it is not every day that you see a rhinoceros, I think you have missed it). But I had not missed, for after following up the tracks for about 100 yards, we found, much to my delight, blood on either side of the track, which proved beyond a doubt that one of my bullets had gone clean through the animal's head.

It was simply marvellous to think that any animal could have received such a shock without staggering or showing any signs of having been hit, and then make off, down a steep hill-side through dense undergrowth, like an express train.

We had not been an hour on the trail before Moung Hpe pointed out the rhino to me, lying down on its stomach amongst some bushes, breathing heavily. I could just see a dirty yellow patch, which I immediately fired at.

On receiving the shot it jumped up and made off again. We came up with him after going about 300 yards further; it was standing broadside on, and I put in a right and left behind the shoulder. I was obliged to use my smooth-bore, as the striker of the Martini carbine was out of order, and would not explode a single cartridge. The last two shots seemed to waken it up pretty considerably, as it travelled some distance before we again overhauled it. We had some difficulty on this occasion in finding the tracks, as it had entered a stream and waded down for a considerable distance. We now found that the rhino was heading back in the direction from which it had come, and we could distinctly hear our camp-followers talking and laughing in the hollow below, and some of them now joined us in trying to find the lost tracks.

While we were thus scattered about up and down stream examining the ground, my attention was suddenly drawn by Moung

Hpe to the rhino, which was standing with heaving flanks on the bank of the stream, within ten paces of a rock upon which I had climbed. I could see from its stertorous breathing that the poor beast was done for and quite helpless, so wading quietly up to within a yard or two I fired behind the shoulder, aiming for the heart. The rhino stood motionless for about four seconds, and then sank dead on to its knees, with a long-drawn sigh.

On examination I found that it was a female, and that only one of my two first shots had taken effect, but this one had gone clean through the head, and only missed the brain by a very narrow margin.

The rhino was 4 feet 6 inches from heel to shoulder, measured between uprights, and had two distinct horns, the front one about 4 inches in length, and the hinder one a mere horny protuberance.

In the male the front horn varies from 6 to 12 inches. The folds were much less marked than in the one-horned species, and commenced behind the shoulders, continuing across the back; there was also a slight overlapping of the skin on the posterior. The colour of the skin approached a light earthy brown; the ears had a long hairy fringe, but were naked inside.

The male rhinoceros, according to Burman hunters, consort with the females about the middle and end of the rain.

On another occasion I saw three animals within the space of an hour and a half on an outlying spur of the Shwe-u-taung hill, two of which I succeeded in bagging. We had been camping in a cave on the top of the hill, and were moving down a spur intending to pitch on new ground as soon as we came across traces of rhino or gaur.

I had been out on a shooting expedition for over a month, having been granted leave from April 3 to May 17, and during that time I had succeeded in getting two elephants, six gaur, one tsine, one sambur, two barking deer, one leopard, and one serow—not a bad bag. Moung Hpe and I left camp early and started walking leisurely down the ridge, leaving the remainder to pack up and follow at a respectable distance.

We had gone about two and a half miles, when Moung Hpe, suddenly stooping down and examining the ground carefully, exclaimed, 'Kyan hnit kaung, mane ga ma net saw zaw thwa ge de' (Two rhinos passed this way early yesterday morning). We immediately called a halt and held a consultation, chose a suitable spot for our camp, and arranged for tracking up the animals. Eventually we agreed that our carriers and camp-followers, numbering in all some twenty men, had better follow our trail in single file, in silence, and at a respectable distance, until we came on fresher tracks, when we could fix on some more convenient spot near a stream for our camp. On this occasion I was armed with a 12-bore rifle, and a double 8-bore burning twelve drams. Moung Hpe carried the 12-bore, whilst I took the heavier weapon. After warning our followers to remain perfectly quiet whilst on the march in our rear, Moung Hpe and I started off and took on the tracks of the rhinos, with every hope of being able to come up with them in some mud-wallow before dark. Moung Hpe as usual took the lead, and whenever he was at fault, which very rarely happened, I assisted him by making wide casts till the trail was recovered. We had greater difficulties to contend with than I had expected to find. The ground was not only covered with a thick layer of leaves, which in places had either been scratched up by jungle fowl and pheasant, or scattered by the wind, thereby obliterating all foot-prints, but at times it was very hard and rocky. After covering some three miles of precipitous but fairly open and rocky country, I felt tempted to give it up in disgust, as we had not as yet come upon tracks which were only an hour or two old. Patience is however usually rewarded in the end, and after many twistings and turnings, backwards and forwards, up-hill and down dale, often going over the same ground twice, we came on fresh tracks, and were suddenly startled by hearing in our immediate neighbourhood the peculiar low, muffled humming sound repeated at intervals; and Moung Hpe at once recognized them as proceeding from the rhinos. To walk noiselessly down the ridge in the direction of the sound was the work of a few moments, but I was afraid our camp-followers would catch us up before we could get sight of the animals. Fortunately, however, they had all seated themselves on the ground for a rest. The peculiar low, buzzing or humming noises now became more distinct, and as we rounded a rocky ridge which overlooked a shallow ravine, wooded with bamboo and an undergrowth of bush and prickly cane, a large mud-wallow, in a small clearing bordering the cane-jungle, came into view, and in this two rhinos were disporting themselves. One animal, the larger of the two, was standing half in and half out of the slushy mud; the other was lying in it half submerged, rolling about from side to side, and uttering the peculiar noises which had attracted our attention.

Telling Moung Hpe to fire at the animal standing on the brink of the wallow, I aimed at the other, and fired both barrels. Moung Hpe's shot rang out simultaneously with my own. There was a terrific commotion for a second or two in the wallow, accompanied by unearthly grunts and screeches, something like the bray of a donkey, and then both animals bolted away in different directions through the cane.

We rushed down and examined the ground and bushes in the neighbourhood of the wallow for blood, several big splashes of which we discovered on the line of flight taken by the rhino which I had fired at. We took on this animal's tracks immediately, and had not gone very far before I caught sight of the beast, limping along with a broken fore-leg; a right and a left behind the shoulder brought it to a standstill, and a third shot completed the business. After marking the spot carefully we returned to the wallow, intending to take on the tracks of the animal fired at by Moung Hpe. We found that his bullet (Moung Hpe having fired only one shot, being unaccustomed to a double rifle, and naturally finding the firing of a second shot with any degree of precision well-nigh impossible) had passed through a creeper about the thickness of a man's calf, which would naturally lessen the penetration of the bullet, although he used a steel-tipped one.

After arranging a suitable spot for camp with our followers, who had in the meantime joined us at the wallow, and who were in great glee at the thought of dining off rhinoceros flesh, Moung Hpe and I started off after the other rhino, which, from a spot of blood found on a cane-bush near the wallow, had evidently been hit.

We were rather ludicrous-looking objects after we had been following the trail for some time, our clothes, hands, and faces being plentifully besmeared with the clayey mud whilst following in the wake of the rhino, which deposited a portion of its coating on the bushes and branches at every step. After puzzling over the tracks for three and a half miles of very rugged country, up and down hill, through cane brakes and ravines, we gave it up in disgust, as not only was it getting late, but the animal had not shown any signs of having been badly wounded. An animal, as a rule, when severely wounded lies down once or twice to rest after the first mile or so: an experienced Burman hunter can often tell by the footprint of an animal such as the gaur or tsine whether it has been wounded or not, the slot of a wounded beast being often deep and irregular, though this very rarely occurs in the case of a wounded elephant or rhino, which, as a rule, only lies down when about to die or unable to stand; though I did once follow a huge 'Muckna' which lay down to rest no fewer than three times within the distance of two miles.

I did not succeed in coming up with this animal, although it had received three 8-bore spherical hardened bullets well placed behind the shoulder.

It was now about 4 p.m., and as we had not eaten any food since 10.30 a.m. we proceeded to dispose of our respective breakfasts, mine consisting of cold salted gaur tongue, biscuits, dried figs, and some cold boiled rice, washed down with beautifully cool, clear hill water. Moung Hpe opened out on boiled rice, cold smoked gaur, and last, but not least, that highly odoriferous national dish, a 'bonne bouche,' amongst Burmans known as 'ngapee.' Whilst in the middle of our well-earned meal we were startled by hearing the peculiar, muffled humming sound already referred to, and which seemed to proceed from no great distance.

We both jumped to our feet with alacrity, sending the remains of our meal flying. I seized my 8-bore and, after pocketing a couple of spare cartridges, moved cautiously in the direction of the sound, Moung Hpe bringing up the rear. After wading through a swampy piece of ground for about 40 yards I caught sight of another wallow, from which the sounds seemed to issue. Arriving within half-a-dozen paces of it I saw a spectacle which made my heart throb at a tremendous pace—a rhinoceros lay submerged in the mud, with its ears and the top of its head occasionally showing as it rolled about from side to side, uttering each time its nostrils and mouth rose above the surface low, peculiar, long-drawn grunts. I cautiously withdrew and beckoned to Maung Hpe to approach a little closer so that he might be of some assistance in case of a charge, and, after seeing him ensconced behind a tree within a few yards of the wallow to my right, I took a steady aim for what I took to be the shoulder of the animal, but which afterwards turned out to be its stomach, and fired.

A tremendous grunting, screaming, snorting, and splashing ensued after my shot, and I was so near to the wallow that several splashes of mud struck my hat and coat.

The rhino, after making several rapid gyrations in the wallow as if trying to bite its own tail, shot out of the pit through mud two or three feet deep, and rushed down the side of the hill as fast as any pig could travel, followed by a second bullet from me and a right and left from Moung Hpe, all of which, as I afterwards found, took effect in various parts of his body.

Notwithstanding all this we had a long, stern chase, the rhinoceros keeping up a tremendous pace for nearly three miles, and leading us through some of the most awful jungle which it has ever been my fortune to travel over. All things must, however, come to an end some time, and we eventually came up with him standing stock still on the side of a deep ravine looking very sick. I was very thankful that we had come up with him, and I am certain that I should not have been able to keep up the pace another mile over such country as that we had traversed, encumbered as I was with the heavy 8-bore. I was literally dripping from head to foot, and almost blinded with perspiration from the violent exercise we had just undergone, enhanced by copious draughts of icy water *en route*.

As the rhino was standing facing away from me, I worked cautiously round till I could obtain a good shoulder shot. I succeeded in getting partially round, but was discovered by the brute, which wheeled round with a loud grunt and walked quickly with lowered head towards me. This was a most unusual proceeding, and entirely unexpected, for I was always under the impression that a rhino was a harmless beast. As may be imagined, I did not wait to see any more, but delivered a quick right and left. On the smoke clearing I saw that the rhino had not only come to a standstill, but was about to fall. In fact, after a preliminary roll or two from side to side, a loud gasping sigh as it collapsed slowly on to its knees proclaimed its decease. A kill of two rhinos in one day was not bad work, but to crown all we found that we, or rather the rhino, had been travelling round in a circle, and that instead of being five or six miles from camp we were only some two or three hundred yards from it. I returned highly elated, but feeling a little knocked up, a dip in the stream which flowed past our encampment (an erection of leaf huts, 'taungzin pet' leaves), and a good dinner brought me up to par once more. Many a long yarn did my *shikaries* and I spin over the flickering camp fire far into the night, before I dropped off into a well-earned slumber.

With regard to a doubt expressed by some naturalists as to rhinos being able to swim, I can say, from what I have seen and heard myself, that they are undoubtedly good swimmers.

Sport in Upper Burma: Sumatran Rhino

I remember very well on one occasion, whilst out with a party of police tracking a gang of dacoits, coming upon some fresh rhino tracks leading into the Kin river and emerging on the opposite bank at a crossing where the water, it being then well on in the rainy season, was quite four to five feet deep. One of my hunters informed me that he had once watched two rhinos, a young one and its mother, cross a stream. Before entering the water, however, the mother had to prod up the little one from behind with her snout several times to induce it to venture in. Being unarmed at the time he was unable to shoot them. He noticed particularly that both swam very strongly and swiftly across, the young one in front, and that only a portion of the snout and head of each animal was visible. The young one on arriving at the opposite bank lay down and rolled over and over again on the grass, in the same manner as a horse would do, but the mother walked steadily on after reaching the bank, leaving her young one to follow. It is said that rhinos deposit huge mounds of ordure,[3] visiting the same spot daily. I have not noticed this with regard to the *R. sumatrensis* (I found the mounds at the foot of the hills near Negrais, and in Assam they were very plentiful wherever there were rhinoceros.—F.T.P.), although I have come across their droppings in the ordinary course of my wanderings, and they all seemed as if left on a single occasion.

Excerpted from Wild sports of Burma and Assam *(London: Hurst and Blackett, Ltd., 1900).*

---

[3] Mr Thom, since this was written, has found these mounds in the ArrakanYomahs.—F.T.P

R. P. NORONHA

A Wild Buffalo Hunt in Bastar

Noronha joined the Indian Civil Service in 1938 and served in the Central Provinces and Berar with great distinction. A junior colleague described him as 'a shikari's shikari, who stalked tigers on foot and always said that an IAS officer was unfit for service unless he drank from the same stagnant pool in the jungle as did the tiger'. His stint as Deputy Commissioner of the sprawling district of Bastar gave him ample opportunities for coming in close contact with the wild buffalo, which he photographed and studied. He even shot them when crops or lives needed protection. The buffalo herds in Bastar live not in riverine savannah as in Assam but often in clearings on the forest edge made by tribal cultivators. Noronha, it might be noted, also wrote an early note on how best to conserve the Bastar wild buffalo for the New York Zoological Society. As an author he is widely known for his book, A Tale Told by an Idiot, *a scintillating account of Indian bureaucratic life. But his writings on animals are of no less interest.*

The Debt

Eram the wild buffalo was lucky to escape with his life when at last his sins caught up with him and he was banished from the herd. He was confused and hurt and resentful and his bad temper clouded his judgement. Perhaps that was why he stopped to look at a pair of tiger cubs frolicking on the bank of a nala. A wild buffalo bull can look at a tiger with more impunity than a cat can look at a king. But no sensible buffalo will stare at tiger cubs, because tiger cubs

A Wild Buffalo Hunt in Bastar

mean tigress and a tigress achieves the impossible in defence of her young—even to the death of the one and half ton mass of fighting force which is the wild buffalo. Eram forgot, and almost paid with his life for the lapse. One moment he was staring moodily at the cubs; in the next he was rolling down the nala bank with a growling clawing tigress on his shoulders, a tigress moreover which had got her initial grip on the back of his neck. The strength of a buffalo lies in his massive neck and shoulders. On the rare occasions when a tiger attacks him or on the more frequent occasions when he attacks a tiger, he always pivots so as to receive a charge on his horns. He knows that once the tiger lands on his neck, the fight is over.

This tiger was on his neck. Eram tried to regain his feet but his back seemed to be frozen, he could not move his shoulders under the deadly grip. His eyes glazed, his movements became more automatic as the struggle progressed. Then, suddenly, there was thunder in his ears and a cloud of dust enveloped him. When it cleared, he was free of the clinging weight on his neck, staring at the backs of a small herd of bison in mad pursuit of the tigress.

Perma the bison is, I am afraid, a snob. He can never forget that he is *Bos gaurus*, first cousin to the cow. He can even breed with the domestic cow—this has been successfully done in the Mysore zoo—although in the wild state he makes no such attempt. Eram on the other hand is a different species, *Bubalus*, and Perma looks down on him. The bison considers himself the aristocrat of the jungle and keeps himself to himself, not mixing with the rest of the herbivora. Another reason for this may be that bison are hill animals while buffalo inhabit plain forest and swamp. Also, buffalo frequently pick up foot-and-mouth disease from the village cattle, and this disease is certain death to bison. But the jungle will tell you that the real reason for their keeping apart is Perma's snobbishness.

Perma and his herd had come down from the hills to drink and had drifted down the nala like ghosts on the wind in time to see the tigress spring on Eram. Normally bison are afraid of tiger and will avoid them. Why Perma started the mad charge that saved the buffalo must, therefore, remain a mystery. Perhaps it was some dim memory of racial kinship, perhaps it was instinctive hatred of the tiger; but I prefer to believe that it was a wave of sympathy for the victim, a stirring of the sense of animal justice, which seems to exist only in the herbivora.

By the time the bison bull returned from his chase, Eram had struggled to his feet. They were both magnificent specimens of their kind, the bison high shouldered, deep necked, the buffalo straight backed, massive. They looked each other over and both were embarrassed. No thanks were offered or expected, the jungle does not believe in thanks; it pays its debts. When the herd moved off, Eram followed and they did not prevent him.

Wild buffalo have a well deserved reputation for ferocity and fighting ability. But they have also a great capacity for affection. The tame buffalo shows it when he rescues his herdsman from a tiger. The wild buffalo demonstrates it less frequently, since opportunities are fewer, but every now and again one comes across a case of a bull joining a tame herd and guarding it faithfully. When Eram joined the bison herd he began a new and unique career, that of guardian to the herd bull and Man Friday to the herd. The jungle, which is conservative, and very critical about any departure from accepted practice, was censorious. Monkeys with chital, barasingha with buffalo, sambhur with pig—yes. But bison and buffalo—no, emphatically. Even the wise crows, watching, cawed derisively. The jungle people spoke about it around their campfires, and their sirahas (witch doctors) said it was a thing of no good omen, and another fowl should be sacrificed to Kankalin. If Eram knew of the criticisms, he did not care. The herd soon got accustomed to him and fully realised his value as a general help and guard. He had changed completely from the surly, ill-tempered despot that he was, to a cheerful, good-tempered grandfather. A harassed matron wanting to graze a particularly succulent patch would entrust her calf to Eram and wander off, and the young bison took unheard of liberties with him. His principal loyalty however remained towards Perma. He was always guarding the weak spot of the bison, the area downwind of him, and his fighting powers were fully up to the reputation of the buffalo, as a marauding tiger discovered to his cost.

The rest of the story I draggerd out from Roy, our Forest Officer. It is said of Roy that he once started a sentence when a nursery was being planted and got around to finishing it when the first fellings began. He is not a man of many words but he knows more about the jungle and its inhabitants than anyone else I have met. He was driving down a forest road at a quarter to seven in the morning to assess the damage done by fire in a coupe. He turned a corner, and about three hundred yards away, scattered grazing on both sides of the road, he saw a herd of bison. What caught his eye was the herd bull, a remarkable specimen, in fact our old friend Perma. He accelerated to have a closer look, and then braked hard. About fifty yards ahead of him a bull buffalo stepped out of the undergrowth and stood squarely across the road. People familiar with buffalo have no illusions about what a big bull can or cannot do. Visions of possible explanations to the Accountant General for a destroyed jeep—the property of Government—flashed across Roy's mind as he braked. But the buffalo did not charge or show any signs of charging. He just blocked the road. For ten minutes Roy raced his engine, hooted, shouted—all to no avail. He even backed the jeep away, out of sight, and returned after a quarter of an hour but the buffalo still held his ground. What particularly struck Roy was that he kept only one eye on the jeep; he was continually looking in the direction of the bison herd as if to see whether it had got away.

Poor Roy! He couldn't even get out and walk, because his coupe was thirty miles away. Eventually, when he had made quite sure that the bison had got away, Eram stepped aside and the jeep passed him at ten yards without his showing the least sign of hostility.

He is still paying his debt to the bison. I know—I saw them only the other day.

BUBALUS BUBALIS

Bubalus Bulbalis, the Indian wild buffalo, is own brother to the domestic water buffalo. In fact, when the domestic buffalo reverts to the jungle the second generation becomes the true wild buffalo; in the outback of Australia there are large numbers of wild buffalo which trace their ancestry to tame buffaloes that were turned loose when the cow substituted them as a primary source of milk. They are indistinguishable from *Bubalus bubalis* in size, massiveness, and horn. Today's record head will certainly come from Australia. In India they have almost disappeared from the jungle, being confined to Assam, Madhya Pradesh, and Orissa for all practical purposes, and that too in small numbers. Even in these areas their numbers are diminishing. Man is the chief agent of extermination, not by killing so much as by being. Where man is, there domestic cattle are; and where domestic cattle are—in India at least—there rinderpest and foot-and-mouth disease too are. The buffalo, tame or wild, is particularly susceptible to these diseases. So are bison, but bison generally keep to the hills which are free of domestic cattle, while buffalo are basically plains animals grazing where cattle also graze, where the seeds of foot-and-mouth are thickly sown, where periodic epidemics of rinderpest occur. Rinderpest is fatal to cattle, foot-and-mouth is not; but both are fatal to wild buffalo. Twice I have walked through jungles that were literally stinking to high heaven with buffalo carcasses.

My first experience of wild buffalo dates back to 1940. I was taking a buffalo calf to tie out as a tiger kill when I suddenly found myself alone, except for the calf. The people who were accompanying me disappeared at a speed that suggested magic, and I heard the cause before I saw it; a large and irate solitary bull, bearing down on me at a good thirty miles an hour. He was bent on rescuing the calf. Fortunately, he lowered his head as he crossed a slight depression in the ground, which gave me a better chance at him for a fraction of a second, but the shot that killed him was nevertheless a fluke. Since that day I have killed three more—each time from necessity, not choice—and have watched and photographed dozens of others. There is something rather appealing about *Bubalus* when you get over your fear of him; I have never looked at a dead specimen without regret.

The tales one hears about his ferocity are largely a myth. Let me qualify that statement before I am murdered by people who have been charged, or by relations of others who have been killed. There are certain circumstances in which buffalo will attack, and when they do, nothing except death will stop them. You can turn a tiger's charge if you hit him squarely with a heavy rifle, even though the wound is not immediately fatal. You cannot do that with a buffalo. Either he drops dead or he keeps on coming. His determination and his resistance to shock are quite phenomenal, reminiscent of the attacking Japanese soldier in World War Two.

He starts at a deliberate trot which soon explodes into a gallop, with the head held high so that all you can see of it is the nose. At the last moment, about five yards from you, he drops his nose—and this is the curious part—sidesteps to his own right and hooks you from the left. The sidestepping and coming in from your left are invariable, to enable him to pick you up on one horn. But there are only two circumstances in which he will charge—when he is wounded, and when he is surprised. Wounds come from man, from tiger (usually when the buffalo charges a tiger that has attacked a calf or a pregnant cow—no tiger in its senses would attack an adult male buffalo) or from another buffalo, one that has ambitions. Surprise—like meeting him round the bend of a dry nala, or stumbling on him in a wallow. At other times *Bubalus* is good-natured enough. Quite often he will walk towards you, being shortsighted and inquisitive. There is nothing to be alarmed at, unless you run. If you do that, the walk becomes a gallop and untoward things may happen. If you stay put he will satisfy his curiosity and amble off. I was once driving along a cart road when I saw a solitary bull standing across the tracks and making like Horatio at the bridge. I refrained from hooting at him because I did not want to replace my car. Instead, I switched off the engine and walked slowly up to within thirty yards of the buffalo. I took a photograph. He turned and walked towards me. I took another picture. Then he suddenly decided there was nothing to worry about and trotted off. I took a third picture... So much for the viciousness of the wild buffalo.

Or take another example. For years I had been trying to photograph *Bubalus* in a wallow. This is even more dangerous than deficit financing because the wallow is usually surrounded by very tall grass at a distance of six to eight feet from it, and taking a picture from this distance (assuming you could get there without alarming the animal) demands qualities of heroism which I do not possess. At last I found a wallow which had tall grass behind, but a clear space of fifteen yards in front. I made my stalk at noon and pressed the release as the buffalo began to rise. What I like about my little Robot cameras is that it does not impede my movements, which are apt to be somewhat rapid on occasions like these. Having got my picture I made off before the buffalo was on its feet. It did not chase me—far—and such an intrusion on privacy *does* justify a charge. I have now accepted it as an axiom that attacks are only caused by wounds or surprise. It is of course a little difficult to define precisely what will surprise a buffalo...

A full grown wild buffalo stands five feet at the shoulder and weighs well over a ton. He is not as tall as his cousin, the Gour, but the spread of his horns is remarkable. Sir Samuel Baker writes of heads more than ten feet from tip to tip, measured along the curve and across the forehead, as being common in Sri Lanka in the 1880s. The wild buffalo of today is not in this class. Ninety inches would be a good head by present-day standards.

The really big bulls are always solitary. Apparently, the master bull of the herd becomes a dictatorial bully at a certain age; when this happens, the other bulls unite and kill or drive him out. More rarely, a single bull with his eyes on the leadership of the herd will take him on in a duel. If he loses, he is banished. The origin of the solitary bull is always loss of leadership. Thereafter he lives alone, with two exceptions. Sometimes another bull which has also been driven out of a herd, joins him. Whenever this happens, the two get along quite amicably, the smaller acting as the chela of the bigger, but the occurrence is rare. What is more frequent is the intrusion of the solitary bull into a domestic herd which has been shifted into the jungle in the dry season to graze. He is not exactly welcomed by the graziers, but there is little they can do to express their disapproval. A lathi is not an adequate means of argument with a wild buffalo.

At the same time one must admit the grazier has justice on his side. *Bubalus* begins by killing all the stud bulls, or at least such of them as he can catch. He has an unreasonable prejudice against allowing any of the cows to be milked. And when his children are born they are so big that nearly half the mothers die. Having generously conceded the case against him, let me also list the points in his favour. No tiger dares to attack the herd; and there are at least ten hours in the day—or night—when the wild buffalo takes himself off for his own grazing. He does not relish the plain fare which is good enough for the tame animals. He has to have the choicest grass and will cover long distances to find it.

The wild buffalo which joins a tame herd is a fascinating subject for study. There was one which rejoiced in the name Banda—tailless; wild buffalo have ridiculously short tails. He grew so domesticated that the children had to coax him to chase them. Once they even got him into a sufficiently good mood to allow a rope to be tied round his neck, and tied him to a tree. After they had tied him, he gave a lazy jerk and snapped the one inch rope casually. But even then he lumbered after the urchins with the utmost good humour, and stopped dead when a three-year-old tripped and fell. He did not allow the adults in the grazing camp to take the same liberties, if he allowed any liberties at all. With cows, the camp dogs, and even the chickens, he was friendly and unassuming. I have never known a wild buffalo to become so tame.

His fate illustrates the dangers of ambition. The herd he joined was a large one, with six stud bulls in it. He killed two, and the rest disappeared into the jungle. He took over the duties of all six, but after a year he was reduced to a shadow, and decided he had made a mistake. When he left the herd for the last time, some of them tried to follow him, an idea that was effectively discouraged with a jerk of his horns. He wanted out—they never saw him again. The oldest woman in the encampment sniffed caustically as she told me the story. She said, 'Just like a man!'

FOOD FOR WORMS

The wild buffalo was huge and old and hungry. He came out of the forested plateau behind Singanpur and eased himself through the thorn fence that protected Nada's field, one of ten or fifteen little paddy fields. In the unirrigated hilly tracts of Bastar, paddy can only grow in natural depressions that hold the run-off during the monsoon, and such depressions are rare. Using every available inch of suitable land, the Murias barely manage to grow enough rice to keep them alive till the next crop. That is why the paddy fields are so important. But young rice in the milky stage is very nourishing for buffaloes, and particularly for old buffaloes who have less than the normal quota of teeth. The wild buffalo grazed contentedly.

There is a saying amongst the aboriginals. 'Old women are the servants of all and the mothers of none.' In the jungle villages old women work from dawn to dusk for just enough food to keep them alive, and their sons have long since forgotten that they exist. Primitive economics is brutal—to each according to the return. What return can there be from very old women? Pakri was old, so old that she could no longer remember the nights of the harvest dancing, nor the wet lips of an infant at her breast. She was half blind too. That big black object in Nada's field—What was it? She decided it was one of the village cattle, and, as the servant of all, she entered the field to drive it away. The buffalo saw her coming and shook his massive head, warningly, irritably. But that she could not see. She went on. He advanced to meet her. She died with surprising gentleness, impaled on one huge horn. To her death came truly as the half brother of sleep.

And that was the wild buffalo's third murder in Singanpur. The first had been a young man, over-optimistic about his speed and strength. He had time to use his lathi—once—before he died. The second was a grazier boy, accustomed to domestic buffaloes, and knowing them to be lovable and placid animals, not given to tantrums. He knew nothing about wild buffaloes, particularly old wild buffaloes and made the blunder of getting too close. That was the second killing. Death is a fact of life in the jungle, but the third killing disturbed the aboriginals, not because Pakri had died, but because it meant that the buffalo would not go away now until he had finished the entire paddy crop of the village. He had bought his right to the crop with blood.

The men took counsel amongst themselves and stopped me the next day at Kondagaon when I was going to Kanker to conduct the annual excise auctions. Kondagaon is fourteen miles from Singanpur, and there is no road between the two places, only a twisty footpath that snakes through the jungle. Nada told me the story without expression, but as he spoke the picture of Pakri's death grew before me, the dark green paddy plants, the dark earth of the field, the darkness of the buffalo, and the bundle of off-white rags that had been Pakri. Before me, too, grew the picture of a tiny Muria village, deprived of its paddy crop, the glum listless men, the quiet women, the pot-bellied children, pot-bellied of hunger. The only weapon I had available was a .275 Mauser, inadequate for buffalo, and that is putting it mildly. Worse still, the cartridges I carried were soft-nosed, the bullets would expand on the thick hide of the buffalo and there would be little penetration, not enough to reach the vitals. In the end it was the picture of the pot-bellied children that exhausted all my options. I got the little Francis Barnet motor cycle lifted out of my pick-up and we started for Singanpur with Nada on the pillion. I left Bonzo, the bull terrier, behind with strict instructions to behave. He promptly curled up and went to sleep on the rest-house veranda; a bull terrier is on his best behaviour when he is asleep.

The motor cycle carried us gallantly through the hills and on to the plateau of Singanpur, although it was we who had to carry it on two occasions, once when we crossed a nala which had too much slush and too much water for the little two-stroke engine to cope with, and once when we had to get across a steep embankment. The fourteen miles took a little over two hours to cover, good going in the circumstances. By the time we reached Singanpur the sun was setting. If Nada had not been an aboriginal he would have taken me first to the village which was quite half a mile from the paddy fields and wasted an hour on pointless hospitality. Being an aboriginal he stopped me at the edge of the cultivated area and said briefly, 'There is a path from here to the fields. If we follow it we will be under cover most of the way. Come.' I followed, the .275 cradled in my arms in the woodsman's carry. It felt ridiculously light, particularly when I though of the two-ton buffalo with which it would have to deal. The footpath ran through half cleared scrub jungle and stopped abruptly at the edge of the fields. I noticed with approval that it had brought us to the village side of the fields, opposite to the jungle from which the buffalo would emerge. There was no animal in any of the ten or fifteen little fields, so we sat down quietly and waited. A very few minutes later the wild buffalo walked out of the jungle, brushed through the thorn fence and entered the paddy. He began to graze. I noticed that his general direction was towards us and made no move to lift the .275. Nada noticed too, and nodded in endorsement of my lack of action. He demonstrated one of the million qualities for which I love the tribals, the ability to understand quickly, and, more important, the ability to refrain from asking unnecessary questions. We watched the animal grazing for all of twenty minutes and I realized why the tribals are so allergic to wild buffalo. He did not graze as domestic cattle do, from patch to patch. Instead, he took mouthfuls of carefully selected paddy, only the very best, and trampled beyond redemption the remainder. The wastage was horrifying.

Now the buffalo was within a hundred yards. Unfortunately, he was facing us and no .275, even with solid bullets, will rake a buffalo from stem to stern, which is the objective of a facing shot at thick-skinned game. Two and a half feet of bone and muscle protected his vitals from in front. I looked at Nada, and again he understood without words having to pass between us. He got up and melted away. A minute later his voice shouted hoarsely from the right of the buffalo and a couple of hundred yards off. The buffalo lifted his head and turned to face the sound, thereby giving me a perfect broad-side shot. I put the bead of the .275 on the line of the buffalo's left shoulder, on the horizontal third, where the head is, and fired. I fired twice more, very quickly, the three flat reports merging into each other. In animals there is a peculiar bend in the foreleg, just about one-third of the body up. A bullet placed there has a sporting chance of missing the massive shoulder bone and reaching the heart. But the bulls-eye is four inches in diameter and I was shooting fast and at a hundred yards. With luck, one of the three bullets might get through the gap and reach the heart. On the third shot the buffalo pecked, and moved off towards the jungle at a slow trot. I did not fire again. If you can't do what you want to do with three shots, you can't do it at all.

A miss sounds hollow. The sound of a hit is solid. I was satisfied that all three shots had been hits; apart from the matter of sound, it was hardly likely that I had missed an animal as large as a buffalo at a hundred yards. But had I hit the jackpot, the only four inches through which the .275 could reach the heart? Nada materialised beside me and said, 'Hit. Badly. His head was down when he passed and he was having trouble with his breathing.' Looking for blood stains in a waterlogged paddy field is an exercise in futility. So neither of us wasted time on the paddy field. Instead, we picked up the trail on the other side of the thorn fence, and after four or five paces we found three, exactly three, drops of blood. We squatted and looked at them carefully. The blood was just ordinary blood, ordinary blood coloured, not the bright frothy pink of lung blood nor the dark dull red of arterial bleeding from close to the heart. The only thing encouraging was that there *were* only three drops. Surface wounds bleed profusely. Perhaps the bullets or at least one of them had penetrated deep enough to reach somewhere close to the heart, if not the heart itself. Perhaps I had got him in the heart. A large animal, heart shot, will sometimes travel an unbelievable distance before dropping dead. We followed the tracks cautiously for another hundred yards but there was no more blood, and it was getting dark. The hoof marks did not suggest an injured limb, the weight was evenly distributed on all four legs, and I cheered up a bit. I may have found the vital four inches after all.

Nada did not offer any suggestions but I know what he wanted to say, and I said it for him. 'This is a job for the dog. I'll have to bring him. There is no blood now.' Nada nodded gravely, 'Yes, we'll need your dog. It is very difficult, further on, in the rocky place.' Rather than walk another half mile to the village—the motor cycle could not get there through the slushy cultivated area—I decided to sleep where I was. Nada brought me a blanket and something to eat, so it was not too bad, except for the mosquitoes.

A Wild Buffalo Hunt in Bastar

At earliest dawn I was on my way back to Kondagaon to fetch Bonzo.

This is the place to tell you something about Bonzo. He was a brindled bull terrier, all of fifty pounds in weight, and compactly built, with the typical heavy chest and jaws of the bull terrier. Like all the bull terriers I have owned, he had brainwashed himself into believing that it would be a pleasure to die for me. But even for a bull terrier he was exceptional. Given a single drop of blood, he would follow a trail from the Himalayas to Kanyakumari if need be. In the jungle at least, he was obedient, answering perfectly to hand or voice commands. He was gentle too, and loved children, even my own who submitted him to the most horrible indignities including the pulling of marrow bones from his mouth. But two words of command in particular he had learned thoroughly. One was 'Get him!' and the other 'kill!' 'Get him!' was meant primarily for thieves, and on receipt of the order he pulled down whatever was pointed at, without doing damage. My children made him practise on them! On the world 'Leave!' he left. 'Kill!' meant just that, and there was no recall. I taught him the 'Kill' command because he was brought up for the jungle. A dog that will fling himself at a charging tiger may just possibly buy you the time to reload—with his life. Bonzo had other qualities, too. Once five pounds of candied peel disappeared from the meat safe in which it had been stored, and my wife gave the servants a hard time until Bonzo reluctantly surrendered to *force majeure* and vomited the stolen property. He had taught himself to unlatch the meat safe with his nose.

Within half an hour Bonzo was on his way back with me, perched on a rubber pad strapped to the petrol tank. That was his normal mode of travel when I took him on the Frances Barnet, and he loved it. When we reached Singanpur, Nada and two men were waiting for us. The standard practice when following a wounded buffalo is for the job to be done by the shikari and three men, one to watch the right rear, one to watch the left rear, one to track, and the shikari to be ready to shoot at all times. I have arranged the actors in order of importance. There is logic behind this traditional arrangement. Like a wounded leopard, the wounded buffalo invariably ambushes his pursuers. The attack, when it comes, will never be from in front. It will always be from the right or left rear, and once it starts nothing but the buffalo's death will stop it. We took

Bonzo to the place where we had found the first blood, and I said, 'Find!' Bonzo sniffed for a few seconds, memorising the scent, and moved forward, head down, tail wagging intermittently as it always does when he is on a trail. Every twenty or thirty yards he looked back to make sure that I was within sight; that was part of his training. A dog which tracks too far ahead of the shikari is almost useless, because even if he finds the animal the shikari cannot come up in time to shoot. There was no blood after the first two hundred yards but Bonzo was never at fault. Another hundred, and he found the place where the buffalo had stopped to rest for the first time. There was a trace of blood on a high leaf, and nothing more. He had not sat down. Nada looked at me and nodded. I knew what he meant. The next stop would be for the ambush. His stopping so soon after the shot meant that he was badly hurt. He would not waste his strength in going too far. I turned the safety off and we proceeded.

Bonzo halted, tested the air with his nose held high, and walked back to me slowly. He stood stock still at my side after a single flicker of his tail, looking to the rear. 'This is it,' I thought. Bonzo always came back to me if a charge was imminent. The tracker who was watching the left rear clicked his tongue, and I half turned to see the buffalo emerging from the undergrowth with the inevitability of Fate, silently, implacably. There was none of the flamboyance of a tiger charge about his onset, and for some reason that made it more impressive. He came at a trot, not galloping as yet, his head held high so that only his chest and his nose were visible. With a really heavy rifle, I would have slugged him in the chest and dropped him. With a reasonably heavy rifle I would have waited till he lowered his head for the typical buffalo hook at the last moment of the charge, and then put the bullet in his forehead. But with the .275 I had neither choice. A chest shot would not penetrate more than a couple of inches, and there was no hope that the soft-nosed bullets with which I was loaded would pierce the heavy bones of his skull. I fired at his nose, and at the same instant shouted, 'Bonzo, kill!' The fifty-pound bull terrier took off like a streak of lightning to meet the charging two-ton buffalo, leaped into the air and fastened on his nose. That is the most courageous thing I have ever seen, and it was not done in hot blood either. The buffalo stopped in mid stride with the dog dangling from his bullet-torn nose, shook his head violently to break the grip, and failed. The

A Wild Buffalo Hunt in Bastar

next thing I saw was his rump as he fled, with Bonzo still dangling from his nose. I thought sadly, 'Goodbye, Bonzo. It was lovely to have had you.'

The crashing in the undergrowth passed beyond earshot, and I cursed my habit of always following any animal I wounded with the rifle that had done the wounding. It is entirely stupid, this giving of hostages to fortune but I have to do it. If I am foolish enough to use an inadequate rifle, then I owe it to the animal to carry my foolishness to its logical conclusion. I loaded another cartridge into the magazine to replace the one I had used, and we got back on the trail. The buffalo was moving in a curious fashion now, wobbling from side to side and bouncing off small trees, like a drunken man, but this was because he had been trying to throw Bonzo off. Once he got him off—I closed my mind to the thought of the broken brindled body that we would discover.

We went a couple of hundred yards more and Nada stopped suddenly. The .275 came to my shoulder of its own accord. Nada said in a low voice, 'The dog—but he is alive!' He sounded surprised. I followed the direction of his stare and made out Bonzo seated in the shadow of a bush, panting, with half a yard of tongue hanging out, very much alive and very much at his ease. He growled when we approached him as a warning to the others to keep their distance because he was guarding something. The Murias halted. I went to him and his tail wagged furiously. He had his paws on a large lump of the buffalo's nose. I said, 'Leave!' and he moved obediently aside. Then I said, 'Come!' and he leaped on me and licked my face and we loved each other. He had fulfilled the first duty of a bull terrier, to hang on. He had hung on to the buffalo's nose until it ripped away in his mouth, and that was what had saved his life. The buffalo's agony must have been tremendous, diverting him from all thoughts of Bonzo. He wanted out, to hell with the dog, and he got out. The Murias were almost as pleased as I was at seeing Bonzo alive but Nada said doubtfully, 'Will he follow again?' 'Follow again? He'll follow to the end of the world!' I replied, scornfully. I turned to Bonzo, pointed at the plate sized hoof print of the buffalo, and said, 'Find!' He put his nose down promptly, and we were off again. After a while it became apparent that now the buffalo was climbing and veering east at the same time. I knew the terrain. There was water beyond the range of hills we were ascending, about five or six miles away, a dried up river except for scattered pools which would be ideal for wounded buffalo to wallow in as a protection from the flies.

Silently from the jungle, from our left rear, the buffalo emerged. Once again it came at us with the inevitability and the implacability of Fate. Bonzo turned and saw him. This time he did not wait for orders. The buffalo was an unfinished job, and bull terriers hate to leave a job unfinished. He streaked for the buffalo. And the buffalo—this no one will believe—turned and ran with Bonzo hot on his heels. Spontaneously the Murias and I broke into laughter and we were still laughing when Bonzo returned, disappointed. Bull terriers are not fast, their legs are too short. Nada said in a relieved voice, 'Now he won't try again, he'll make straight for the water. There's no hurry, let's have a drink ourselves.' The drinking of water to a Muria in the jungle does not quite mean drinking in the normal sense. It means taking a large mouthful of water and letting it trickle down your throat, drop by drop, for who knows when you will be able to replenish your supplies? And that mouthful is only permissible when a pebble in the mouth no longer works. Each of us had our mouthful from the single hollow gourd that carried our total supplies. When it was finished, I put Bonzo back on the scent and we resumed our climb. An hour later we were at the top of the hills. Far down below was a deep gash in the jungle that told of the river. We stopped to breathe ourselves before we began the steep descent, for panting is bad for concentration and the jungle is no place for absent-minded professors.

I was proud of Bonzo that day. In spite of the rocky ground—which dissipates scent—he was never at fault and there were no checks. He even showed off by taking short cuts, leaving the scent where the buffalo had gone in a curve and going straight to the point where it resumed the general direction of travel. Now we were close to the river after eight miles of tracking. Nada said, 'Leave the dog behind. He will go for the buffalo as soon as he sees him.' I made Bonzo sit and gave him the command 'Stay!' That was all that was needed to keep him there until I returned, after two hours or two days.

We began the final approach, what is technically known as the close stalk. It *had* to be absolutely silent. If a patch of dried leaves was to be crossed, the foot was not placed *on* them but wriggled *under* them slowly, without sound, and weight lowered on to it ounce by ounce. If a tall rock stood in the way, we went round it, not over it—climbing over a rock increases height and therefore visibility. That last furlong took us nearly an hour. Nada had steered so accurately that we came out of the river at the place of the pools, where, with any luck, we could find the buffalo.

There was a high, wooded bank on our side, and below it was the river bed, dry and littered with huge rocks. About twenty yards from our side and almost overhung by the bank was the most likely pool, Nada whispered to me. We could not see it from where we were. We left the other two Murias behind and crawled along the bank until a slight pressure on my leg from Nada stopped me. I eased myself forward, inch by inch, until I could look over the bank. There, almost directly under me, was the pool, small and rock encrusted, and in the centre of the pool was the buffalo. He was immersed in water up his ears and every now and again he lowered his head into the pool, leaving only his horns exposed.

After an incredibly long time he raised it to breathe. I waited. The next time his head came up the bead foresight of the .275 centred on the hollow behind his ear and I fired. His head dropped into the water, two horns protruding. I waited for a minute or two—there was no movement. Nada and the Murias came up and we stood looking at the buffalo. I thought, 'For everything that lives, something has to die. It was not my fault. The Murias have to live so this poor devil had to die.' But I was only making excuses. I told Nada to send the head to Kondagaon and soon afterwards Bonzo and I were on the Francis Barnet and on our way.

I was in Kondagaon two hours later. After a hasty bath and change I set out for Kanker and spent the next four days auctioning liquor vends, a boring business. I stopped at Kondagaon on my return journey to headquarters and asked the Tahsildar if the buffalo head had arrived from Singanpur. He said, 'What head? Nothing has come from Singanpur.' I cursed silently. Those damned Murias, they had gorged themselves on the buffalo and washed it down with their home-brewed liquor and were still recovering from the hangover. The footpath to Singanpur is visible from the rest-house where I was talking to the Tahsildar. At that very moment a line of Murias emerged from the jungle along the footpath, eight of them carrying a huge buffalo head tied to poles, Nada leading. They saw me and came to the rest-house, and the first thing I noticed was that the head was absolutely fresh—it had certainly not been removed five days ago.

'When I came back from the village with men and pool was empty, he had gone,' said Nada, wearily. 'Then we followed him for days. He was weak, but not weak enough for us to kill, so we waited for him to die. He died this morning. Neither your bullets nor your dog killed him. Worms did that. The last bullet flattened out against his neck bone without breaking it, and none of the first three reached his heart—they broke up before getting there. But a piece of one of them nicked the covering of the heart and the worms got in and did the rest. His nose and mouth were all smashed up. Of course he would have died of starvation but that would have taken a month or more. No, the worms killed him.'

Excerpted from Animals and Other Animals *(Delhi: Sanchar Publishing House, 1992).*

PART II

On Hill and Mountain

During one beat for Barasingha (Kashmir stag or hangul), I sat in a tree on the crest of one of the lower slopes towards the Jhelum valley. Behind me gleamed the long line of snow mountains; all around and below was every shade of autumn tint; beyond and stretching away for a hundred miles, was the opalescent blue of the vale of Kashmir...It did not really matter whether the stag was in the beat or not.

CLAUDE H. HILL, c. 1929

ISABEL SAVORY

Himalayan Black Bears

The world of shikar was pre-eminently a man's world. Hunting anecdotes are replete with references to 'manly pursuits', to pig-sticking as a means for 'the purgation of lusts' and to the need to hunt to overcome 'effeminacy'. British writers often claimed that only white men and a few Indian communities like the martial races (Gurkha, Sikh) and forest tribals (Toda, Santhal) were brave enough to hunt dangerous game on foot.

Isabel Savory was an exception, an English woman whose book A Sportswoman in India recounts her exploits. The valley of Kashmir, already a favourite destination for vacations, camps and big-game shoots, was among the places she visited. The Himalayan black bear was a prize trophy to get, a reminder that up in the mountains, there was another whole world of wildlife, in lands beyond the tropics and tigers. Savory shared a big-game hunter's view of India: she disliked Indian trappers and fowlers even as she extolled the glories of European sport. She hunted bears in summer, when they emerge from hibernation sites and greedily feed on insects, grubs and fruits.

> But you've no remorseful qualms or pangs
> When you kneel by the grizzly's lair:
> On that conical bullet your sole chance hangs,
> 'Tis the weak one's advantage fair,
> And the shaggy giant's terrific fangs
> Are ready to crush and tear;
> Should you miss, one vision of home and friends,
> Five words of unfinished prayer,

> Three savage knife stabs; so your sport ends
> In the worrying grapple that chokes and rends:—
> Rare sport, at least, for the bear.
>
> LINDSEY GORDON

The next day was spent in marching to Keypoor, still down in the Valley of the Pohru; and we had a restful afternoon wandering about its banks. Our tents were again put up in a grassy orchard; and towards evening a whole troop of monkeys came out of the forest, walked across the shallow river, picking their way over the stones, and invaded our orchard. They swarmed up the apple—and mulberry-trees not far from us, shaking the boughs and tumbling down fruit into the arms of the leery old ones waiting underneath.

Khubr was brought us that evening, and very encouraging *khubr* too. Lalla had secured plenty of coolies for the next morning, and we went early to roost with renewed hope.

Damp and chilly dawn saw the faithful band leaving their camp shrouded in white river mist, picking their way across pastures and through a field of Indian corn dripping with dew, and thence disappearing into the jungle. Lalla and S. were walking first, myself on the *tat* came third, and the *chota shikari* brought up the rear. The coolies had been divided into three bands, and sent on some miles from where we were eventually to take our stand.

The dim light of the morning as yet barely penetrated the gloom of the jungle; deodars towered over our heads, and on either hand formed masses of impenetrable depths, suggestive of containing much of the unknown. Silently we trod, avoiding stepping on a branch or anything which might betray our presence.

I soon had to leave the *tat* behind, tied up to a tree, where I hope he enjoyed himself more than I should have done under similar circumstances. To be left behind alone was one of the last things one wished, and I crept and climbed and scrambled assiduously, keeping my breath for the time when it would be wanted, and often gratefully accepting a tow from the *chota shikari* with his stick.

It was very hot as well as very silent work. We were following a small stream, and coming to a deepish pool in one place, Lalla, after peering over the wet soil round it, whispered in an awed and triumphant voice, with many gesticulations, '*Harpat's* bath'—pointing to the large, unmistakable track ending in a sharp claw-mark; there were even splashes all over the rocks and drops on the ground, showing where the bear had shaken himself. So there in that dark hollow, among the thick jungle grass and in the little, clear tank, a bear must actually have been within the last half-hour. This was distinctly encouraging; after all we had gone through, 'fought for, and wrought for,' a tangible prize really lay somewhere at the end of the struggle.

More silently than ever we crept on, the two *shikaris* moving with the greatest care and very slowly, 'tracking' on either side of the path for more traces of the *harpat*. Every sign pointed to his being still ahead of us. We walked and climbed in this way for some time, until the jungle began to open a little and show a clearer space, where one could at least see to shoot between the tree trunks, the undergrowth having disappeared. By gesticulations Lalla showed us we were to stop; and he and the *chota shikari* then proceeded to break off and bend back any twigs which they seemed to consider it was possible might come in the way across the line of fire.

Here, then, we were to take up our stand; and here was to be—success or failure. The inevitable cold tea bottle was brought up to the fore, and we both had a drink. It sounds uninviting, but anything is better than stimulants in a hot climate; and after all, it was just as well on these occasions to be cool and clear-headed. And so we stuck to cold tea.

The *shikaris* chose a bank sparsely covered with deodars to stand on, to the left of us a small nullah, in front of us the jungle sloped gradually away into the distance.

The beaters, as I said, had been sent on some miles ahead—in fact, to the extreme limit of this patch of jungle. One party was to start from there in line, beating towards us; the second contingent was to act as stops, standing at intervals, on one side lining the country between ourselves and the beginning of the beat, in order to try to turn any bear which should try to break sideways, instead of coming straight on; the third party was distributed as stops opposite the second party.

We had a long wait—long enough to cool the most 'jumpy' nerves when once they had grown accustomed to that unbroken silence, which is apt to become almost painful.

The slightest rustle! one sees visions and dreams dreams! Once a great red fox stole by, with a brush almost as big as himself. And a picture he was! In the mind of the fox-hunting sahib rose recollections of a marshy ride, of a whip at the corner of the cover waving his cap, of the long mournful, and withal heart-stirring cry, 'For-r-ad—aw-aaay.' We saw a pig with a pair of curling tushes; and then after that again dead silence.

At last, far away, the well-known vibration of tomtoms, the yells of the natives; the beaters are coming, the beat has begun. Now is the time to harden the heart against disappointment, or to be ready to face any emergency. Two of the big deodar trunks hid us all four completely. Alert, straining every nerve to see and to hear, curbing every wish to move and rustle, to breathe almost, we crouched, kneeling, waiting.

The tom-toms and cries gradually got nearer and nearer, louder and louder; it seemed a long time—too long—to be in such a state of tension. I was just beginning sadly to think that this *honk* had nothing in it either, when unexpectedly came, as it always comes, the supreme moment, and the present was alive with reality. Lalla's sinewy back in front of me was stiffened with excitement; I caught the gleam of his eye as he turned, and the *chota shikari* hissed '*Harpat*' into my ear.

We were all ready to move in a second, and there, coming quickly towards us, was a great black form, which now emerged into the open, now disappeared behind a tree. The bear suddenly turned off to the left, and was passing our stand without seeing us, when at that moment the *chota shikari* moved and cracked a stick under his foot. The bear turned and growled savagely; but a shot rang out almost simultaneously, and a thick smoke hid what followed. Hit or not, the bear had gone on.

We rushed after him. Whatever happened, I knew that I must keep up with the rifle, and I did my best, hauled along by the *chota shikari*. But through such tangled jungle it was no joke moving fast, and a difficult matter to get through some places at all. Left behind once or twice, at last, to my joy, we caught Lalla and S. up, hurrying down into a nullah. In the gloomy shadows it was difficult to see far; there seemed no traces whatever of the bear.

By this time the shouting and the tom-toms sounded quite close, the sticks rattled, the beaters could be heard pushing through the bushes. Suddenly, above the yells of the coolies, rang out another cry—a shriek of alarm—which was instantly taken up on all sides; the jungle echoed with shout after shout.

We turned round and saw at once, on the other side of the nullah, another and much larger bear coming along the top. He turned off sideways; and we saw, to our horror, that one of the 'stops' was right in his path. He was unfortunately an old man too, and could not spring quickly aside. Lalla was an idiot to have allowed him to beat.

What followed was the work of an instant. To shoot would have been even chances of killing the bear or the native. With furious growls the bear sprang towards the half-naked coolie, and springing up at once erect on his hind-feet, he hit the man on the top of the skull a buffet with one great forepaw, with the other he struck the man's up-raised arm, and at the same time bit him in the chest. With agonising shrieks the poor native was thrown upon the ground; the bear left him, and as he moved away from him S. fired. It was a long and, I suppose, difficult shot...Gracious heavens—he has missed him!

The bear turned in an instant, saw us, and rushed down the nullah straight at us. It was a moment to turn one a little pale. This 'glorious hour of crowded life' seemed likely to be overcrowded. Lalla and the *chota shikari* shouted and yelled for all they were worth, in the last extremity of terror, but no cries would deter this *harpat* from his purpose—he must be on us in a moment.

I remember thinking of getting hold of the spare gun, but saw with the corner of my eye that the *chota shikari* was 'shinning' up the nearest tree, and the eight-bore lay at the bottom. I stuck to S., which was the only thing to do, and hoped for the best.

On the old bear came, in far less time than it takes to read this, growling with rage; S., with his last barrel, waited to make it a dead certainty; but the *harpat* was most appallingly close, not farther than four yards at the longest estimation, when his gallant charge was ended by a bullet behind his ear. His body rolled over and over to the bottom of the nullah. It was certainly with a sense of relief that we looked at each other, for a bear who will charge in this way is *not* usual, and we were unprepared for quite such an emergency.

We ran across to the poor native, who was soon surrounded by a sobbing throng; his scalp was lacerated, his wrist broken, and his chest mauled. A native doctor was either fetched or appeared from somewhere. We asked him what he would do. He replied, 'Give him medicine.' We asked, 'What medicine?' 'Oh! he would buy something down in the village bazaar,' he said. Any one who has ever walked through a filthy little native bazaar will understand why we took the patient out of the doctor's hands, got him into a *dhoolie*, and sent him off to Soper hospital, two days' journey, under the care of Sala Bux and relays of coolies.

Lalla said to S., 'We'll carry him to the camp, and when we get there, we must get some pins, and pin his head up.' So much for the primitive Kashmiri surgery! It turned out that the man's skull was slightly fractured; but he recovered, and left the hospital in six weeks' time. S. went to visit him there, after a letter from the surgeon-in-charge, in odd English, in which he said, 'If you will come and visit him here, his sickness will be half.'

Meanwhile the great black bear, an old male, measuring six feet three inches, with worn-down teeth, and a rusty coat, was worth examining. He had the usual white crescent-shaped mark on the chest, and a white lower lip. His head was particularly short and round, but very broad and massive, with its cruel, piggy little eyes. His claws were unusually long and sharp, better adapted for climbing than for digging. He must have stood up to the beater seven feet in height quite.

It still remained to have a last look for Bear No. 1. Searching carefully, blood was seen on the ground; and, tracking his marks warily, we at last came on him, stone dead, in the open, shot through the side. He was a smaller bear, with a better coat, a deep, glossy black.

I left S. still arranging about the poor native, and with a coolie as a guide I went back and found my *tat* and rode to the camp. Later on a tremendous noise resounded up the valley, and by-and-by into our orchard wound a triumphal procession, headed by tom-toms thumping wildly and natives dancing and capering, while around the two bodies of the bears, borne upon poles, a shouting throng jumped and cheered.

The paying of the beaters followed. At last they all cleared off, alternating between weeping like children over the recollection of the accident, and screaming and dancing with joy over the successful issue of the *honk*.

We sat down to a soothing meal, while Lalla and Co. skinned the two *harpats*; after which the skins were carefully stretched and secured by numerous pegs along their margins, powdered alum was thickly sprinkled on and thoroughly rubbed in, and some parts were well anointed with arsenical soap. The next day, rolling the skins up with the hair inwards, we sent them straight off on the back of a coolie to Srinagar to be roughly cured by our skin-man, before despatching them home to Rowland Ward. On skinning the big bear they found he was literally stuffed with mulberries.

As an example of what natives are, after having skinned the bears they never troubled to bury the bodies, but left them both in the sun, and in the wind's quarter. Though late in the afternoon, the smell which blew into the camp was quite unbearable; however, the servants were unconscious of it, and would, I am sure, have lived next to the carcases cheerfully for the rest of the autumn.

Having had such luck with bears, we thought that we could not do better than march away farther afield and higher up, and try for a *bara singh* (twelve horns), by which name the Kashmir deer is generally known among *shikaris*. He is almost identical with the red deer in Scotland; but he is, alas! gradually being banished from many hills where he once abounded by the vastly increased herds of cattle, especially buffaloes. Not only so, but the deer are massacred almost wholesale by the natives, with dogs, in the snow in winter. Gurais was once a favourite locality, and in this direction we turned our faces.

From Keypoor we marched by degrees to Imbresilwara, where we were in camp all among fir-trees. From there we marched to Alsoa, along a ridge, with a magnificent view of the Lolab Valley stretched out below us on our left. We saw Nagmerg in the distance; and then descending from this high ground, we came down—down—to Alsoa, almost on the borders of the Wular Lake.

Excerpted from A Sportswoman in India: Personal Adventures and Experiences in Known and Unknown India *(London: Hutchison & Co., 1900).*

F. W. F. FLETCHER

The Nilgiri Wild Goat

The Nilgiri hills in southern India which include the famous hill station of Ooty and picturesque tea gardens that are a get-away from the plains, had a spectrum of fauna reminiscent of the Himalayas: pine martens and the Nilgiri tahr among them. The latter, often called 'wild goat' or 'ibex' is found only in the highlands of southern India. With age the bucks' coats become almost black with a distinctive light patch on the back which looks like a saddle, leading to the name 'saddleback'. The Nilgiri tahr is a highly elusive animal, able to climb craggy slopes and take refuge on cliff tops. Yet a thirst for trophies may have wiped it out had a group of sport-hunters not formed the Nilgiri Game Association in the 1870s and enforced strict quotas. Fletcher wrote at a time the tahr numbers had begun to recover. A few select males were shot for another half century. His own account is memorable for his references to how different groups of hunters, European and tribal, hunted the ibex of the cliffs.

I chace the Wild Goats o'er Summits of Rocks.

DRYDEN

The Nilgiri wild goat is found on the Nilgiri and Anamallai Hills, and on the higher spurs of the Western Ghats as they traverse Travancore and Cochin. It does not descend below four thousand feet, save in a few localities where the lower slopes are rugged and broken. North of the Nilgiris, the grand western mountain chain affords in many places (*e.g.* the higher peaks of the Vellarimallais) an ideal habitat for this cliff-loving goat; but I have never been able to obtain evidence of its occurrence in such localities. In view of this long stretch of suitable country trending northwards, the present isolated habitat of the Nilgiri ibex is curious, more especially as the relationship to its Himalayan congener is sufficiently close to warrant the assumption that at one period a single species inhabited the whole of Western and North-Western India; the two existing species being merely local races whose distinctions are due to their present environment.

In describing the habits of the Himalayan tahr Blanford writes: 'Col. Kinloch's account is excellent. He says, "The tahr is, like the markhor, a forest-loving animal, and although it sometimes resorts to the rocky summits of the hills, it generally prefers the steep slopes which are more or less clothed with trees... Old males hide a great deal in the thickest jungle."' Here a wide divergence occurs between the habits of the two species, for the Nilgiri ibex is the very reverse of a 'forest-loving animal.' It frequents the beetling crags and towering precipices of the Kundahs, far above the forest line, coming up to the grass slopes which border the cliffs to feed. These grass slopes usually hold *sholas* in the folds between the hills, but save when wounded, I have never known an ibex take refuge in them. About April ibex frequently leave the cliff-line, and roam a considerable distance inland, attracted by the fresh sweet grass which springs up after the annual fires; but if disturbed, they retreat at once to the inaccessible cliffs.

In former days—the halcyon days of sport on the Nilgiris—ibex were found in very large herds, an assembly of even one hundred being not uncommon according to the accounts of old-time sportsmen. But owing to incessant persecution the numbers were thinned at such a rapid rate that at one time the ibex stood in imminent danger of extermination. The introduction of the Nilgiri Game Act in 1879, which prescribed a close time for all game, did much to avert this calamity: the absolute prohibition of ibex shooting, which followed a few years later, did more. Under this salutary legislation there was such a steady increase in the herds, that in 1908 it was found possible to permit the shooting of one saddle-back under each licence issued in a season, and this rule still obtains, though, I need hardly add, a saddleback does not fall to the lot of every sportsman who goes on a shooting trip to the Kundahs. The largest herd I ever saw was at Bettmund on a glorious January morning in 1890, and numbered twenty-nine individuals. Curiously enough, this large herd did not contain a single warrantable buck; and, ensconced behind a rock, I contented myself with watching them for a couple of hours, while they fed up to within a hundred yards of my post.

Ibex begin to feed at sunrise and continue feeding till about nine or ten o'clock. If the country is quiet, they then lie down in some warm nook, sheltered from the wind, near the cliff-line, rising to feed again in the early afternoon. But if they have recently been disturbed, they descend some distance down the cliffs before couching. As said before, I have never known them seek the shelter of a *shola* for their midday siesta, or at any time save when wounded: in fact they may rightly be described as open-loving animals. A sentinel is invariably posted to watch over the slumbers of the herd, usually a doe, and an extremely wary sentinel she is. Perching herself on some dizzy eminence, for an hour or more she makes the most minute survey of the surrounding country; then, if satisfied that no danger threatens, she lies down, being careful to place herself in such a position that she can still maintain a vigilant watch. Frequently, for a part of the year at least, an old buck leads a solitary life; and while a bachelor, he goes through all the above precautions every morning himself. But, in common with most wild animals, ibex are never very suspicious of danger from above; and it is their want of caution in this respect that sometimes gives the sportsman his opportunity, for if he can get above them, a stalk is an easy matter provided the wind is right. I say 'sometimes' advisedly, for as ibex usually select the highest ground for their midday couch, it is not often that a chance of an approach from above occurs; while so wary and keen sighted are they, that a stalk from below, no matter how carefully conducted, is almost always hopeless. It is this extreme caution on the part of the quarry, coupled with the grand country they frequent, that makes ibex stalking the cream, the poetry, of Nilgiri sport.

As young kids run with the herds all the year round, ibex do not appear to have a regular breeding season; but so far as my observation goes, more kids are dropped in the spring months than at any other period of the year. Some writers have stated that the doe always has two kids; but though I have seen two kids of apparently the same age with one mother, more frequently I have seen only one; and I do not think it is the rule that two are produced at a birth.

Every phase of South Indian sport has its own peculiar charm; but from one standpoint—that of mountain scenery—ibex stalking o'ertops them all. When following the Nilgiri wild goat over the beetling crags on which he loves to dwell, you meet Dame Nature in her grandest aspects; and there is ever present too that spice of danger without which any sport loses its attraction. In the words of Lindsay Gordon (slightly altered),

> No sport was ever yet worth a rap,
> For a rational man to pursue,
> In which no accident, no mishap,
> Had need to be kept in view.

One needs a cool head and a sure foot amongst the tremendous cliffs which form the western face of the Nilgiri plateau.

In December, 188—I was out in camp with J. near T—mund, on the Kundahs. We sent on our baggage and tent from 'Ooty,' and shot our way out to camp, getting a fair bag of small game, including two wood-cock, on the road. The next morning J. started with his *shikari* to work up the long valley in front of the *mund*, while I paid a visit to the B—cliffs, perhaps the best ibex ground on the Kundahs, with my gun cooly. I reached my destination just as day was breaking, and the view that was unfolded as the sun rose over the rocky ridge behind me I can only call sublime. In front of, and round me, in a semi-circular sweep, the cliffs dropped down to the low country in a sheer unbroken wall. A carpet of green turf ran along the edge, while in every valley and ravine nestled a *shola* of beautiful indigenous trees, running through every shade of colour from dark green to brilliant red. The rhododendrons were in full flower, and the masses of carmine blossom turned each hillside into a garden. Far away below me the plains stretched out to the sky line in an emerald carpet, through which the hill streams wound in bands of silver, sparkling and flashing in the rays of the morning sun. At the foot of the cliffs lay the dense forest which clothes the foothills along the whole western face of the Nilgiris—the home of elephant and bison. To my left, miles away, the needle-like cone of Mukarti shot up into the blue sky, and further still the bold ridge of Nilgiri Peak ran out into the plain, its summit broken into fantastic pillars and cupolas of granite. I know of no sensation to be compared with the feeling of awe that comes over one in the presence of such mighty works of Nature as these. The towering heights: the awful depths: the vast gloomy forest bring home to a man his own insignificance with overwhelming force. And over all broods that tremendous silence; broken only by a stream rippling over the cliffs in a veil of silver, or the swish of a bird's wing as it darts down the sheer drop with a velocity that makes one shudder. I very much doubt whether any mountain range in India, save of course the Himalayas, possesses such marvellous scenery as this part of the Nilgiris.

I took up my post at the edge of a dizzy cliff, close to a waterfall which fell perhaps two hundred feet into a large pool. I had not been watching long when I saw a sambur stag ascending the opposite hill. Through the glasses I made out that he carried a decent head. After traversing a small *shola*, he lay down under a clump of rhododendrons on the summit of the hill; but ibex were what I wanted, so I left the stag alone. An hour passed, but nothing else appeared, and I was on the point of starting after the stag, when my gun cooly spied an ibex below. I stretched myself at full length, and peeped over the edge of the cliff. There, sure enough, far down, were three ibex, and shortly they were joined by three more. I made out that two at least were bucks, but they were so far below that I could not with any certainty judge the size of their horns. For another hour we watched them as they sauntered along, nibbling at the bushes which here and there grew out of the rock, until they reached the fall. Just here the cliff curved inwards, the water falling clear of the rock. A narrow ledge ran round the face of the cliff, and on this, under the waterfall, the herd lay down. I could now see them clearly through my glass. Two were good brown bucks, one carrying a fine head: the others were does and kids. In the hope that later on they would ascend the cliff to browse on the grass, I lay still; but though I watched for a couple of hours they did not move, and it was evident they had couched for the day. If I was to get a shot the only way was to go down after them, but the prospect was scarcely pleasant. Above the ledge on which the ibex were lying, the cliff was a sheer wall of rock, down which it was impossible to climb. But some distance to my right a narrow ravine ran down the cliff, and the soil in this cleft supported a scanty growth of scrub jungle. From the peculiar formation of the ground I could not see where this ravine led, or how far it descended, but as it afforded the only chance of reaching the ibex, I determined to follow it down. For some four hundred yards the descent was easy, as the trees gave us support; but then the ravine stopped and the jungle with it. I now found myself on the ledge along which the ibex had passed in the morning. Below was a precipice of unknown depth. Looking ahead, I could see the waterfall, but the ibex were hidden behind a curve in the cliff wall. The only way round this was by the ledge, and the mere thought of the journey made my flesh creep. The cooly was too frightened to come further, and I told him to wait in the ravine for my return. I sat down to recover my breath, and to wait till the feeling of awe had passed off a little: then I started along the edge. Curiously enough, once out on the cliff my nerve retuned, and borne up by the excitement, I lost all sense of danger or risk. Inch by inch I crept on, and soon I had rounded the curving face of the cliff. The ledge here was broader, and peeping round I saw the ibex about sixty yards away under the fall. The one big buck was nearest to me and on his legs; the others were lying down. All unconscious of danger, the buck half turned to reach a clump of *Strobilanthes* over his head. Taking careful aim at his neck I pressed the trigger: he dropped at once, and lay kicking with his legs in the air. I made sure he was mine, but alas! one convulsive kick took him over the ledge, and I saw him drop through space until he fell into a pool far below, sending the water up in a cascade. At sound of my shot the other ibex rushed back towards me along the ledge. Seeing me they swerved, and ran along the cliff below, over ground that I would not have believed could afford foothold for a fly. The second buck was amongst the last to cross, and as he passed I fired. The bullet caught him in the flank and he staggered. Here the rock was smooth and slippery: he began to slide: made a desperate effort to recover himself: and went headlong over the precipice. Where he fell I could not see, but as I heard no sound, he must have gone right to the bottom. My disappointment, and the names I called myself for having followed ibex over such ground, I leave to be imagined. There was nothing left but to go back, and that return journey was the most 'skeery' experience I have ever had. With the ibex in front of me, I had lost all sense of danger, but now that the excitement was over, the thought of the road I had to traverse made me shudder. As I crept back each step made me sick, and once or twice the feeling came over me that I could not face that perilous ledge; but I kept my eyes steadily in front, and at last surmounted the dangerous bit. How long it took me, I cannot say, and the sense of relief that came over me when I joined the gun cooly can be better imagined than expressed. I have a strong head, and ordinarily dangerous ground does not flurry me, but I freely confess that nothing, not even the 'biggest saddleback wot ever was seen,' would tempt me to repeat my journey down that cliff.

Next morning J. and I returned to the cliffs, to see if there was any chance of recovering my lost ibex. At the point where we struck them, a little north of the scene of my adventure on the previous day, the ground slopes up in a long incline, so that the cliffs are hidden till you reach the edge. J. had not visited the Kundahs before, and as I was keen to see how the awful grandeur of this face of the plateau would strike him, I did not give him warning of what was in store. We reached the last valley, crossed the stream, and climbed the last ascent, J. unconscious of what a marvellous vision, lay beyond the summit. Suddenly, as suddenly as if it had been a fresh slide in a magic lantern, the scene changed. The grassy upland lay behind, and we stood on the brink of the precipice, with the whole panorama of rugged spires, yawning chasms, giddy heights and bare walls of rock streaked with silver ribbons spread out before us. 'Well?' I asked. J. took one look round, and simply said, 'My God'; and were I to write pages I could not describe with half the force of those two words the feeling of mingled pleasure and awe that seizes and holds one in presence of such wondrous works of Nature as these. Mark Twain, speaking of the Alps, says, 'There was something subduing in the influence of that silent and awful presence (the Jungfrau); one seemed to meet the immutable, the indestructible, the eternal, face to face, and to feel the trivial and fleeting nature of his own existence more sharply by the contrast. While I was feeling these things I was groping, without knowing it, toward an understanding of what the spell is which people find in the Alps and in no other mountains' (M. T. is wrong there) 'that strange, deep, nameless influence which once felt cannot be forgotten—once felt always leaves behind it a restless longing to feel it again—a longing which is like home-sickness, a grieving haunting yearning, which will plead, implore, and persecute, till it has its will.' This is a long digression; but Mark Twain's words apply as forcibly to the Kundahs as to the Alps, and those who have been under the spell of mountain scenery themselves will forgive me for lingering on these reminiscences.

We reached the waterfall and peered over. Far below, a mere speck, we could see my first buck lying in the stream, but alas! to reach him was an impossibility, and his trophy was lost to me for ever. How I wished then that I had left him in peace on the chance of finding him on better ground at some future time. After a last peep, we separated, J. and his *shikari* turning to the right, and I going in the opposite direction. I sauntered on for a mile, but though marks of ibex were plentiful, I saw nothing, and about midday I sat down under a tree on the open hillside for an *al fresco* meal. Here the strong wind disturbed my cooly's equanimity, and he went over the knoll in front to get out of it. While I was considering what line to take on my road home, the cooly came running back to say three ibex were lying down on the further side of the hill. I climbed up, and through the glasses made them out to be a buck and two does. The ground was dotted with stunted trees, and I got to within sixty yards without difficulty. The buck was lying broadside on, lazily munching a mouthful of grass, and I fired for his shoulder. I heard the bullet tell loudly, but he jumped to his feet, and, with the does, rushed straight down the hill, which was as steep as the side of a house. I had a flying shot at him with my second barrel, but missed. All three ibex plunged into a strip of *shola* that fringed the stream at the foot of the hill, but to my satisfaction only the does reappeared on the other side. We followed as fast as we could, though caution was necessary in going down the slippery hillside, and lying on his back in the stream we found the buck, stone dead. My cooly, an Ooty 'beater' with a smattering of English, was first up, and he called out 'saddleback sar,' but on coming up, which I did with such undue haste that I contrived to fall down a bank and hurt my hip severely, I found him to be a good sized brown buck, with thirteen inch horns. The bullet had blown his lungs to a jelly, but so great was the impetus he got from the steep ground, that he had travelled full two hundred yards before falling. This was luck with a vengeance. Many a time I have fagged for days without even a sight of an ibex, but during this trip the fates were propitious, for here on two successive days I had come across a herd, and we saw another before we left the neighbourhood.

On the way back to camp, I was witness of the most striking effect of light and shade I have ever seen. The morning had been fine, but towards the afternoon the sky became overcast, and a dense bank of clouds gathered in the West, which rapidly spread upwards. In a short space they had covered the whole dome of the sky with an ink-black mantle, and by the time I had secured the head and skin of the ibex it was so dark and threatening that I decided to make for camp direct. Just as I reached the path along the cliffs, the sable pall split down the centre in front of the sun, and through the rift came a great beam of golden light which bathed everything in its path in glory while all else remained in deepest shadow. 'Heaven peeped through the blanket of the dark.' The effect in such sublime surroundings was magical, but I did not stay long to contemplate it, for it was evident a big storm was brewing. And sure enough, before I had covered half the distance to camp, down came the rain in sheets.

The next day was a blank, and that evening we decided to shift camp to Mukarti the following morning. It was a longish tramp, but this is one of the lions of the Nilgiris, and J. had not seen it. If the visitor to the Blue Mountains climbs the hill behind the old church, or better still if he strolls along the Kotagiri road to the point where it overlooks the Botanic Gardens, he will see, far away over the waving sea of blue gums that girdles Ooty and over the outlying spurs of the Kundahs, outlined clearly as a silhouette against the setting sun, a needle-like peak with a double apex, which by its peculiar formation stands out prominently about mid-way along the serrated line of peaks that bounds the horizon. This is famed Mukarti.[1] Long ago some kind but—from my point of view—misguided philanthropist cut a bridle-path to the summit of the peak. The distance is only some seventeen miles, so many visitors 'do' Mukarti now, and they are amply repaid for their trouble, for the view from the summit is amongst the best that can be obtained anywhere on a range renowned for glorious views. But the constant intrusion of sight-seers is not an unmixed blessing to the sportsman, for Mukarti and its neighbourhood is a favourite haunt of the ibex.

We pitched camp on the stream at the foot of the peak, and early the next morning were on the cliffs. These we followed in a gentle curve until we reached a place where, near a stream which went

---

[1] 'Mukarti Mallai' signifies in Kanarese 'the Peak of the severed nose.' According to Metz (a well known authority on the hill tribes) the local legend is that Ravana, incensed at the greater reverence paid to his enemy Rama by the hillmen, cursed them with a plague of vermin (a lasting curse, by the way!). Rama in revenge cut off the nose of Ravana's sister, and set it up—transmogrified into Mukarti Peak—as a proof of his superior power.

The Todas believe that from Mukarti's dizzy height the souls of dead men and buffaloes take their last plunge into Amnordr, the World of the Shades below this World.

thundering over the cliff, there had recently been an extensive landslip. We had not been scanning the broken country below very long, when far beneath us the *shikari* spotted a herd of five ibex, two of which through the glasses seemed to be bucks. We could not by any possibility get down to them, so had no choice but to possess our souls in patience on the chance of the herd coming up to feed on the grass above. They sauntered up the cliff, and at last lay down some five hundred yards below us. The *shikari* Selvia said that from where they were lying two tracks led up to the summit of the cliff, one of which had been swept away by the landslip. He therefore thought our best plan was to wait where we were, as the other, and—as he supposed—the only practicable path debouched close by. Between us and the landslip lay a narrow *shola* which extended to the verge of the cliff-line, scattered shrubs running some way down. We made ourselves comfortable under a bush, taking occasional peeps over the edge of the precipice at the ibex, which still maintained their position. Time passed, and as we had a long tramp before us to camp, I was beginning to think we should have to leave the ibex for another day, when suddenly, and as if moved by one impulse, all five sprang to their feet and ran a short way up the cliff. There they stopped, wheeled round, and gazed intently at something below. This manoeuvre was repeated several times, until they reached the landslip. Under this they huddled for a minute, and then—with the peculiar whistle which is the alarm note of the ibex—began to climb rapidly up the broken ground. It was now evident they did not mean to take the track which led past us. Their dark hides showed up clearly against the red earth of the landslip, and the distance could not have been more than two hundred and fifty yards. 'Is it good enough?' whispered J., but I shook my head, for I felt sure we should be able to work round the head of the *shola* in ample time to meet them when they reached the summit of the cliff. With this intention we jumped up and were on the point of starting, when the *shikari* seized my coat, and pointing down said, '*pillee, pillee.*' The erratic movements of the ibex were now explained, for sure enough, far below us, was a tiger creeping up the face of the cliff, on the line the ibex had taken. These of course were forgotten at once, and we held a hurried consultation. Selvia urged us to stay where we were, as the tiger would perhaps take the easier path up the precipice. But a glance at his face showed me that this counsel was dictated by 'funk,' and as it seemed most probable that the tiger would follow the track the ibex had taken, I advised a general move for the landslip. J. agreed; but the *shikari* begged earnestly that one gun should remain to guard the other path. 'All right,' said J., 'I'll stay here and you run round.' So far all the luck had been with me during the trip, and I would gladly have given J. the chance; but there was no time to argue, so I hurriedly made my way along the edge of the cover. On reaching the landslip I looked over, and saw the ibex huddled on a grass slope about a hundred yards away. It was a strong temptation, but remembering the nobler game in view, I refrained. From my position the *shola* hid the face of the cliff where I had last seen the tiger; but J. signalled he was coming up in my direction. With my heart going like a steam pump, I crouched at the edge, expecting every moment to see the tiger's round face appear. But I was doomed to bitter disappointment, for suddenly J. raised his rifle, and two rapid shots followed. I jumped up, and far away down the cliff I saw the tiger going at breakneck speed. How he went over such ground at that pace was a marvel. In desperation I sent a couple of bullets after him, but I have no doubt both were misses. I looked round for the ibex, but scared by the firing they too had vanished. Then slowly and sadly I rejoined J. He told me the tiger had crept steadily up until below the landslip, when the wretched tiffin cooly was so overcome with fright that he had run along to a big tree at the edge of the *shola*, up which he began to climb. The movement above at once attracted the tiger's attention: he stopped, gave one look at the cooly, then turned and ran back down the cliff. Seeing my chance of a shot had gone, J. had fired at a range of some three hundred yards. His first shot, he said, was a palpable miss, but the tiger had seemed to respond to the second. However, it was impossible to follow, so we were obliged to leave without even the satisfaction of knowing that the tiger had carried away a souvenir of his visit to Mukarti. What we said to the cooly was, I fear, unfit for publication. We did not reach camp till 9 P.M. and as it was a dark night, and the ground was very much cut up by deep buffalo tracks, our tramp was the reverse of agreeable. Altogether the day, with its chapter of accidents, had been too much or us, and when we got back neither of us was in a very enviable frame of mind. But our factotum had not been idle, and over the wonderful repast we found ready, we recovered our tempers. Truly the Indian 'boy' is a marvel.

The Nilgiri Wild Goat

> There is a realm of magic sable,
> Sable monarch he of it;
> One wave of his kitchen ladle,
> And *ex nihilo* dinner *fit*!

In April 188—I was staying with H., who owned an estate on the Kundahs in the neighbourhood of some of the finest ibex ground on the whole range. A few miles from his *tote* the plateau ended in a wall of cliffs that overhung the low country in a sheer drop of perhaps three thousand feet, and these cliffs were, when the fresh grass had sprung up after the spring showers, a sure find for ibex.

We were sitting in the verandah of his bungalow, indulging as usual in yarns of *shikar*, when I remembered that the next day would be my birthday. 'What would you like best?' asked H. 'A sixteen-incher above everything in the world,' I replied, for this had long been the summit of my ambition. 'Well,' said H., 'it's rather early for ibex about here, but we'll try the cliffs'; and at 5 A.M. the following morning we were well on our way thither.

We struck the cliffs at what H. called the Waterfall just as day was breaking, and sat down to enjoy the view. To our right the cliffs swept round in a wide curve broken into the most fantastic shapes. Pinnacles of bare grey rock shot up from the dense *shola* which here extended to the verge of the cliff-line, the view on that side being closed by a bluff capped with a helmet of rock, which towered like a giant above the rest. Far beneath us the plains spread out to the horizon in a sheet of green, through which a stream wound in a silver thread. On our left the wall of cliffs was not so rugged, and grass took the place of jungle along the edge. A couple of hundred yards lower down on this side was the Waterfall, a stream which bounded over the cliff in a veil of gossamer, sparkling in the morning sun like burnished steel.

We had carefully swept every nook and cranny with our glasses, but had not seen a living thing of any kind. H. had just remarked that apparently we were in for a blank day, when I chanced to look below the fall. A narrow belt of jungle crept up to the basin into which this discharged, clinging to the bare rock in a way that set one wondering where and how the roots of the rhododendrons of which it was mainly composed found support. I was taking stock of this, when out stepped an ibex on the side furthest from us, followed by her kid. Straggling after her came the rest of the herd, until we counted nine. Two were fair bucks, the others does and kids. The wind was blowing straight down the cliff, and there was nothing for it but to remain where we were until the herd fed up to the level ground above. For a full hour we watched them as they gradually ascended, nibbling at the bushes amongst the rocks, till they reached the fall. Here a ledge ran round the cliff, and one by one we lost them as they turned the angle. Then we held a council of war. The track which gave access to the summit of the cliff ran up a cleft in the rock, and was hidden from our view. I thought we might with safety creep a hundred yards further down, to be within range directly the ibex appeared, but H. considered it would be more prudent not to move till we saw exactly what they meant to do, and I gave in to his judgment.

Half an hour passed without a sign of the ibex, and we were all getting impatient—Bill, H.'s *shikari*, was with us—when at last the leader, the same old doe I had first seen, appeared on the grassy plateau above the cliff. In another five minutes the other members of the herd had joined her, and my heart bounded when I saw that they had brought a grand saddleback with them. Ye gods! how I gloated on his curving head and grey saddle. The herd were quite unsuspicious of danger, and began to feed at once. Very soon they reached the stream, and we watched anxiously to see if they would cross. If they did, we had but to wait till they fed within range; if not, to get a shot would be a more difficult matter. Another half hour passed, and still the ibex persistently kept on the further side of the stream. The wind meanwhile had shifted, and was blowing directly from the ibex to us. H. and I agreed that the Mountain did not mean to come to Muhammed; but how was Muhammed to get to the Mountain—the saddleback—who had kept religiously in the rear the whole time? We were then hidden behind some boulders on the summit of a knoll. Round the foot of this ran a brook, parallel with the cliffs, which fell into the main stream just above where the ibex were feeding; and we settled that I should creep down the line of bushes which fringed this brook—H., with his usual good nature, resigning the shot to me. 'It's your birthday, my boy,' he said, 'and there's the sixteen incher. Go ahead, and luck go with you.'

Gradually I crawled backwards, until the ibex were out of sight, and then made tracks for the brook. Silently I crept along, until I reached the larger stream, on the further side of which the ibex should be. I half raised myself and peeped over the bushes in front. Within twenty yards of me was the old doe, with her kid frisking round her, the rest being some thirty yards further back and to the left; but where—great heavens where—was the saddleback? I subsided and looked back to H. I could just see the top of his brown hat over a rock, and in dumb show I asked him what had become of the big one. A wave of his hand towards the cliff told me the saddleback had fed lower down than the rest. Here was a dilemma with a vengeance. I had been over the ground before, and knew that a hundred yards in front, just at the edge of the cliff, was a depression due to an old landslip, and in this the big buck must have ensconced himself. But I dared not advance another step. As it was, the doe and kid were so close that my heart was in my mouth. A slight slant of wind, an incautious rustle, and good-bye to any chance of a shot at the saddleback. Either of the other bucks I could have secured from where I knelt. They had decent heads, and in other circumstances I would have considered one of them an ample reward for the stalk; but the grand old fellow I had seen made them appear insignificant, and better a blank day than any head but his. In this position, *vis-à-vis* with the doe, I remained till I had almost lost patience, and thought of creeping back to H.; but on taking another peep, I saw the doe sauntering in the direction of the others, who in turn were feeding up the side of the hill in front.

Breathing a fervent prayer that they would go over the crest, I resumed my vigil. The Fates for once were propitious, and shortly I saw the last ibex disappear over the summit. The next instant I had crossed the stream, and was stealing along the hillside. The depression I have mentioned was of considerable depth; and, on hands and knees, with the utmost caution I approached the edge. Imagine my delight when, peering over, I saw the saddleback lying in the sun, with his head turned from me, not fifty yards away. My heart was going like a steam pump as I brought my Express forward. But it would not do to risk a miss through excitement, so I waited till my hand got steadier: then, with both elbows planted firmly on the ground, I took a careful aim well behind between spine and shoulder, and pulled the trigger. The buck made a desperate effort to rise; but I had heard the unmistakable 'thud'—pleasantest of all sounds to a *shikari's* ear—and knew that he was mine. He had just strength to stagger to his feet, and then fell back with all four legs in the air. My reward for days of fruitless toil had come at last! H. soon joined me, and together we gloated over my prize. He was a magnificent buck, with a well developed saddle, and horns which—measured fairly along the outside curve—were just fifteen and three-quarter inches in length. My over-night wish had been fulfilled almost to the letter, and a more acceptable birthday gift than the one Fortune had bestowed, I could not possibly have had.

The day was still young, so we sauntered down the cliffs for half a mile, to a point where the grass had been burnt over a wide stretch of country some time before. We found the whole area covered with tender shoots of grass springing up amongst the cinders; and the recent footmarks on all sides showed that both sambur and ibex had been busy. Here we sat down to an *al fresco* breakfast under a rhododendron, whose gnarled trunk and branches were festooned with long beards of grey moss. The cliffs at this point were rugged and grand in the extreme, and we had a superb view of the foothills. Bill, the *shikari*, squatted on the very edge of an awful precipice, and peered down from that dizzy height in the most unconcerned fashion imaginable. Every time I looked at him I felt a creepy sensation down my spine; but H. said he was used to it, and gifted with an unusually strong head. We had just finished the contents of the tiffin basket, and were lighting our pipes, when Bill crept a yard or two backwards, and excitedly signalled to us to come up. On our joining him, he told us he had seen two ibex below. I stretched myself full length on the ground and craned my neck out over Eternity. At first I could distinguish nothing, but shortly I saw the ibex on a ledge that jutted out from the face of the cliff, far down below us. While I looked they were joined by four others which had been hidden by the bushes, and the herd were evidently coming up for a banquet on the fresh grass. H., who knew every inch of the country, told me that the track reached the summit of the cliff some three hundred yards further down, and we started at once to wait for them there. The ibex were in no hurry, and a full hour elapsed before they rounded the sweep of the cliff and came into view. We waited till the herd—which consisted of two good bucks and four does—had fed some distance away from the edge, and then began our stalk. There was nothing to choose between the bucks, and H. decided to take the one on the right, nearest the cliff, while I was to go for the other directly he had fired. The stalk was an easy one as the ground was strewn with large boulders which afforded us ample cover, and in ten minutes we had crept up to within fifty yards. H. had a perfect broadside, and hit his buck fairly behind the shoulder. He staggered forward, but dropped almost at once. At sound of the rifle the rest of the ibex made for the cliff at top speed. I got a fair shot at the other buck, and hit him hard; but the pace at which he was racing down the steep slope gave him an impetus which carried him over the cliff. Luckily he did not fall far, and Bill was not long in bringing up his head. The horns of H.'s buck measured thirteen and a quarter inches, and of mine twelve and three-quarter inches—both well up to the average. It was now late in the afternoon, so we made tracks for home, well satisfied, as may be imagined, with our day's sport.

This is the record of an exceptionally fortunate day. Ibex are scarce now owing to the indiscriminate slaughter in the past of bucks and does, aye and even kids. I have heard of one sportsman(?) who years ago got a herd of nine ibex into a *cul-de-sac*, and squatted himself at the entrance and blazed away until he had finished the lot. But for several years ibex shooting on the Nilgiris was absolutely prohibited, and under this salutary regulation the herds have increased. At present, as already mentioned, one saddleback can be shot in a season under each licence issued, and this relaxation of the embargo can do no harm, for only a couple or so are bagged each year, and these patriarchs can scarcely be of much use for breeding purposes. But the difficulty is to enforce a rule of this kind, for the supervision of the game on a distant and quite uninhabited range like the Kundahs must necessarily be superficial, and it is impossible to stop the constant poaching by the Kurumbas living on the lower slopes. Both the tiger and the leopard take their toll of the ibex; but it is the two-legged poacher who does the damage. That ibex may again increase and multiply is a consummation devoutly to be wished, for—from the sportsman's standpoint—it is ibex shooting that lends to the Nilgiris their chiefest charm. Personally, I know that whether Fortune smiled or whether she frowned, I never regretted a jaunt to the Kundahs, for the wild grandeur of that mountain chain had for me a fascination that never palled. The sense of boundless freedom that thrilled through every nerve with each draught of the keen mountain air was in itself ample reward for a day's toil; and I never look back to the halcyon days

I spent with H. after ibex without regret. The Kundahs and the ibex are there, and to them I can return at any time; but H., true sportsman and truer friend, sleeps the sleep that knows no waking on the Blue Mountains he knew and loved so well.

Excerpted from Sport in the Nilgiris and in Wynad *(London: Macmillan, 1911).*

C. H. STOCKLEY

The Kashmir Stag

By the 1930s, there was a heightened awareness of the need to protect rare fauna from the threat of extinction. The Kashmir stag or hangul, a sub-species of the red deer, was among the handful of species the British Indian government was keen to help protect. But its only habitat was in the princely state of Kashmir where it had long been a privilege of the Dogra ruler and a few nobles to lay claim to the stags. The maharaja's game department charged a hefty fee from sportsmen who wanted to 'bag' a stag. Guests, including VIPs were exempt. C. H. Stockley, the big-game hunter, was among the most experienced sportsmen to have shot in the Himalayas. And he was one of the most knowledgeable of those who stalked the Kashmir stag.

There is no animal in the world so closely associated with British history as the red deer, for many of our laws, the badges of our great families, and outstanding historical episodes had their origin in the chase of the stag or preservation of the wild red deer for royal sport.

In England, and particularly in Scotland, we are apt to think of the red deer as almost exclusively an insular possession, whereas in Central Europe its hunting and preservation, with, above all, critical estimation of trophies, has been carried to the highest pitch for many centuries; with the resultant evolution of a hunting etiquette stricter than any of the drawing-room, and a set of laws of feudal derivation which, even up to the Great War, remained of almost feudal severity.

Farther east the red deer persists through Asia Minor and the

Caucasus, where his habitat grows rougher and wilder, and farther east again culminates and ends in the glorious country of the mountains bordering the lovely valley of Kashmir.

There can be no grander surroundings to enhance the joys of stalking than amongst the pine and birch forests which clothe the rocky ridges, deep ravines between filled with lush feed for the deer, climbing to open slopes of short green turf below granite cliffs and patched with low clumps of dark-green juniper.

There can be no finer quarry than the Barasingh, as he is called by the British sportsman, or *hangul* as known to the Kashmiri, and there surely can be no more invigorating climate than that of those forest-clad hills in September and October.

But first, that we may the more enjoy our sport, let us have knowledge of his ways and life history.

The Barasingh is distinguished from the other species of his group by his horns, which have the brow tine always shorter than the bez and never develop a cup at the crown; and by the light yellowish caudal disc which, fringed with dark brown, is often the most conspicuous thing to catch the eye in a bad light.

The coat is variable from dark grey to a dark brown, and I have seen dark liver-coloured stags, their hair so sleeked that it looked as though it were watered, and these stags invariably carry fine heads. The hinds do not vary as much in colour as the stags, but appear to be lighter in youth than in their prime, then turn greyer.

The fawns are heavily spotted and hinds sometimes carry spots in their second year.

Stags run 50 to 54 inches at the shoulder, hinds 44 to 46.

The horns carry 10 points normally, and the brow tine is shorter than the bez, but I have seen as many as 16 true points, not counting 'snags' or 'offers,' but a royal of 12 points (6 on each horn) is hard to get, and symmetrical heads with points above that in number are rare in the extreme; I shot the only absolutely symmetrical 14-pointer I have seen, in the Liddar Valley in 1919.

It is impossible to lay down the law as to what constitutes a good head, but length is far from being the sole dominating factor. The record head of 51½ inches is a straggly 11-pointer, narrow and with a horrid kink in the right beam. One of the finest 'wild' heads I have seen was a 48½ inch 15-pointer in the mess of the 4/6th Rajputana Rifles, massive and well-pearled; while a 46½ inch royal, shot by Capt. Leese, 60th Rifles, in the Liddar Valley in 1905 remains in my memory as the most symmetrical head with long and heavy points. Both these heads were very heavy, and a good girth and strong points are essential to establish a trophy's claim to excellence.

Spread is an added glory, and much to be desired.

The finest stag I have ever seen was at the top of the Liddar Valley in October 1933: a royal, about 48 inches on the beam, with a spread of over 40 inches, every point on the massive beam was long and matched its opposite. The three glorious tops on each crown were so level, that I believe a board placed across them would have lain even on their tips. This stag lived on the face of a projecting bluff between two deep nullahs, and on three days I tried to stalk him for a photograph, but failed. On the third day he was sent off by a stupid village shikari trying to be clever, and I never saw him again, but he probably still lives to furnish sport, a grand adjunct to grand scenery. He will soon be past his prime, so let us hope for a merciful bullet for him, and another fine trophy for someone's walls.*

Among the many fine heads I have seen there has been none with a cup in the crown, and, to my mind, this peculiarity does not add beauty to the horns of a British red deer, so its absence in the Kashmir species is not to be lamented.

I have myself shot a 40-inch 8-pointer, and seen two more of 42 inches and 44 inches; in all of these the trez tine was missing. They were from Kishtwar, the Kishenganga and the Sind Valley respectively, so it is not a local aberration.

Often a master stag will stamp his pattern on the horns of a wide area, and such a one must have dominated much of the Liddar-Tral divide during the war, for in the succeeding years several fine heads were shot with abnormally long trez tines; one in particular, shot by Major-General Pitt-Taylor had trez tines of 17 inches in length, and I saw one stag, which I failed to get up to, whose trez's must have been longer.

The best heads used to be fairly evenly distributed over the whole area of the barasingh's habitat, which comprises from the Kishenganga

---

* I told H. H. Holkar of Indore of this stag, and sent my shikari with him next season (1934). He bagged the stag, which turned out to be a 49½ inch royal.

Valley to Kishtwar, and the south-east flanks of the Pir Panjal Range; but nowadays it is rare to see a good head from the Kishenganga or Bandipur direction; they are too much poached by Gujars. The Maharaja's State rukh of Dachigam has acted as a nursery for fine stags, and the finest heads of recent years have come from adjacent areas. Within my knowledge three 50-inch heads have been shot on the east side of the Sind Valley in the last ten years, the last being killed in September 1931. I saw a 10-pointer in October 1933, up the Aru (W.) branch of the Liddar, which must have been just about 50 inches. Then farther south-east, barasingh have increased greatly in the lower half of the Wardwan Valley, and fine heads are to be got from this area, which is little shot.

Farther south again the deer are spreading into Chamba State and increasing steadily owing to strict protection.

The first horns of a stag, when it is termed a 'pricket,' appear in the spring of the year following its birth, and are usually short and stick-like, only about 8 to 10 inches long, and show no sign of points; but I have seen them with a slight fork at the top and indications of a brow tine.

The second pair of horns, the stag being then termed a 'brocket,' have well-developed points, usually six in number and are 20 to 25 inches in length. In March 1934 I saw four 6-pointers together at the south-east end of the Kashmir Valley, and with them was a well-developed 8-pointer which looked to me also to be a brocket, and whose horns were 28 to 30 inches in length. These five youngsters were evidently the progeny of two different sires, the formation and similarity of horn being most conspicuous in three of them and in contrast to the remaining two.

I have several times seen young 8-pointers with horns of 20 to 25 inches in length, very thin in the beam and with very short points, which I have taken to be brockets.

The trez tine would seem to be the last developed.

The third or fourth pair of horns will normally have the full 5 points on each side, and a stag will carry 10 points for several years in succession before reaching his prime at nine or ten years of age, when he will begin to throw out extra points, usually beginning with the bifurcation of one inner top. The number of 11-pointers met with amongst old stags is remarkable, and I find that of the fifteen stags that I have shot, all over 40 inches on the beam, there have been one 8-pointer, six 10-pointers, five 11-pointers, two royals, and one 14-pointer: these are all true points without counting snags or offers.

The stags will fight furiously, and points are often broken to the detriment of a head.

Their enemies are numerous. Black bears are destroyers of new-born calves, and will work along a hill-side trying the upward wind for the scent of hind and young.

Leopards are a terror to hinds and young deer, but the stupidity with which hinds treat a leopard makes it a marvel that any survive. Kashmiris say that a leopard will spring on a calf and lie on it without killing it, until its bleatings draw the mother near enough for the leopard to seize her, and the calf is also then killed. Two reliable observers have told me of coming on a scene which would bear out this Kashmiri story, and in each case the interruption sent off the leopard and the calf was quite unhurt, although the leopard had been lying on it.

Such behaviour by the hind is easily explained on the score of maternal affection, but in the Liddar Valley in October 1927, being attracted by the noise, I saw five hinds advancing in an open semicircle barking and steadily drawing nearer to a leopard, which was lying at the foot of a pine tree; unfortunately a stray eddy gave them my wind and dispersed them all when the hinds were within fifteen yards of the leopard. Another observer, who has spent many years in Kashmir, told me of a similar happening, when the leopard was lying openly on a rock, one of the hinds eventually advancing to within five yards and then being seized by the leopard, which my informant drove off by firing his shotgun in the air. A full-grown stag is a match for a leopard, and will face the cat with lowered horns swept menacingly from side to side.

By far the worst enemies of the barasingh are the Gujars and the shepherds; and within the last few years barasingh have almost disappeared from many of their western haunts in the Kishenganga Valley, where Gujars are particularly numerous. In the Sind and Liddar Valleys, Gujar villages have increased and the deer have vanished from their vicinity. Twenty years ago the Lang Nai in the Liddar was excellent ground for deer, but the Gujars settled at the mouth have increased and in 1927 I saw but one solitary stag, which was on passage.

Most damage is done amongst the hinds, and many are shot in summer by the shepherds high up above the tree line.

Foot and mouth disease has also taken terrible toll of the deer in the last ten years, and the cause of this must be put down to errors in preservation. Every winter large numbers of deer crowd into the safety of the State rukhs, of which the principal, and far the largest, is Dachigam. In February and early March 500 to 600 deer are allowed to assemble on the low ground at the mouth of this rukh, which is much fouled by village cattle, and allowed to remain there, instead of being driven to the ample protection of the rukhs immediately adjoining on the east where there is plenty of clean grazing. These deer contract the fatal disease and carry it up with them to the high grounds when the snows melt, infecting others and spreading the disease over a wide area. Fortunately barasingh are fairly conservative as to their summer haunts, and return regularly to the same areas, so that the south-eastern districts seem to have escaped the infection so far.

Yet there are plenty of barasingh left in most of their haunts, and this is ascribable to their wariness and partiality for cover.

In April or May, when her time comes, a hind will slip off in a dense thicket to calve; usually amongst the undergrowth of *skimmia* or in dense hazel at lower elevations. For forty-eight hours the calf is in great danger from bears and leopards, but then staggers to its feet and quickly acquires strength to out-distance the first and take advantage of its dam's warnings when there is danger of the second. Usually the hinds will make up small parties; possibly five or six in number, perhaps three of them with calves at foot, a couple more which calved the previous year (for they usually calve in alternate years), and a wary old yeld hind in charge of the party.

Young stags usually form parties of their own, from three to six in number, as do old stags.

As the snow melts they greedily follow the fresh green grass upwards, until by midsummer they may be found on open moors at 12,000 to 13,000 feet. Some of them penetrate far back over the passes, and they are then to be found on the hills beyond the Zoji La, in the vicinity of the Kamri Pass, and in Tilel. In early September they begin to descend, the hinds first, while the stags mostly remain and clear their horns of velvet in the birches, thrashing many a sapling to shreds.

Usually the spring and autumn lines of migration are well marked, the deer using the same levels along main ridges and rarely crossing a big valley. There is no main migration as used to be supposed, the movements of the deer being entirely regulated by the feed available. Thus the deer of the Liddar-Wardwan divide spend the summer on the uplands on the east side of the crest and, while the deer which winter on the east side of the Liddar Valley travel north-westward to their mating grounds, those of the country round the Marbal Pass arrive there from the north and east. The Kishenganga Valley deer mostly summer west of the Tragbal Pass and in Tilel, then converge in September towards Gurez and Bandipur.

By mid-September most of the stags are clean, but even then few of the big ones come down below the birches until forced down by hard frosts or an early fall of snow, and, though the shooting season opens on September 15th, it is best to wait another month before trying for a really fine trophy.

In the first week of October the stags begin to roar, though in 1905, after 101 hours unceasing rain, I heard four stags call on September 4th, and never another sound for six weeks. The young stag has a high whistling note, like a police siren, but one in his prime a long crescendo bellow ending in a high note, and sudden drop; 'Aunghr-r-r-r-r-e-e-oh,' while a really big fellow will seldom roar, but utters a deep low moan, particularly when lying down.

Many a good stag never quits the forest, and never joins up with a party of hinds; only cutting out an occasional one and consorting with her for a day or two. I once saw a fine 13-pointer in Kishtwar which had twenty-three hinds for his harem and two young stags as outliers and sentries, though usually the number of hinds to a stag is three to ten. These parties almost invariably feed in the open spaces, either at the top of the tree-line, or between tree-clad ridges, and it is from them that the sportsman under the guidance of a Kashmiri shikari usually takes his toll; for the Kashmiri is helpless in forest.

As a first essay in Himalayan stalking an attempt on the barasingh has a great deal to recommend it. The ground is easy of access, for three days at the most from a motor road will take one to the most favoured nullahs, while much of the ground in the Liddar Valley and towards the Hoksar Pass is within a short day's march. Then supplies are easily obtainable and shikaris innumerable. The only objection, from the point of view of the serving officer, is that leave is hard to get at the best time of year—October 15th to November 15th.

The deer are much lower in the winter, but remain much in the shelter of the forests, feeding on chestnuts and browsing on bushes, while they also invade the higher fields for stubble.

Let it be assumed that six week's leave has been obtained from October 1st, and the sportsman has arrived in Srinagar on that date by the adventitious aid of a complaisant C.O. and a 'flying start.'

The first thing is the choice of a shikari, and this needs great care, for it is not to be governed by the same consideration as—say—a trip to Baltistan.

It must be born in mind that the meat of a barasingh represents a definite pecuniary asset, sufficient to overbalance any shikari's scruples with regard to a shootable head, and the Kashmiri shikari, whose avarice is beyond computation and who is one of the world's most plausible liars, will outdo all previous performances in his efforts to secure a sahib to hunt barasingh.

It is essential to select a shikari who is well acquainted with the best ground, and one who proposes hunting ground west of the Sind Valley should be discarded immediately; he is almost certainly lying about the number of deer and backing his luck in the hope of meat and good pay for a couple of months. If a man applies for the job and proposes the Sind or Liddar Valleys, or country still further east, make him discuss his proposals with the map. The deer do not come down the main ridges and enter the nullahs lower down the valleys until forced by the weather, and the average shikari is apt to disregard all such conditions and go for any nullah where he has ever seen a stag before. Thus, if a shikari were to propose to take me to Versirran nullah in the Liddar in the first week in October of a rainless autumn, I would never engage him, for it is extremely sunny and the feed dries up easily; but if he were to propose Mamal ten miles higher up I would certainly question him further, for Mamal has much high and broken ground and the big peak behind it attracts rain, so in a wet autumn it can be unpleasantly trying, but in a dry one there is nearly always a good stag to be had by taking a bivouac high up.

Having engaged a shikari make him send out a man immediately to arrange pony transport, and follow next day in a lorry to the nearest possible point to the nullah it is proposed to occupy. There the pony transport should meet one, and it frequently pays to engage it for a whole month, as coolie transport is expensive, and ponies often most difficult to get, when on stag ground; and, if the ponies are present at hand, it is possible to take immediate advantage of weather conditions which entail a change of plan.

Take the camp well up the nullah, but conceal it as thoroughly as possible; a grassy flat is often to be found by the bend of a stream, well hidden by pine forest and, a most important point, invisible from the feeding grounds.

Having pitched camp, find out the best vantage points from which to view the feeding grounds, and watch morning and evening for a couple of days before actually walking over the ground, at the same time listening for calls.

If no sign of deer is seen or heard, climb up higher and try the small secluded ravines, particularly those holding water and a thick growth of such vegetation as wild angelica, which is much loved by the deer.

When traversing the upper ground, move with the greatest care and never hurry, paying particular attention to the wind. Barasingh are far from conspicuous, their grey-brown coats blend amazingly well with lichened pine trunks, and even in the open on a shady slope their resemblance to a fallen log is often misleading to the point of disaster, and seeing a stag without his seeing you is essential to success.

Do not expect to see monarchs of the glen on the skyline; if barasingh cross an open skyline it is almost invariably at night or so late in the evening as to be valueless for stalking purposes; careful inspection of shady places, slopes criss-crossed with fallen trees, and hidden dips at the top of the tree-line will eventually reveal a great dark-grey stag moving quietly and warily, keeping always within easy reach of cover as he crops the grass.

If hinds are seen, make absolutely sure that there is no good stag with them before moving on; it may take hours, but it is worth it. I once watched a dozen hinds for three hours without spotting a good stag, which I was sure was somewhere there, and eventually detected his horns, sticking up among the broken boughs of a fallen pine on the far side of which he was lying, 70 yards from his nearest hind.

I have located a calling stag within an area 80 yards square, and taken an hour to spot him with the glasses.

Once seen the head has to be judged, and in this matter put little faith in the Kashmiri shikaris; the trashy heads brought back to Srinagar are sufficient evidence of their untrustworthiness, and, at the height of the stag season, I have seen head after head pass through Pahlgam (at the top of the Liddar) not one of which should ever have been shot.

The first thing to look for in a head is the number of points; normally 10 are essential to a shootable head, as 8-pointers over 40 inches are rare in the extreme. Then look at the tops; which should be the longest points by far, except in the case of abnormal trez tines; little 6-inch forks at the top of the horns are sure indications of a poor head in a 10-pointer.

Any head of over 10 points is shootable, though it may be a poor one 'going back'; I once shot a royal of only 34 inches on the beam.

Having made up your mind that a stag is a good one, due consideration must be given to the question of time. If the spying position is a good one and there is any doubt of being able to get up to the beast before he retires to lie down for the day, it is far better to watch his movements, noting the route by which he returns to cover, and will then come out to feed again any time after 4 p.m.

If the feeding ground has a hot afternoon sun on it the stag may delay his emergence till sunset or dusk, or may feed on shadier ground on the other side of the cover in which he lies up for the day.

An early fall of snow will usually make the stags call, and they are then given to coming out on the upper margs, as soon as the weather clears, and roaring defiance at each other. I once found three good stags making a deafening noise, bellowing at each other on a snow-covered marg up the Aru branch of the Liddar, and they were a grand sight on the sunlit snow, each with a group of hinds around him.

In some years stags hardly call at all, especially when it is very dry, and in October 1927, I hunted in four nullahs of the Liddar and the Trisangam (E.) branch of the Bandipur nullah and only heard three calls in the Liddar and none at Trisangam.

The stags were very high that season, as there was no snow up to 14,000 feet, and on November 1st I shot a good stag at that height, which I had watched lie down on snow in the shade of a big rock, although the cold was bitter and 2,000 feet lower down every rock in the main streams was heavily wreathed with ice.

It is very rarely that barasingh are seen on bad ground, and then only when making a change of ground, but there is often a long plug up to the high ground, while a stalk may entail traversing slopes in small forest where soft mould is held up at such a steep angle that a footing is most difficult to obtain, and the return journey makes the latter half of the old tag: '...*sed retrograre difficile est,*' come forcibly to mind.

So far little mention has been made of the stag which never quits the forest to feed in the open, but browses on bushes or grazes in secluded ravines and clearings. Almost invariably solitary and carrying a grand head, these stags provide the best sport, which involves considerable knowledge of the art of still-hunting. A stag of this type is usually first located by his tracks, often discovered at some soiling pit, where he comes to wallow in the cool mud. The general direction of the tracks being ascertained, it is better to try and locate them again in likely haunts rather than to keep steadily on them, unless they are very fresh. It may take two or three days to puzzle out the habits of the quarry, and throughout, the greatest care must be taken not to alarm him; moving slowly, avoiding noise such as breaking branches, and above all never talking except in very low, just audible, tones. A constant watch must be kept for the slightest movement in the forest ahead and *above* the direction of the tracks, for the stag will almost invariably turn uphill for a little before lying down. If a good stag be come upon unexpectedly in the forest, it is very unlikely that he will give a chance of a shot before bolting, and it is usually far better to leave him alone and try for him next day. He will keep a very sharp look-out for some hours, but, seeing and hearing nothing, will assume that the disturbance was accidental and relapse into normal behaviour.

One habit these forest stags have is of the greatest assistance to the hunter. Being solitary, yet often desirous of the company of a hind, they call at any time of the day, though rarely twice in the same hour. The stags which feed in the open will often call frequently in the early morning and late evening, although with several hinds, but very rarely at other times of the day.

When a call is heard from a forest stag it is best to wait ten minutes or so on the chance of his calling again, then make one's way carefully and quietly to a point estimated to be about three hundred yards from where the last call was made, and sit down to listen. If another call is heard, and the place of origin located, then creep in slowly and watchfully, lifting dead sticks to avoid treading on them, never allowing a branch to spring back suddenly when pushed aside, testing each foothold where there is any likelihood of a slip, and doing everything possible to avoid startling pheasants into departing with raucous outcry.

If no other call is heard search for fresh tracks, moving in the same cautious way. It is better to spend five minutes getting over a fallen tree without noise than have a dead branch crack loudly through a hasty movement, and so spoil the work of many hours, or even days.

Perhaps the first glimpse of the stag will be his feet moving below the level of the pine branches, or the top points of a horn, or merely the violent shaking of a sapling as he rubs his horns on the stem; but whichever it may be, stop, get ready to cover the nearest gaps, and let him do the rest of the moving. Pushing further in will almost certainly alarm him; there will be a sudden cessation of movement, a tense moment as he listens, then a sudden plunge downhill and he is gone.

But if he is not alarmed and the wind is right (and, if possible, always keep a little above him to get the benefit of the uphill trend of nearly every breeze), then he is sure to move quietly and slowly, sooner or later, to cross some open gap at 40 or 50 yards.

This forest hunting is a most fascinating game, but one that never meets with the approval of the Kashmiri shikari, who is clumsy in the extreme in cover, an inept tracker, and impatient of slow movement. Keep him behind and do the job yourself and, if he is careless and treads on cracking sticks, clears his throat, or talks unnecessarily, or loudly, tell him he can go back to Srinagar and then carry on yourself with local help: the sport will probably be greater if he does go back.

If the stalk is successful and a good stag falls, then supervise the skinning of the head with particular care, and on no account have it mounted in Srinagar. A mounted stag's head is very bulky, a costly thing to cart about, and difficult to keep in good condition in the plains of India; but it may be that a particularly fine trophy or one that has given memorable sport is thought worthy of preservation complete with mask. If the work is done in Srinagar the modelling will be anything from poor to execrable, the mask will be filled with an insanitary collection of old rags, paper, and uncleaned wool, and the curing so bad that the hair will fall off in large patches in the first monsoon. If a stag's head is to be mounted it is worthy of being done really well, and England is the place to have it done.

But, fully mounted or not, the trophy is bound to conjure up memories of grand sport amid glorious surroundings, and inspire the hunter to go back again and again to feel the keen breeze at the top of the tree-line, thrill to the long crescendo of a big stag's call, and feel his pulse hammering as he crawls in to the final act as a great grey beast stands doubtful, sensing danger, yet not knowing surely whence it comes.

Excerpted from Stalking in the Himalayas and Northern India *(London: Herbert Jenkins Ltd., 1936).*

JAMES FORSYTH

Stalking the Gaur

The gaur, the great wild ox of the Indian hill forests, has retreated in recent times not only due to over-hunting or denudation but also due to its vulnerability to rinderpest contracted from domestic livestock. Forsyth's memoir of central India shows how the gaur was highly sensitive to the use by humans and cattle of forests even in the late nineteenth century.

In after days I spent many a long day in the chace of the bison on these splendid hills; and have also made the acquaintance of the mountain bull in many other parts of the province. Some account of his habits may, therefore, not be out of place here, particularly as they are frequently a good deal misrepresented. And first as to his name. The latest scientific name for him is *Gavæus Gaurus* [now *Bos gaurus*] but what he is to be called in English is not so easily settled. Sportsmen have unanimously agreed to call him the 'Indian Bison,' which naturalists object to, as he does not properly belong to the same group of bovines as the bisons of Europe and America. They would have us call him the *Gaur,* which appears to be his vernacular name in the Nepalese forests. I would, however, put in a plea for the retention, by sportsmen at least, of the name 'Indian Bison.' In the first place it fully accomplishes the object of all names in distinctly denoting the animal meant. Ever since he became known to Europeans he has been so called, and no other animal has ever shared the name. Then his structural distinction from the true bisontine group appears to consist chiefly, if not solely, in his having thirteen instead of fourteen or fifteen pairs of ribs, and somewhat flattened instead of cylindrical horns (Jerdon). Lastly, there is no vernacular name universally applicable to him, 'Gaur' being unknown in Central India; while his occasional Central Indian name of *Bhinsa* (with *Bun* or 'wild' prefixed to it) is almost identical in sound with 'bison,' and is no doubt derived from the same root. If you ask for 'bison' in these forests where he is known (and speak a little through your nose at the same time), you will certainly be shown *Gavæus Gaurus* and no other animal.

The respective ranges of this animal and the wild buffalo (*Bubalus*) have sometimes been defined by sportsmen in the saying that the bison is not found north, nor the buffalo south, of the Narbada river. Like most apophthegms, however, this contains little more than a flavour of the truth. Not only does the bison inhabit many parts of the Vindhya Mountains, directly to the north of the Narbada, but he also stretches round the source of that river and penetrates into the hills of Chota-Nagpur and Mindapur, and crosses over to the Nepalese Terae, and the hilly regions in the east of Bengal. The wild buffalo also covers the whole of the eastern part of the Central Provinces far to the south of the latitude of the Narbada, and also the plateau of Mandla and the Godavari forests, directly to the south of that river. In fact, the bison appears to inhabit every part of India where he can find suitable conditions. These appear to be, firstly, the close proximity of hills, for though he is sometimes found on level ground, he is essentially a lover of hills, and always retreats to them when disturbed; extensive ranges of forest little disturbed by man or tame cattle, for, unlike the buffalo, he cannot tolerate the proximity of man and his works; a plentiful supply of water and green herbage; and lastly, so far as I have observed, the presence of the bamboo, on which he constantly browses. In the Central Provinces of India all these conditions are unfortunately still present over enormous tracts of country. Thousands of square milles in the central range, much of which will one day be reclaimed to the uses of the plough, are now the very perfection of a preserve for the bison.

Perhaps he is nowhere more completely at home than in the Mahadeo hills. There, as a general rule, he will be found to frequent at any season the highest elevation at which he can then find food and water. During the cold season succeeding the monsoon they remain much about the higher plateaux, at an elevation of 2000 to 3000 feet, where they graze all night on the bamboos that clothe their sides, and on the short succulent grasses fringing the springs and streams usually found in the intervening hollows. They generally pass the day on the tops of the plateaux, lying down in secure positions under the shade of small trees, where they chew the cud and sleep. Their object in lying under trees seems more the concealment thus afforded to their large and dark-coloured bodies than shelter from the sun, as the shade is seldom dense, and a secure windy position is always secured irrespective of the sun. I have observed that single animals always lie looking down wind, leaving the up wind direction to be guarded by their keen sense of smell; and, in my experience, it is far easier to baffle their sense of vision in a direct approach, than to stalk them down wind, however, carefully the approach may be covered. It is extraordinary how difficult it often is to distinguish so strongly coloured an object as a bull bison when thus lying down in the flickering shadow of a tree.

The colour of the cows is a light chestnut brown in the cold weather, becoming darker as the season advances. The young bulls are a deeper tint of the same colour, becoming, however, much darker as they advance in age, the mature bull being almost black on the back and sides, and showing a rich chestnut shade only on the lower parts of the body and inside of the thighs. The colour of both bulls and cows varies a good deal in different localities. The lightest coloured are those of the open grass jungles in the west, the darkest those of the deep bamboo forests of Puchmurree and the east. The white stockings, which are so characteristic a marking of this species, also change with advancing age, assuming a much dingier colour in the old bulls. A singular change also occurs in the growth of the horns. A young bull has horns approximating to that of cows which have horns that are slender and much recurved at the points. A slightly older bull has horns considerably increased in girth at the base, which and have assumed a more outward sweep, with less incurvature at the points. In a full-grown bull the horns are still thicker and more horizontal, with some signs of wear at the tips. A very old and solitary bull has horns which are extremely rugged and massive, with scarcely any curve, and are considerably worn and blunted at the points. I once measured a bull in the Puchmurree hills which was five feet eleven inches at the shoulder, measuring fairly the right line between two pegs held in the line of the foreleg, and I am convinced that this is about the extreme height attained by them in this part of India. I strongly suspect that the much greater heights often given have been taken from unfair measurements. A common way is to take an oblique line from the forefoot to the top of the dorsal ridge, and follow the curvatures of the body besides. In this way twenty-two hands may doubtless be made out, but we might as well measure the distance from nose to tail for the height as this.

At this season of the year (the winter months) the bison are rutting, and they will be found collected in herds numbering ten or twelve cows, with one bull in the prime of life, and a few immature males, the remaining old bulls being expelled to wander in pairs, or as solitary bachelors, in sullen and disappointed mood. Very old bulls with worn horns are almost always found alone, never apparently rejoining the herd after being once beaten by a younger rival. These solitary gentlemen wander about a great deal; while the herd, if undisturbed, will constantly be found in the same neighbourhood. Each herd appears to possess a tract of country tabooed to other herds; and in this are always included more than one stronghold, where the density of the cover renders pursuit of them hopeless. When frequently disturbed in and about one of these, they make off at once to one of the others.

As the hot season advances, and the springs in the higher ranges dry up, the bison come lower down the hills; and may even, if compelled by want of water, come out into the forest on the plains, drinking from the large rivers like other animals at that season. But they are always ready to retreat to their mountain fastnesses when much disturbed; and as soon as the fall of the rains has renewed the supply of water, and freshened the grass in the higher hills, they retire again to their favourite plateaux. At this season the cows begin to calve, and separate a good deal, remaining for two or three months secluded in some spot where grazing and water are plentiful. The bulls and young cows are then often found together in herds of six to ten, the oldest bulls, however, always remaining alone. During the lulls in the monsoon a species of gadfly appears in the jungles, which is exceedingly troublesome to all animals. At such times the bison seek the high open tops of the mountains; and I have then seen a solitary bull standing for hours like a statue on the top of the highest peak in the Puchmurree range.

Though at first sight a clumsy-looking animal, which is chiefly due to his immensely massive dorsal ridge, the bison is one of the best rock climbers among animals. His short legs, and small game-like hoofs, the enormous power of the muscles of the shoulder, with their high dorsal attachment, and the preponderance of weight in the fore part of the body, all eminently qualify him for the ascent of steep and rocky hills. For rapid descent, however, they are not so well adapted; and I have known cases of their breaking a leg when pushed to take rapidly a steep declivity; a bull with one foreleg broken is at once brought to a standstill.

Terrible tales are told of the relentless ferocity of the bison by the class of writers who aim rather at sensational description than at sober truth. I have myself always found them to be extremely timid, and have never been charged by a bison, though frequently in a position where any animal at all ferocious would certainly have done so. In all my experience I have only heard of one or two cases of charging which I consider fully authentic, and in all these cases the animal had previously been attacked and wounded. Captain Pearson was once treed by a wounded bull in the Puchmurree hills, which charged and upset his gun-bearer; and an officer was killed by one some years ago near Asirgarh. Often the blind rush of an animal bent on escape is put down by excited sportsmen as a deliberate charge. Much, too, of the romance attached to the animal must be attributed to his formidable appearance; for the sullen air of a mighty bull just roused is very impressive; and much to the wild tales of the people in whose neighbourhood they live, who always dilate on their general ferocity, but can seldom point to an instance of its effects, and who are, moreover, frequently from religious prejudice, desirous of withholding the sportsman from their pursuit. Still there is sufficient evidence on record of the occasional fierce retaliation of the bull bison when wounded and closely followed up, in some resulting even in the death of the sportsman, to invest their pursuit with the flavour of danger so attractive to many persons, and render caution in attacking them highly advisable. The ground on which they are usually met is fortunately favourable for escape if the sportsman be attacked, trees and large rocks being seldom far distant.

Although a closely-allied bovine, the Gayal of trans-Brahmaputra India, has for ages been domesticated and used to till the land, all attempts to do so with the subject of my remarks, or even to raise them to maturity in a state of captivity, have failed. After a certain point the wild and retiring nature of the forest race asserts itself, and the young bison pines and dies. It has always struck me as curious why the most difficult of all animals to reclaim from a wild state are precisely those whose congeners have been already domesticated. The so-called wild horses, and the wild asses, are almost untamable; so also with the wild sheep and goat, the wild dog and the jungle-fowl. A young tiger or hyena is infinitely easier to bring up and tame than any of these.

This unconquerable antipathy of the Indian bison to the propinquity of man is slowly but surely contracting its range, and probably diminishing its numbers. Gradually cultivation is extending into the valleys that everywhere penetrate these hills; and the grazing of cattle, which extends far ahead of the regularly settled tracts, is pushing the wild bull before it into the remotest depths of the hills. I have, in a comparatively brief acquaintance with these hills, myself known considerable areas where bison used to be plentiful almost entirely cleared of these animals. Other wild beasts retire more slowly before the incursions of man, partly subsisting as they do on the products of his labour. The tiger who finds himself suddenly in the middle of herds of cattle merely changes his diet to meet the situation, and preys on cattle instead of wild pigs and deer. Even deer seldom live entirely in the deep forest, but hang on the outskirts of cultivation, and, mainly subsisting on it, need not materially decrease in numbers so long as there remain uncleared tracts to furnish a retreat when pressed. But the bison admits of no compromise. I have never heard of his visiting fields even when he lives within reach; he never interbreeds with tame cattle; and the axe of the clearer and the low of domestic cattle are a sign to him, as to the traditional backwoodsman, to move 'further West.' It may be that the time is not far distant when the tracts now being marked out, to remain for all time as reserves for the supply of forest produce to the country, will be the only refuge for these wild cattle, as has been the case with the bison and the wild bull of Europe.

On the day appointed for our grand hunt I started early, with the young Thakur had and a few of the Korkus, by a way that led right over the top of Dhupgarh. After walking along the open plateau for about three miles we commenced the ascent of the hill, which is close on 1000 feet above the plateau. The zigzag track was hardly distinguishable among the grass and bamboos that clothe the hill; and every here and there a road had to be cleared with the axe, no one having passed that way since the preceding rainy season, when all vestiges of paths in these hills become obliterated. We were amply regarded, however, for the climb by the magnificent prospect that awaited us when we gained the summit—the finest by far in all this range of hills. The further slope of Dhupgarh was not nearly so precipitous as that we had come up, but fell, by steps as it were, to the bottom of a deep and extensive glen, which was the one we were about to beat. Beyond this again rose the mural cliff that buttresses the whole of this block to the south; and far past this, to the left, stretched out below us the wilderness of forest-clad hills, that reaches with scarcely a break to the Tapti river—a distance, as the crow flies of sixty or seventy miles. All this immense waste is the chosen home of the bison; and beyond it, on either side of the Tapti, on the elevated Chikalda range, and in the wild hills of Kalibhit, lies another tract of equally wide extent, where, too, the mountain bull roams, as yet scarcely troubled with the presence of man or cattle. This is region of the Teak tree *par excellence* in this central range of mountains, to which I will have the pleasure of conducting the reader in a future chapter.

Tracks of bison and sambar were numerous on the top of the hill, which is covered with bamboo clumps and with a low thicket of the bastard date. I have frequently, on other occasions, found both bison and sambar on the very top of Dhupgarh in the early morning. The descent of the farther side of the hill, over long slopes of crumbled sandstone, and the curious vitrified pipes of ironstone that exfoliate from the decomposed surface of these hills, was fully more tiresome than the ascent. Many a time after this did I tread the same path to reach this valley, where bison were nearly always to be found, and many an effort did I make to discover a shorter and less precipitous road. But all in vain; for the sheer ravines that everywhere else hem in the flanks of the Dhupgarh mountain render a passage round it a matter of infinitely greater time and toil than the way over the top. At the bottom of the valley, below a shady grove of wild mango trees, where the stream that drains the large valley has formed a considerable pool in a rocky basin, I found assembled three or four of the Raj-Gond chiefs whose possessions lie in the hills to the south of Puchmurree. They differed not at all from him of Puchmurree, unless that they were somewhat more intelligent and polished in manner. Each had brought his small retinue of matchlock men, and a large gang of common Gonds and Korkus to beat; so that altogether we mustered some twenty guns, and between two and three hundred beaters. The people were well acquainted with all the beats and passes, having always several great hunts of this sort during the year; and everything had been arranged before I came. The bulk of the beaters had gone on hours before to surround the valley, and, as we were a little later than was expected, it was likely that they would already have commenced to beat. We lost no time, therefore, in taking up our posts, which stretched in a long line right across the lower end of the valley. First, however, I had to furnish powder to load the whole of the matchlocks of my native friends; and had I not guessed that such would be the case, as usual, I would certainly not have had sufficient in my flask. Six fingers deep is the rule for these weapons, and it is of no avail to point out the superior strength of our powder. They will have six fingers of Hall's No. 2, whatever the consequence. As they put generally two bullets, a leaden and an iron one, on the top of this charge, and wad with a handful of dry leaves, the result often is the bursting of the barrel, and always considerable contusion of the user's shoulder.

This was to be a silent beat; that is, the people were to advance without noise, beyond the rapping of their axes against the trees, as there was another dense cover lower down which usually held bison, and sometimes a tiger, and which was to be beaten also in the afternoon. I had sat an hour at least behind the screen of leaves that had been put up for me when the first sign of the beat appeared, and for another half-hour nothing was heard but the occasional knock of an axe-handle on a tree. Presently a shot rang from the extreme flank of the line of guns, then another, and a clatter of hoofs inside showed that a herd of something had been repulsed in an attempt to escape. As the beat advanced more shots were heard on either side, and the galloping about of the imprisoned animals, now and then met by a shout from behind when they attempted to break back, became productive of considerable excitement on my part. At last a rush of animals advanced down the side of the steam where I was posted, and eight or ten sambar clattered past within half a stone's throw. I had just fired both barrels of my rifle at a couple of the stags, dropping one of them in his tracks, and had advanced a few paces towards it, when I heard a shot on my immediate right, and a fine bull bison, with two cows and a small calf, trotted past almost in the same line as the sambar had taken. Those were not the days of breech-loaders, and though I had another rifle it was a little behind, leaning against the tree, and before I could get hold of it nothing but the sterns of the 'beeves' (as a friend used to call them) were to be seen. When I got it I favoured the bull with both barrels *a posteriori*, but there was no result. The young Thakur, who occupied the post on my right, had been more successful; and when the beaters came up immediately afterwards I found a fine four-year-old bull lying dead, with two of his bullets through the centre of his neck. All the guns now came dropping in, and gathered in a group round the slain bison. One had seen a bear, another a couple of sambar, and so on. All had fired, and of course hit hard, but the net result was the thakur's beeve, my sambar, and two little 'jungle sheep,' as they are called, the proper name being the four-horned antelope.

I had never seen a bison before, and though this was only a young chestnut-coloured bull with small horns I was much struck with the bulk and expression of power belonging to the animal. Such was the width of the chest that when lying on the side the upper fore leg projected stiff and straight out from the body, without any tendency towards the ground. The head in particular has a fine highbred, and withal solemn appearance, which is still more noticeable in old bulls. From the eye of a newly slain bison, turned up to the sunlight, comes such a wonderful beam of emerald light as I have seen in the eye of no other animal; and the skin emits a faint sweet odour as of herbs.

We tracked the wounded sambar and bison a little way down the valley, the former showing signs of being hard hit, and a little blood was found also on the track of the bull. We left a few of the best trackers to follow up their trail with the next beat, and went round to take up our places about a mile further down, and close to my camp at Rorighat. The same process was repeated here, and this time with much shouting and hammering of drums, as a tiger was usually somewhere in this part of the valley, and his tracks had been seen in the morning. I did not get a shot on this occasion. One of the Gond Thakurs shot another sambar; and my wounded stag was found and killed with their axes by the Gonds. The wounded bull was in the beat, and broke near one of the Thakurs retainers, who was too astonished to fire. The rest of the bison, or another herd, broke through the side of the beat, and plunged down a very steep and rocky descent, which the people said they had never attempted but once before, when one of them had broken a leg. Certainly I should not have thought that any animal so large as a bison could go down that place and live.

Nothing had been seen of the tiger, and had I known him as well as I afterwards did, I would not have been surprised. I knew that tiger intimately for many months after this, and yet I never once saw him. He was a very large animal indeed, but entirely a *jungle* tiger, that is, preying solely on wild animals, and keeping during the day to the most inaccessible ravines and thickets. He frequented the bison ground round Dhupgarh, and hung on the traces of the herds, apparently with an eye to the young beeves. I never came across evidence of his killing any of them, though I once saw a place on the plateau where the whole night long he had evidently baited an unfortunate cow with a calf. Within a space of some twenty yards in diameter the grass had been closely trampled down and paddled into the moist ground by their feet, the footprints of the calf being in the centre, while the tiger's mighty paw went round outside, and the poor cow had evidently circled round and round between the monster and her little one. I am glad to say that I tracked the tiger off in one direction, and the courageous mother and her calf safe in another. The tiger cannot, I believe, kill even a cow bison, unless taken at a disadvantage; and with a bull he could have no chance whatever. I seldom went out without meeting the tracks of this tiger; and often followed him through his whole night's wanderings, which were laid out as on a map in the clean sand of the stream beds; but I always lost him in the end, though I believe he often let me pass within a few yards of him without saying anything. He came at rare intervals, like the bison, on to the plateau; but his regular beat was round the bottom of Dhupgarh, a thousand feet lower down. Once, long ago, a tiger took up his post on the plateau, and became a man-eater, almost stopping the pilgrimage to Mahadeo, till he was shot by the uncle of the Thakur.

I followed the wounded bison bull for about a mile from where he was last seen; but he was moving fast, and the blood had ceased to drop. He would never stop, the people said, till he got to a stronghold of the bison of these hills, about five miles off, a hill called the Buri-Ma (Old Mother); and so I reluctantly gave up the pursuit. When I returned all the beaters were assembled; and a more wild and uncouth set it never before had been my lot to see. Entirely naked, with the exception of a very dingy and often terribly scanty strip of cloth round the middle, there was no difficulty in detecting the points that mark the aborigine. They were all of low stature, the Korkus perhaps averaging an inch or two higher than the

Gonds, who seldom exceed five feet two inches; the colour generally a very dark brown, almost black in many individuals, though never reaching the sooty blackness of the negro. Among the Gonds a lighter-brown tint was not uncommon. In features both races are almost identical, the face being flat, forehead low, nose flat on the bridge, with open protuberant nostrils; lips heavy and large, but the jaw usually well formed and not prominent like that of the negro; the hair on the face generally very scanty, but made up for by a bushy shock of straight black hair. In form they are generally well made, muscular about the shoulders and thighs, with lean sinewy forearm and lower leg. The expression of face is rather stolid, though good humoured. Some of the younger men might almost be called handsome after their pattern; but the elders have generally a coarse weather-beaten aspect which is not atractive. All the men present carried the little axe, without which they never stir into the forest, and many had spears besides. During the beat they had killed a good many peafowl and hares, and one little deer, by throwing their axes at them, in which they are very expert.

The Korkus, I found, were prevented by prejudice acquired from the Hindus from eating the flesh of the slain bison; so the Gonds from Almod, and a number of a tribe called Bharyas, who had come from the Motur hills, had him all to themselves, while the Korkus set to work on the sambar with their sharp little axes, which are all that is wanted for skinning and cutting up the carcass of the largest animal. My servant secured the tongues and marrow-bones, and a steak out of the undercut of the bison—all delicacies of the first water for the table of the forest sportsman; and the remainder of the flesh was given up to the hungry multitude. As night fell, they lit fires where the bison had fallen, and near the village where they had brought the deer; and for hours after continued carrying about gobbets of the raw meat, which they hung up on the surrounding trees, broiling and swallowing the titbits during leisure moments. This was only the preliminary to the great feast, however—the dozen of oysters to whet the appetite for turtle and venison. Soon the trees were fully decorated with bloody festoons, and the savages set to work in earnest to gorge themselves with the half-cooked meat. The entrails were evidently the great delicacies, and were eaten in long lengths, as Italians do maccaroni. The gorging seemed to be endless, and I sat outside my little tent for hours looking on in wonder at the bloody orgie. The bonfires they had lighted threw a ruddy glow over the open glade, and on the crimson junks of flesh hanging on the trees, bringing the dusky forms of the revellers into every variety of picturesque relief, and forming a wild and Rembrandt-like picture which I shall not soon forget. Till a late hour many new arrivals continued to add to their numbers, winding down the steep path that leads over the Rorighat, with lighted torches and loud shouts to show the way and scare wild beasts. All were welcome to a raw steak and a pull at the pot of Mhowa spirit that stood beside every group. Ere long they began to sing, and then to dance to a shrill music piped from half-a-dozen bamboo flutes. The scene was getting uproarious as I turned in; and my slumber was broken through the greater part of the night by the noise and the glare of the great fires through the thin canvas of my tent.

Next morning I was roused by the crow of the red jungle-fowl, which swarm in the bamboo cover of this little valley, and by the unremitting 'hammer, hammer' of the little 'coppersmith' barbet, of which there seemed to be more in this valley of Rorighat than in all the rest of the country. I found the revellers lying like logs just where they had been sitting; and it was no small labour to rouse and get them together. A couple of days' supply of flour was served out to each, as remuneration for their labour in the drive; and plenty more was promised if they would come and help to build the lodge at Puchmurree. I also gratified the Chiefs by presenting them with sundry canisters of powder and all my spare bullets; and we parted, I believe, mutually pleased with each other, and with promises of plenty more hunting-meets of the same sort. I had enough of that sort of sport, however; and, excepting once with the Thakur of Almod, never again drove the hills for game. It is poor sport in my opinion, and is seldom very successful even in making a bag.

Excerpted from The Highlands of Central India, Notes on Their Forests and Wild Tribes, Natural History and Sports *(London: Chapman and Hall, 1879).*

PART III

Princes and Sahibs

Programme of a viceregal visit to Kathiawar and Bombay, 1924.
Col. R. B. Worgan, Military Secretary to the Viceroy, 6 October 1924.

Wednesday, 26 November

7.15 a.m.	Departure from Veraval
8 a.m.	Arrival at Talala
8.15 a.m.	Departure by motor car
9.15 a.m.	Arrival at Sasan, Gir Forest
11.30 a.m.	Beat for Lion
1.30 p.m.	Luncheon at Sasan
3.30 p.m.	Beat for Lion (if required)
5.00 p.m.	Afternoon Tea
8.30 p.m.	Quiet dinner

Thursday, 27 November

9.30 a.m.	Breakfast
11.30 a.m.	Beat for Lion
1.30 a.m.	Lunch
3.30 a.m.	Beat if required
5 p.m.	Afternoon tea
8.30 p.m.	Dinner

Friday, 28 November

10.30 a.m.	Departure from Sasan

LOUIS ROUSSELLET

Govindgurh

The French writer and traveller Louis Roussellet travelled to several princely states in the 1870s. As a guest, he was given a red-carpet welcome in Gwalior, Udaipur and Rewah. The last named princely state had well-stocked hunting reserves in the highlands in central India. Roussellet's is a story of the hunt not as a means of conquest or of survival, but as a ritual in which Indian princes honed their war-like skills against the creatures of the forest.

On the 28th of March, accepting the Maharajah's invitation, we set out for Govindgurh, eleven miles distant from Rewah, at the foot of the Kairmoors.

The country to the south of the capital presented a far more cheerful appearance than towards the north and the west. The soil seemed covered with a fertile mould, and we no longer met with those beds of sandstone whose level surfaces form small deserts in the midst of the plains of Tons. The villages, too, displayed themselves coquettishly on little heights, always adjoining a pretty jheel, and some cool groves of mango-trees.

About half-way on our road we passed through the small town on Mukunpore which was for a short time the capital of the kingdom, before the occupation of Rewah. The Holi was still being celebrated here; the streets were filled with noisy crowds; and on the banks of a magnificent pool of water was raised the obscene idol of Holica. The festival was held around it; and, to hear the cries of the mountebanks and the priests, the uproar of voices, and the piercing sound of the reed flutes, which almost resemble the *merliton* (or reed-pipes), one could imagine oneself at some fête in the environs of Paris.

On leaving Mukunpore the country changed its aspect; the ground rising in sudden undulations, broken by ravines which ran surging on to the foot of the Kairmoors, whose declivities now came into sight. Forests succeeded to the cultivated ground, but they are almost entirely composed of productive trees, dyewood, fig-trees, and mhowahs.

I have not yet spoken of the *mhowah*, the tree pre-eminently belonging to Central India, and having the same connection with these wild regions as the cocoa-tree to the banks of the Indian Ocean. Providence has endowed it with such wonderful properties that it supplies the primitive inhabitants of these plateaux with all that the most industrious nations have obtained from the whole united vegetable world.

The mhowah or mahwah is one of the finest trees of the Indian forests. Its straight trunk, of immense diameter, bears its branches arranged with regularity, and gracefully raised like the sconces of candelabra: and its dark green foliage spreads itself in dome-shaped storeys, casting a thick shade all around it. Towards the end of February its leaves fall almost suddenly, leaving the tree completely bare. The natives pick up these leaves, which they use for many purposes, such as bedding, roofing, and head-coverings. Within a few days of shedding their leaves, the candelabra become covered with astonishing rapidity with masses of flowers, resembling small round fruit, and arranged in clusters. These flowers are the heavenly manna of the jungle, and on their greater or lesser abundance depends the prosperity or the misery of the whole country. The petal, of a pale yellow colour, forms a thick fleshy berry, of the size of the grape, which leaves room for the stamen to pass through a small aperture; and, when fully ripe, this petal falls naturally. The Indians simply attend to removing the brushwood from around the tree, and every evening the fallen flowers form a thick bed, which is carefully collected. This shower continues several days; and each tree produces on an average a hundred and twenty-five pounds weight of flowers.

When fresh, this flower-fruit has a sweetish flavour, rather pleasant to the taste; but to this is added a musky, acrid, and almost sickening odour. The natives nevertheless consume great quantities of it in this state, and they also employ it in the manufacture of cakes, and in different sorts of nourishing food. But the greater part of the crop is dried on osier screens. This operation causes the fruit to lose its unpleasant flavour; it is afterwards made up into loaves, or reduced to flour. By fermentation the mhowah flower produces a wine of an agreeable taste, that must, however, be drunk while new; and by distillation a strong brandy is obtained from it, which is considered by the Indians as the most precious production of the tree, and which, when old, may challenge comparison with good Scotch whiskey. From the residue of the flowers a good vinegar also is derived.

As soon as the flowers have disappeared, the foliage returns, and rapidly covers the tree again; and then, in the month of April, comes the fruit to replace the flowers. The fruit of the mhowah is of the same shape, but a little larger than the fruit of our almond-tree. The shell is of a violet-tinted colour, covering a smooth, hard, and woody envelope, in which is found a fine almond; which is of a milk-white colour, with a delicate and rather oily taste. The Indians use it for cakes and pastes, and by a simple pressure extract from it an excellent eating oil, while the refuse serves for fattening buffaloes. This oil is already in large demand in the commerce of Bombay, and promises to become a fruitful branch of the export trade of the country. Finally, to wind up this enumeration of the wonderful properties of the mhowah, let me add that its bark yields a woody fibre used for making rough ropes, and its wood—easy to cleave, though uneven in the grain—is invaluable in the construction of huts, as it resists the attacks of the white ant.

Making a rapid summary of the preceding lines, we see that the mhowah supplies a nutritious food in its flowers and fruit, besides yielding wine, brandy, vinegar, oil, a textile material, and valuable timber for building. It will not, therefore, be a matter of surprise that in the Vindhyas and the Aravalis it should be ranked by the inhabitants as equal to the Divinity. The Gounds, Bheels, Mhairs, and Mynas owe their existence to it. They hold their meetings beneath its shade, and under it they celebrate all the important events of life. On its branches they suspend their rude votive offerings, lance-heads or ploughshares; and around its roots they spread those mysterious circles of stones which supply the place of idols to them. They will fight, therefore, with the energy of despair in defence of their mhowahs; and where the mhowahs disappear, the Bheel and the Gound are seen no more. This precious tree is occasionally planted and cultivated in the plain, but it grows naturally in the mountains.

At four o'clock we reached a fine mukkam, about half a mile distant from Govindgurh; where our camp was spread over a narrow glade, above which mango-trees and mhowahs formed a dome of verdure. A little farther on began the slope of the mountain, stretching on an imperceptible angle in a smooth acclivity to its summit; and, on the opposite side, a deep ravine, the bottom of which forms a small lake, separated us from Govindgurh.

Our people and baggage, transported by four elephants, arrived an hour after us; and, while they were unloading the animals, I walked beneath the shade of the mhowahs, whose branches were being pillaged by a tribe of langours. While so engaged, I observed the approach of a wretched-looking, mangy, bald-coated dog. Not caring for his company, I flung a stone at him, which caused him to turn off across the brushwood, towards some neighbouring huts. Shortly, however, loud cries proceeded from the direction, and some armed peasants issued forth, preceded by my 'dog,' which turned out to be nothing less than a big wolf. He had seized a kid, which he still held in his fangs, and, in spite of its weight, passed by us with such rapidity that neither stones nor sticks could intercept him. In fact, it is very hard to distinguish the Indian wolf from the half-tame dog that frequents the native villages. The form is the same; the hide alone is less yellow, in which respect it greatly resembles the tawny wolf of Poland.

The fact of the existence of wolves in India is not generally known. Travellers, being entirely absorbed in tigers and other animals of the same species, have neglected to mention these carnivorous beasts, which, though of a less assuming type, cause quite as extensive ravages. Wolves are extremely numerous over the range of Hindostan from the Vindhyas to the Himalayas. Never attacking men, even when in packs, they penetrate into the villages and farms, and carry off many children, dogs, and kids. They have even been known to get into European dwelling-places. I know a planter in Doab, whose child was carried off before his own eyes by one of these wolves from the verandah of his bungalow; and, in spite of the high premium which the Government has offered for every wolf's head, they are still far from having obtained the mastery over this scourge.

Towards evening, as we were about retiring to rest, we had a slight alarm. A sudden whirlwind rushing down from the Kairmoors passed over our encampment, overthrowing the small tents of our men; and the kulassees of our own tent just had time to throw themselves on the ropes to prevent our being buried beneath the khanats. These gusts of wind are of very frequent occurrence in these regions, as we have shortly to experience.

March 29th.—We received a deputation from the palace, consisting of four nobles who came to pay us the customary compliments; when, as usual, I had to exhibit our stereoscopic views of Paris, and distribute some presents among them. They were particularly delighted with the boxes of Eley's percussion-caps, which I gave to each of them.

In the afternoon we paid our visit to the maharajah; for which purpose we had to cross the entire length of the town, if such a title may be awarded to a confused mass of cane huts of the most provisional appearance. The streets of the future capital are certainly wide and well designed, but the probabilities are that the rajah's project will not outlive him, and Govindgurh will in a short time fall back again to the rank of a mere hunting-station. The principal entrance to the palace is situated at the end of the great street. It is a fine triumphal arch of marble, perforated with three small indented arches in the best Rajpoot style; and beyond this gateway lies a small courtyard, surrounded by fine facades, whose effect is heightened by wide colonnades. It was an agreeable surprise to find all the simplicity and elegance of the buildings of the sixteenth century in a perfectly modern edifice; for throughout the eastern part of India the Rajpoot style of architecture, so well adapted to the climate, has been abandoned in favour of that hybrid kind introduced by the English, in which is found mingled every style— the neo-Grecian of Munich, the Renaissance, and even the Gothic.

The prince received us in the fine verandah which serves as an audience-chamber. Etiquette was still more scrupulously observed than on our first interview; the nobles standing ranged on each side of the throne, and the king radiant with jewels and decorations. Every one rose on our entrance, and the king emphatically bade us welcome to Govindgurh. After an audience of a few minutes, the essence of roses and the betel were brought in, and we retired; but scarcely had we reached the staircase when a moonshee came after us, and conducted us into a small saloon, where the king speedily joined us. He had put aside all his glittering ornaments, and with them all his kingly majesty, for he warmly pressed our hands, and frankly declared the pleasure our visit had afforded him. Having had notice some time previous, through the agent at Nagound, of our approaching arrival, he was impatiently awaiting us, in order to give us the spectacle of a tiger-hunt. One of these animals was now in the ravines of the Kairmoors, watched by the shikarees, who had cunningly decoyed them with some propitiatory victims, and was destined to fall to-morrow by our bullets.

The maharajah then discharged the honours of his palace in person. Its interior corresponds with the simplicity of its exterior; but, to make up for this, the principal room or grand saloon for fêtes, attains to the very acme of bad taste in its over-profusion of glass, gilding, and utterly incongruous ornamentation; of which, however, in our anxiety to humour our amiable host, we feigned the greatest admiration.

March 30th.—The shikarees came in to give us notice to be in readiness for the evening; and, at three o'clock, the hunting-sowari passed by our encampment, the rajah reclining on a litter smoking his hookah, which a young page held by his side, while the nobles, soldiers, and followers of every description crowded around him. Mounting our *mukna*, we rejoined the troop, and soon we are ascending the mountain all together.

I have already remarked that at this point the Kairmoors present a gently inclined slope, extending for a considerable distance without any break. The extremely easy angle of this slope renders its ascent commodious on any part of it; but the naked rock is covered with innumerable detached block of very unequal sizes, through which the elephant can with difficulty open a passage. The nature of the soil seems to me volcanic; for on all sides, between the blocks and the crumbling stones, may be seen small castings of a black substance similar to pitch when it is cooled, brilliant as jet, and which I should not hesitate to pronounce lava, if its very formation did not seem to be, relatively speaking, of recent date. On the whole surface of this slope nothing is to be seen but a few stunted acacias, with here and there a sal growing in the hollows of the rocks, where only a limited quantity of vegetable earth is to be found.

From a very slight elevation, a splendid panorama unfolded itself. The view extended over the whole plateau, and the horizon towards the north appeared uniformly flat; the forests formed a line some miles in depth at the foot of the chain, and on the opposite side stretched the vast cultivated plains which surround Rewah and

Mukunpore; while, on the west, the sharp crest of the Bandair Mountains were distinctly visible.

The king's cortége wound picturesquely up the side of the mountain. The royal palanquin, carried by eight men, in the midst of a group of servants holding up parasols and fly-flaps made of yak-tails, took the lead; then followed a long line of elephants, with their hunting-houdahs and motley trappings; and these were succeeded by men on foot, and the horsemen leading their horses by the bridles, the animals leaping from rock to rock like goats.

We soon reached a fine plateau, covered with a rich vegetation; and on all sides towered high rounded peaks, between which, one thousand feet below, we caught glimpses of the lovely valley of the Sone. This river, which is one of the chief southern tributaries of the Ganges, at this point becomes confined between the Kairmoors and the buttresses of the Burgowa plateau. It descends from the heights of the Amar-Kantak, where it has its source in the same group as the Nerbudda. Its general course is towards the north-east, and, crossing Bogelcund, falls into the Ganges near Dinapore, after a course of four hundred and sixty-five miles.

Continuing our march across the plateau, we halted finally at the foot of one of the cones surrounding us, leaving the hunters to reach the houdi alone and on foot; and, after a laborious ascent of these declivities covered with underwood and young trees, we reached the summit, which presented the appearance of a vast funnel hung with verdure, its depths forming a small pool. (May not this be an ancient crater?) We descended to the verge of the water in the profoundest silence, and there at length we got to the houdi.

The lake, on the borders of which it stood, is the only place in the whole mountain where the animals can find water. It is the rendezvous, therefore, of all the denizens of the forest, and the tigers especially are attracted thither by the double bait of water and an abundant prey. When one of them is signalled, he is permitted the peaceful enjoyment of this paradise up to the day when he becomes the object of an expedition like the present.

The houdi is brought to even greater perfection here than in Meywar. It is quite a small habitation, containing rooms, and surmounted by a terrace. The walls are battlemented, and their loopholes command a full view of the spot where the animals are forced to come to water, the rest of the lake being surrounded by a little wall which prevents all access to it.

In the principal room of the houdi we find a table and chairs, and a basket containing refereshments and some bottles of Moselle, which are to supply us with patience to await the arrival of my lord the tiger; it is, however, strictly forbidden to speak aloud or to smoke. A perfect arsenal of carbines, ranged along the wall, are destined for our use and that of the king and the few nobles who have followed us.

Darkness at length spreads over the little valley; the hours wear on till it is past midnight: as yet nothing has stirred; but, towards one o'clock, the forest seems to become animated; presently some boars arrive, then stags; a little later a solitary sambur halts proudly at about thirty yards distant from us, his graceful head crowned with magnificent antlers, reflected in the mirror of the lake, lit by myriads of stars. But all these temptations do not make us forget the tiger we are expecting.

As is always the case in these hunts by night, the most interesting moments are those of expectation, when the hunter, momentarily unarmed, sees the whole nocturnal life of the forest defile before his eyes. When the tiger appears, there is another interval of excitement. Then the unfortunate animal, fatally condemned beforehand, advances almost without any mistrust. A discharge is sent from the houdi, and the tiger falls with a roar, his body riddled with bullets. This last act, which appears to be the principal one, is not the most to be admired. For my own part, I have always felt a sort of remorse in making one of eight to assassinate a tiger from behind a wall two feet in thickness.

This time also, everything happens as I had foreseen: out of our number of eight, the tiger has received five bullets, which fact does not, however, prevent the courtiers from complimenting the king on his skill, as though he had been the only one to fire.

On hearing the shots, the attendants arrive on the spot, bearing torches; the carcase of the tiger is placed on a stretcher, and, remounting our elephants, we return on our road to Govindgurh. At four in the morning, we found ourselves in our tent, after a fearful run, mounted on our elephant, stumbling by torchlight in the midst of the chaos of rocks I have already described. It is quite a miracle that no accident has occurred, for I have a notion that, while we were drinking the Moselle in the houdi, the king's attendants were doing honour to the new wine of the mhowah.

This expedition is only the prelude to a series of *battues* which we are to make during the following days in the valley of the Sone.

March 31st.—Today, being Sunday, is devoted to rest, which, for that matter, our night-excursion fully entitles us to.

Towards evening, the king honours our camp with a visit of ceremony, pompously carried in a litter ornamented with silver plates, and escorted by a regiment of his regular infantry, and followed by an endless sowari of elephants and horsemen. This unexpected visit throws our mukkam into confusion; our people rush about here and there, hurriedly putting on their turbans and gala dresses. We have scarcely time to arrange our few campaigning-stools and arm-chairs in an imposing line, before the procession issues from the mhowahs. The king alights from his litter at the entrance to the camp, and, leaning on my arm, advances to take his place on one of our iron arm-chairs, making us sit on either side of him, while the nobles and soldiers form themselves into a square.

Feeling really confused at all this display, I again repeat to the prince that we are only simple travellers, and that nothing in our position calls for this avalanche of honours; which gives him the opportunity for putting in another fine speech. We then exhibit all our curiosities, photographs, and water-colour sketches; then the nautchnis execute their dances by torchlight, and finally a few crackers and two fire-balloons are started off, bringing the ceremony to a brilliant termination. The king, on taking leave of us, can imagine no better way of expressing his gratification than to say to us, 'You are my brothers; my kingdom is yours.'

After this departure I perceive that, by his orders, an ample supply of mhowah brandy has been distributed among our people, who are all in a sad state of drunkenness. This state is, however, at the present moment common to the majority of the Baghela and Gound populations, who thus celebrate the harvest of the precious flowers.

Scarcely is our encampment buried in silence, than I am awakened by cries mingled with roars and smothered growls. I spring upon my gun, and the scared servants crowd round the entrance of the tent. Under a tree, twenty paces off, two panthers are tearing one of our dogs to pieces. With my eyes still heavy with sleep, I can only send them a chance shot, which puts them to flight. We find the dog at his last gasp; the unfortunate animal, tied to a tree, was unable to escape. It was a fine spaniel I had given me at Nowgong.

Panthers abound in the environs of Govindgurh, and in the neighbouring plains; the winding irregularities of the rocks, the low thick tangled jungles, affording them their most favourite haunts. The panther feeds almost exclusively on animals of middle size, dogs, goats, and sheep, which it comes in search of in the very midst of the abodes of men. It scarcely ever attacks men or larger animals, but is infinitely more to be dreaded than the tiger, as it unites greater courage to a far superior agility, springing upon the hunter as soon as it perceives itself to be attacked; it also has the advantage of being able to climb up the trees, and many a hunter has been dislodged from his place of ambush by these vindictive animals.

From the 1st to the 4th of April we have grand *battues* among the ravines of the Sone. We are encamped in the middle of the forest with all the Court. The ceremony of our departure took place with great pomp; and we left Govindgurh in solemn procession, each one of us mounted on an elephant, with a numerous retinue of servants carrying palms, besides musicians and singers.

My readers have already accompanied me in several of these hankhs; I shall not therefore stop to particularise the details of this one, which was in no respect inferior to the others. The booty of these four days comprised, besides boar, nilghaus, and stags in abundance, two black bears of a small species, some pretty *chikara* gazelles, and a fine lynx.

The bushrangers employed on this hunting-excursion, twelve hundred in number, belonged for the most part to the Gound race, with some Kolees and Bhoumias of the East. I soon found out that a savage from the high plateau of the Sirgouja was amongst them, his presence having excited the curiosity even of our apathetic companions.

Although situated geographically near to Bogelcund, the Sirgouja and its mountainous groups are still almost entirely unknown to the dwellers on the banks of the Sone and the Ganges. The poverty of the country, and above all the pestilential emanations of its terrible malaria, have prevented any colonisation movement from approaching these regions, which are still surrounded by the mysterious veil of legendary fables.

Many a time since I had entered Bundelcund, I had listened of an evening to our people round the bivouac talking of this frightful country, their fantastic tales representing it to me as infested by the most formidable animals, elephants and tigers of gigantic size, while the human race was only represented by creatures having the appearance of apes, living in trees and shunning the eyes of men. I had often conversed on this topic with Englishmen long settled in Central India, and their opinion had been that these descriptions would seem to apply to some large species of ape—some unknown anthropomorphites, possessing, like the Hunouman ape, a certain degree of social organisation. In support of this hypothesis, some travellers, who had passed through the country, asserted having seen and even pursued some large apes, similar to the ourang of the Malay islands.

It may be imagined with what delight I learned that one of these men-apes or *Bundarlokh*, as the Indians call them, was within my reach, and about to afford me some elucidation of this obscure problem.

One of the maharajah's hulkaras brought this representative of the Bundars to our camp. I was struck at once by his low stature, scarcely five English feet, and, above all, by the length of his arms, which, united to the animal expression of his wrinkled countenance, fully justified the title of ape given to him by the natives. The low forehead disappeared beneath woolly tangled locks; the nose thick at the extremity and flattened at the bridge; broad raised nostrils; small deep-sunken eyes, a fleshless chin; and to complete the ugliness of this mask, on each side of the mouth wrinkles running in parallel lines, covering the cheeks. This face, in spite of its ugliness, bore the impress of a profound sadness, which had nothing of the savage in it. The body itself was of a shocking leanness; the skin, of a reddish black, like tanned leather, hung in creases on the limbs; the abdomen, sunk inwards as though withered up, bore in the middle a shapeless protuberance covering the navel, and doubtless proceeding from the umbilical cord.

The presence of Europeans had considerably embarrassed the unfortunate savage, and it was impossible to extract the slightest information from him. The Gound who accompanied him furnished us with the few details he had obtained from the man himself. It appeared that this savage belonged to a tribe, of a hundred head, inhabiting the forests east of Sirgouja; that the name of the race was Djangal, which is only the derivative of the word jungle, and is applied by the Indians to all savages in general; and that he had left his tribe, driven away by the famine which was decimating the country. These details were, as may be seen, of the vaguest description, and told us nothing.

My comrade made a rapid sketch of the face and profile of the Bundar, and in the hope of attracting him to Govindgurh, where I had left my photographic apparatus, I ordered some rupees to be given him. But the sight of us, and our questions, had so frightened him, that he escaped during the night, and could not be found again. It is probable that he had been banished from his tribe for some crime he had committed, and, after having wandered miserably among the valleys for a long time, had made up his mind to implore the hospitality of the Gounds.

It was a pity that I could not get a more faithful likeness of him than a mere sketch, for I am convinced that chance had brought me face to face with one of the representatives of that interesting Negritto race of India, which, after having at a certain period peopled all the western coasts of the Gulf of Bengal, has now almost entirely disappeared. Some remains of them may certainly be found in the still almost unexplored group which extends between the Sirgouja, Sumbulpore, and Singboum; but it is evident that only some few families of them still remain, who have taken refuge in the most inaccessible places.

On the 5th we returned to encamp again under the mhowahs at Govindgurh. The maharajah had promised us for the 7th an elephant and four waggons drawn by oxen, to conduct us as far as Bhopal.

During the night after our return we experienced one of those terrible hurricanes so common at this season throughout all the mountainous region. The Indians give them the name of *tofan*.

The tempest burst upon us with so much suddenness that our servants had barely time to awake us; the canvas khanats were rent asunder, the stakes flew up in the air, and the wind furiously swelled under our tent. We rushed outside, and at this very moment a typhoon of rain and dust, mingled with pebbles and branches, hurled us to the ground, carrying me to some distance, stifled and bruised. The darkness was so dense that it was only with difficulty, and guided by my companions' cries of terror, that I succeeded in gaining the tree at the foot of which they sought shelter.

Even here the hurricane enveloped us with its whirlwinds of warm rain and stones, which take away our breath. Thunderbolts constantly ploughed the ground, bursting through the darkness with great violet-coloured flashes. The tempest brought us the sound of the cries of the unfortunate inhabitants of Govindgurh, buried under the ruins of their dwelling-places, while, from the mountain the roaring of the torrents reached our ears, accompanied by the crashing of the rocks carried away by the sweep of the storm. We might imagine we were present at the final cataclysm which is to swallow up our world. For a whole hour the hurricane persisted in all its intensity, then suddenly calm succeeded, and we are scarcely recovered from our emotion before the sky appeared glittering with stars. We left the shelter beneath which, masters and servants mingled in one group, we have passed through the tempest. Every one sets to work; the kulassees raised up the tent, great fires were lit, the baggage and furniture was withdrawn from the swamp in which it had sunk, and all was restored to order.

Daylight discovered to us the extent of the ravages of the *tofan*; on all sides trees were to be seen uprooted, and rocks displaced. Govindgurh appeared all in disorder, and the lake, yesterday quite dry, displayed a broad sheet of water.

This terrible night had quite sickened us of Govindgurh, and it was with pleasure we heard that nothing further prevented our departure. We went to pay our last visit at the palace, where we found the king presiding at his kutchery or state council. The ministers and the clerks, squatting round his chair, read the official documents in a nasal tone, or scribbled interminable rolls of paper; while at the end of the hall the nautchnis sang a sleepy tune. This mode of despatching business is not without a degree of originality. The king, while he was talking to us, interrupted himself to make some observation to the moonshee, who continued his report.

At last I presented him with some photographs I had taken of his durbar and of the palace, and in return he offered us a very handsome khillut. We took leave of each other with mutual protestations of friendship and remembrance.

Excerpted from India and its Native Princes, Travels in Central India and the Presidencies of Bombay and Bengal *(London: 1882).*

C. B. FRY

Panther Shoots with Ranji

Ranji or Ranjitsinhji was described as 'a prince among cricketers', but sport in his own lexicon included partridge drives and angling, quail hunts and panther shoots. His own little home state included a stretch of the Barda hills where panthers crossed over from neighbouring territories. C. B. Fry was an old associate whose work includes this account of hunting panthers in Nawanagar state.

Besides princes and politicians, India contains tigers and snakes. This is well known. The snakes are not as easy to find there as politicians, nor the tigers as princes. Some people talk as though there were no tigers left in India; but there are. In the jungles of the Central Provinces and the Sunderbunds of Bengal and all along the foothills of the eastern Himalayas, as well as Southern India, there are still plenty of tigers; but India being a vast country the individual animals are often some distance apart. Anyone who has had the fortune to be a guest of the late Maharajah of Alwar would be unlucky were he not to see a tiger. Alwar is within easy reach of the populous districts of Delhi and Agra. There are tigers in the forests and hills of Alwar quite close to the capital. As for snakes, I saw only two: one not anywhere in this world, and the other puffily swaying in a charmer's basket. Thousands of barefooted villagers die every year from snake-bite, so it is remarkable that I should have spent hundreds of hours in the jungle without ever seeing a snake there.

In Indian India, shikar holds a prominent place when important events bring important visitors, and in quiet times no guest of an

Indian Prince would be considered even fairly entertained without being taken out to shoot.

Ranjitsinhji's State of Nawanagar in Kathiawar, being for the most part flat and open, was not all its ruler would have liked it to be in the matter of shikar. There have been very few keener hunters and fishermen than Ranjitsinghji, and I cannot think of one more skilful in the pursuit of any form of game, great or small, furred, feathered, or scaled.

It must have been a regret to Ranjitsinhji, who in latter days was far keener on fishing than on shooting, that the few rivers of his State were not of the sort to provide sport. Nor did he himself pursue the gentle craft in India; he did not even seek out the rivers of India where the fiercely running mahseer can be caught. He would have liked, I expect, to frequent the trout-streams of Kashmir, and his friend the Maharajah would have been very glad to see him. Ranji, however, found that the high country of Kashmir was bad for his asthma, and I believe he only went there once or twice...

But as for shooting, two exceptional examples he had in his own State. One was an island near his port of Rozi, where, contrary to all advice and all expectations, he had succeeded in raising an abundant stock of partridges. There are various kinds of partridges all over India, but Ranji's was the only partridge shoot in the English sense of the term. No walking up; proper drives with hundreds of birds whizzing over the guns, and a bag at the end of the day worthy of Six Mile Bottom.

Then the panther shooting. There were panthers in nearly all the outlying districts of his State, but his show spot was around his shooting bungalow about seventy miles from the capital in the southern corner along the borders of the State of Porbandar; his famous panther jungles in the Burdar Hills. These hills form a wide circle, heavily jungled with bamboo and all the common smaller trees of Western India; a huge saucer, in the centre of which was his bungalow of Kaleshwar. The circle of hills in that outlying part are sparsely inhabited by the wandering herdsmen tribe of the Robaris, magnificent people with Greek features and terra-cotta skin; but there are no villagers within the circle of hills; an immense, silent, and lonely amphitheater.

At Kaleshwar there was an ancient temple a stone's throw outside the garden of the bungalow, surrounded by tall pipal trees, with here and there an aged and expansive banyan tree, under one of which the Jam Sahib used to hold informal audiences of cultivators from the distant villages. Outside the gates of the garden was a row of low buildings, inhabited only when the far countryside made pilgrimage to the temple.

It happened that the first time I was with Ranjitsinhji in Jamnagar, the capital of his State, he had no other guests, and after a busy fortnight, in which I was learning about Indian India from the Jam Sahib and his Ministers, my host suddenly announced after dinner that to-morrow we would go to the Burdar hills. This meant that at six o'clock three Ford cars of the old type and a lorry awaited us under the portico of the white Jam bungalow. We started with the sudden decision characteristic of Ranji, who rarely told anybody what he was going to do till the last moment; he always knew perfectly well days before. Away we went, past the two broad lakes with their ancient round fortresses, and past the high outer walls and the frowning west gate of the city into the sea of plain beyond. For some miles there was a proper road. Then we forked off on what was called a fair-weather road, which meant an earthen track across the plain. Soon we seemed to be far from human habitation, seeing only occasional cultivation on one side or the other, and distant villages, few and far between.

Within twelve miles of the city we began to see herds of blackbuck dotted here and there, and once, quite close to the track, a herd of little, brown, gazelle-like chinkara. To one who had had no experience of bigger game, except fugitive springbok in the Transvaal, the quantity of deer was exciting.

Within two hours we had traversed the remainder of the yellow plain to the foothills of the Burdar range, which seemed to spring from the earth like a sudden island at sea. We passed close to the site of the ancient and obliterated city of Ghoomli. Scarcely a fragment remains, but close to the hillside there are some deep wells, in which, according to the legend, the inhabitants deposited their treasure. Attempts to extract the water have failed. One can descend by steep steps almost to the water's edge, and the water looks dark and deep enough for anything. We climbed a rough track over a series of shoulders and wound down towards the green saucer in the centre.

There was the pink-washed bungalow, square-faced and unobtrusive, rather like the country houses one sees down the Rhone valley. The rough but well-tended compound was brilliant with a dozen giant clumps of bougainvillea. There was shade in the wide verandah and under the big pipal trees. A grey stone fountain was splashing in front. All round was the wide circle of the wooded hills, complete silence, and a brazen blue sky. The quiet experience of arrival was more than thrilling to one who only knew the woods of Europe and the treeless stretches of the high veldt. Were there not sambur and cheetal somewhere in those mysterious jungles, and panthers sleeping in caves or under shelves where the outcrop of rock showed on the ridges of the round hills?

That afternoon a score of white-clad cultivators appeared by magic. Under an immense banyan tree, whose wide branches were supported by pendent pillars of itself, they sat and awaited Bapu's pleasure.

The Jam Sahib, impassively patriarchal, sat in a camp-chair and heard their case. It was that they were being oppressed by a moneylender. Shylock was there with all the signed documents and other bundles of legality in the fold of his cotton homespun. Everybody stated his case except Shylock. He merely exhibited his papers with a laconic gesture. There was no doubt that the cultivators had signed themselves into trouble.

The Jam Sahib explained to me that the man had all the law on his side, and the protestants had been very foolish not to take advantage of the Land Bank which he had established. But the *banya* had been outrageously extortionate; and what would I do if I were he? He was always testing me in small matters of practical wisdom in India, and, I think, took sly pleasure in showing me that the ruler of an Indian State had no sinecure in his capacity of Bapu—the daddy of his people.

I knew better than to hesitate. I said that whether the *banya* had the law or not in this case, did not His Highness think that in equity the *banya* could be severely called upon to justify so flagrant a breach of the spirit of His Highness's intention in establishing Land Banks? Why not give the *banya* the verdict and then, on the count of equity, fine him enough to recoup his victims?

To my surprise, the Jam Sahib, in measured vernacular, and in the style of Haroun Al Raschid, delivered his two judgments accordingly. Afterwards he told me that the *banya* had paid up with alacrity and on the nail, and had been glad to escape so lightly. I believe I went up in Ranjitsinhji's opinion quite several pegs over this small incident. And so to tea in the verandah, and then an early dinner, including some excellent fresh fish rather like mullet, which had been brought in by runners from the sea coast in the morning and conveyed in ice on the lorry. Early because we had to be away and hidden in an old tower in the jungle before nightfall.

Directly the sudden night falls with scarcely any interval of twilight a panther which has been lying up in the hills, choosing a coign of wide observation, wakes up and begins to think of coming down to find his dinner. The cunning beast knows very well that the goats will be driven home before nightfall, and that, like himself, the stray dogs, if any, will be prowling round for derelicts of daytime near the habitations of man.

The Ford car drove down a jungle track to within half a mile of the tower and overtook the party of shikaris and half a dozen goatherds. The goatherds had one unfortunate youthful goat on a short tether, and walked ahead calling out their musical summons. 'Arree! Arree!' they go, not loud, but audible for miles. This would be a sign to all the denizens of the jungle that the goats were being called home.

When we got to the little clearing with the small two-storied tower of old stone, called a *kotah*, the guns and two servants slipped in by a back door. The theory is that the panther cannot count. Ranji and I mounted to the upper room, where the soda-water and refreshment for the night were laid out, and sat down in two large wicker chairs. Meanwhile the outside party walked on down the jungle path, still calling their goat-call. At last they came back, tied up their goat (which as yet had said nothing) on a large artificial boulder resembling an ancient pagan altar.

Such *kotahs* with the little clearing and its artfully simulated boulder in the middle were disposed at various spots in the jungle. Formerly, when younger and less *rusé*, Ranji used to sit up for panthers in *machans*—that is, shallow Indian wooden beds plaited across and arranged as platforms in the branches of convenient trees. But when he became the ruler and could get things done, he either adopted the existing little old towers or had new ones built at convenient spots where goats could be grazed in the daytime. The panthers, watching from the hills and keenly alive to noting where they would be likely to find a dinner, inferred (with due respect to the psychologists) that a clearing and a *kotah* meant a chance of a random goat after dark.

So the party moved casually around, still calling, and then disappeared, their voices diminishingly audible afar off. Just before leaving the clearing one of them had lit a big lamp hung up on a high pole. Panthers do not mind a light slung up at a fair height, so there was no need to wait for a full moon. The carbide arc-light made no sound, and sound is what frightens a panther away. Perhaps, too, the panther has learned to interpret the hanging lamp as a nearer and a better moon.

There we sat in silence. The arrangement may seem rather artificial, but when the Indian night falls over the jungle, followed by a spell of silence, and then gradually the distant honk of the sambur is heard from the hillside, and the little various night sounds of the jungle begin, the impression is one of complete loneliness in a mysterious savage world. There we sat for hour after hour until well past midnight. Not easy to keep awake, peering out of the narrow windows. But, of course, the goat, as soon as he found himself deserted, began to bleat; and he bleated on through the night hours with rhythmical monotony.

Suddenly Ranji touched me on the knee and whispered, 'He has come.' He pointed across to the bushes at the corner of the clearing and handed me his binoculars. Good glasses disentangle and magnify objects even in the darkness.

I could not see anything at first. Then suddenly a pair of phosphorescent lights switched on in the dark. The panther's eyes. There he sat for half an hour in the blackness of the bushes, himself as black. Then the eyes were switched off. Ranji made a pass with his hand to indicate that their owner was circling round us in the bushes. The ground was covered with dry leaves and dry sticks; the bushes themselves, as one knew from moving through them in the daytime, might well have been designed to produce a stage crackling and rustling. But not the least sound.

Presently the brilliant green eyes appeared at another corner; then they disappeared again. Meanwhile the goat had stopped bleating and was standing up. It kept facing round with its nose like a compass-needle in the direction of its circling enemy. Very slowly, but not always correctly. Suddenly from the side of the clearing behind it a grey form, as of a giant tabby on very long legs, moved out into the open. Just before it stalked from the bushes the form looked dead black, but the moment it came into the artificial moonlight it was grey as the ground—so grey and similar as to be invisible when still. The goat jumped round to face it as it stood for some minutes as still as stone. Then the panther moved forward like a grey ghost and leaped with marvellous ease and grace on to the edge of the boulder. There it sat like a cat on a doorstep, with its forepaws between its hind legs, and contemplated the goat. The goat stamped its feet and blew a succession of snorts through its nostrils.

Ranji had an electric pencil with a light at the end, such as hospital nurses use. He wrote on a small block: 'It is a lady, but you can shoot her if you like.'

Whether from innate gallantry or from sympathy with the goat—probably the latter—I took the block and wrote: 'No. Let her go.' I lowered my rifle on the cushions provided for the very purpose, and Ranji knocked with his knuckles. There was a grey streak, and the lady was gone.

Ranji then wrote: 'She or the old man will be back in twenty minutes.' So I drank a bottle of soda-water and relaxed.

Sure enough, the lady came back, but only to sit half-way between the goat and the edge of the clearing. Again, after an interval, she did the same from the other side, but this time she went away for good, and Rajni whispered—he could whisper without any sound of breath—'The old man is watching in the bushes. I saw his eyes last time she came.'

But the old man did not show that night. Ranji and I took it in turn to lie down for a snooze on the charpoy, but I was not good at this, because I kept verging on a snore, and every time Ranji leaned across and pinched my nose.

Finally the grey morning washed over the hills and the jungle. Ranji blew a whistle, and presently our overnight party began threading the jungle path towards us, going through the same routine as before. You must not let a panther into the know. Their 'Arree! Arree!' echoed nearer and nearer, we came down, and the whole party, the reprieved goat now cheerful, returned to breakfast and to bed till noon.

The old man panther had seen us home. The shikari reported that he had followed on the edge of the jungle fifty yards behind us to within a stone's throw of the house.

On our undisturbed way home Ranji told me of a goat which defeated a panther at the very spot where we had spent the night. The panther had jumped on to the boulder. The goat, after pulling back to the full length of its tether, launched itself like a bullet at the panther, which was so surprised that it fell off backwards, made an acrobatic landing, and shot away into the jungle to think it over.

'What is more,' Ranji added, 'you can see the white goat any time you like in the Zoo at Jamnagar. His compound is labelled *Caper heroicus*.'

For six successive nights I sat up in *kotahs*; different goats but the same performance. Every time either no panther or else a lady.

Ranji was not too pleased; he wanted me to get a panther. He told me next time it did not matter whether it was a lady or not.

But I would not shoot a lady; I had found out that there were not too many panthers in the Burdar jungle. If you shot a gentleman, his successor would soon arrive from the jungles over the Porbandar boundary; if you shot a lady, her gentleman would depart over the Porbandar boundary to find another wife.

Nevertheless, the whole experience was exciting. It is no feat to shoot a panther in the strong artificial moonlight at a distance of twenty yards, but it is intensely interesting to wait while interpreting the mysterious drama of sounds in the jungle at night.

My self-restraint was well rewarded. At the end of the week Ranji announced casually at lunch, 'Charlo, you are in luck. There is a panther down on the plain by Moti Gop. We will go and get it.' He had known this all morning.

So in the brazen afternoon three Ford cars with canvas hoods—the kind we now call Lizzies—threaded up the hills, out of the Burdar saucer and down over the other side, a few miles out into the plain, where there was a big, single, rounded hill like a giant ant-heap and the little village of Moti Gop.

Directly we got there we saw a ring of villagers in their white cotton *dhotis* on the side of the hill, ringing a mass of jumbled rock. The panther had come down from the hill in the night and had been seen in the early morning creeping up the hill to lie up for the day, no doubt mediating another try for a stray dog near the village. We began to climb. Ranji was then quite stout, but he was almost as good a walker as in the old days over twenty years before, when we used to tramp the big fields after partridges at Duxford, near Cambridge.

The panther beat us. He did a bolt when we were about half-way up. We saw the upper arc of villagers scattering like white ants over the brow of the hill. They were not going to lose sight of Bapu's panther.

So we climbed down again, and the head shikari came to report that the panther had gone over the hill and into a cave at the head of a dry watercourse in the flats on the other side. Away we went after him by car. How they got there in the time was amazing, but the villagers were already standing at intervals in a big ring round the curious underground cavity at the head of the deep narrow nullah, which was like a single-line railway cutting. The villagers were opened out in a long line on either side, and we found a little thorn zareba arranged for us on a spur overlooking the nullah. There I sat with Ranji, while about a hundred yards away the shikaris insulted the panther with stones and epithets.

Suddenly there was a silence. The panther was under way. View halloos might have caused him to break back. I saw a creature like a yellow maggot threading the bed of the nullah towards us. He crept slowly along, with his white belly scraping the ground. Very different from the tall, ghostlike panthers of the night jungle. They were feline greyhounds. This fellow, until he stood up fifty yards away and began to lope, was a huge dachshund-mastiff. As he went by, just after he was past me, I had an easy shot and bowled him over. One should never shoot while the animal is approaching; if he is wounded he may up and at you like lightning. A panther can flash from immobility into electric speed of spring.

Shooting panthers on foot is a very different thing from sitting up in a tree or in a *kotah*. A wounded panther is dangerous. Some say more dangerous than a wounded tiger, because of its savage swiftness. Nor can one readily find an emplacement in the open where a panther cannot climb or spring; both of which evolutions he performs a good deal faster than anything most animals can manage on the flat. But with Bapu's servants one was almost safe. Any one of them would have jumped in front and taken the charge, and even jammed his arm in the panther's mouth, rather than that Bapu's guest should be clawed. That was my first panther.

It happened that two years later, when we were in the Burdar hills again, another panther was reported at Moti Gop. I had been out with a party in the jungle, where a drive was being attempted for General Sir Archibald Montgomery-Massingberd, to give him a shot at a sambur. He was Chief of Staff to the Commander-in-Chief. I got separated from the party in the maze of jungle hillside, and went home.

Ranji, sitting in his shirt-sleeves in the verandah, welcomed me with the words: 'Lucky fellow, Charlo. There is another panther at Moti Gop.'

This time it happened there was a largish party at the Kaleshwar bungalow, including four ladies, all of whom, as always, wanted to come and see the show. So this time we went away in four Ford cars.

When we arrived at Moti Gop we found that the panther had eluded the villagers from the start and gone away to the nullah cave at the first intention. We all pushed off again and found everything curiously as before. If Sir Archibald had not been high on the Burdar hills hoping for a sambur, he, being a short-time and very important guest, would have been the man to shoot. As it was, Charlo came in for the luck.

But when the ladies arrived with Ranji, and he saw the frail zareba on the top of the spur as before, he decided that this was much too dangerous a spot for them. We were accordingly all moved about a hundred yards down the nullah to a higher, but, for the rifle, by no means as convenient a spot. The chief shikari shook his head and indicated to His Highness that the panther was likely to break back towards the hill up a grass slope almost opposite our former station.

Ranji brushed this aside in a word, so the noise of dislodgment began at the cave, as before. The panther came out all right, but he did exactly as the shikari had said. He broke back up the grass slope at a gallop. At the top was a long row of villagers who immediately chased him back as if he were a sheep. One youth with nothing on except a loin-cloth ran after him, beating him on the back with a little stick. Meanwhile, four unfortunates had been badly clawed. The villagers were not going to let Bapu's panther escape under the eyes of Bapu. They treated him with positive indignity; and he streaked back to his holt in the cave. He was too far away from me to shoot; and even if I had I might have only wounded him, which would have been bad for the villagers.

Again after a lot of insults he was dislodged. Again he went up the side of the nullah and clawed three more villagers. Then he returned to earth. Ranji sent my gun-bearer, Bhimji with a bottle of neat brandy to pour into the scratches. Bhimji, also took a double-barrelled gun loaded with slug.

This time the panther refused to budge, so in the end the head shikari, a little man with a face exactly like a panther, crept down to the mouth of the cave with a rifle and shot the panther at about five yards. He did this bold feat as though he were merely going into the backyard to unchain a dog. This man had an instinctive knowledge of panthers. If there was one within half a day's march he would find it; he would know everything it had done in the last twelve hours, and everything it would do in the next. He was a quiet little fellow of few words, but he was a panther-man.

The panther-man was not the only interesting character among the cotton-clad shikaris. There was the man with the red beard. These shikaris were scattered all over the State; in general, one to each village, but wherever you happened a party of a dozen or so could always be relied upon to turn up by dawn full of the required information, whether of birds or of beasts.

Excerpted from Life Worth Living: Some Phases of an Englishman *(London: The Pavilion Library 1939).*

LORD HARDINGE

A Viceroy's Game Diary

As viceroy, Lord Hardinge continued the time-honoured tradition of hunting in various princely states as a guest of the Indian rulers. In six years beginning in 1910, he visited a range of states, each of which 'served up' its own special game. It could be grouse in Bikaner, gaur in Mysore or geese in Bharatpur: the game diary tells us all. The custom continued till the British left India: the Viceroy on tour got the best 'game' in the land.

BIKANER

Of all the Provinces and States of India the State of Bikaner is the most sporting. There is no big game in Bikaner, but the State provides the most varied list of smaller game such as black buck, chinkara, sand-grouse, duck, demoiselle crane, wild boar, smaller bustard and partridge.

On arrival in India in 1910 the Maharajas of Bikaner and Gwalior were two of the first to greet me at Calcutta, and these two were my first guests in Government House and honoured me by their visits many times later. They also extended to me on many occasions the most generous hospitality in their States and provided me with the most wonderful sport. I always regarded them and the Maharaja of Mysore as my greatest friends amongst the Indian Princes, and though, sad to relate, Maharaja Scindia of Gwalior is no more, I still cherish a very warm feeling of friendship for the Maharaja of Bikaner, a progressive ruler of his own State, and a loyal supporter of the British Raj in India.

It was in November 1912 that I paid my first visit to Bikaner.

I was met by the Maharaja and his principal Sirdars at Sujangarh, the frontier of Bikaner and Jaipur, at the close of my visit to the Maharaja of Jaipur, during the course of which I had paid a visit to the plague camp, the plague being rather bad at that time in Jaipur, owing to the refusal of the Maharaja, on religious grounds, to sanction the destruction of rats, and to other reasons.

My train halted at the frontier at 8 a.m. on the morning of 22nd November, and from my carriage I could see the Maharaja and his Sirdars assembled on the platform to receive me. But unfortunately I had been poisoned by something I had eaten at a banquet in Jaipur on the previous evening, and as I had a high temperature my doctor, Sir James Roberts, refused to allow me to get out of bed. It was decided that my train was to remain in a siding of the station, and no reception took place. When it was known that I was confined to my bed, the word went round that I had got the plague! The Maharaja sent me word that there was a very fine black buck that he wished me to shoot a few miles away, and I replied that I would do so the next day. My doctor ridiculed the idea, saying that I would be at least three days in bed, but I was right, and I shot the buck next day, though really hardly equal to the effort.

It was the first black buck I had ever shot, and I was greatly interested by the *modus operandi*. The country was desert, with patches of undergrowth here and there. I was driven out in a carriage for some miles to a spot where a watcher pointed out the buck in the open a long distance away. Now black buck are quite accustomed to see village carts drawn by oxen and are not afraid of them, provided that they keep a certain distance away. A village cart with oxen was there awaiting me, and I got in at the tail of the cart. The peasant leading the oxen went round the black buck in a large circle, which gradually grew smaller, and when the cart was about only two hundred yards away and the buck was looking in another direction, I slipped quietly off the back of the cart and laid flat on the ground with my rifle in front of me. The cart in the meantime went on quietly, the centre of attention from the buck, while I was able to take a steady shot and killed the buck, to the joy of the Maharaja. It was a fine buck, but during the course of my visits to the Maharaja I killed several others with much finer heads, some measuring over twenty-seven inches.

This is a very useful way to shoot a buck that is known to have some special merit, such as an exceptionally good head, or something unusual about it, but it is much more sporting and interesting to endeavour to select and shoot the best buck in a herd, particularly when they are moving in line, as is usually the case. The black buck moves very fast with great springs in the air and, owing to the considerable variation in the height of its body from the ground, presents a fairly difficult shot.

The chinkara is a small gazelle with a very pretty head. They are not found in large herds and are fairly easy to stalk, as they like to frequent places with undergrowth. A good head is about twelve inches or more. The best I shot was thirteen inches. During my recent visit to Bikaner I shot two or three chinkara by pursuing them from a motor-car, but on the whole I do not think that a very sporting way of shooting them.

The sport for which the State of Bikaner is perhaps best known is its sand-grouse shooting. There are two kinds of sand-grouse: the imperial sand-grouse, a very strong big bird; and the common sand-grouse, which is considerably smaller. At certain seasons of the year the sand-grouse come in great flocks in the morning to drink at certain pools. These pools are well known, and butts are erected near the pools. These birds are curiously regular in their flight and arrive exactly at a certain hour. I remember that the first time I shot sand-grouse I was urged by the Maharaja to be very careful to be in my butt at 7.30 a.m. exactly, and the first flight of grouse passed over me at that hour to the very minute. They fly very high and strongly, and can carry a good deal of shot.

After the first shot into a pack of grouse they zigzag worse than a snipe and make a most difficult shot for the second barrel. They continue arriving for about two hours, and they are followed by the common sand-grouse in smaller numbers for about another half-hour. All these sand-grouse having failed to obtain water at the pools fly away into the desert, possibly to other pools. Thus for about two and a half hours, when the sand-grouse are plentiful in numbers, shooting goes on very rapidly, and the bags are sometimes very large. The best morning I ever had was one thousand four hundred grouse to thirty guns, but I have heard that sometimes as many as four thousand have been shot in one morning. These grouse are very good to eat if properly cooked.

The Maharaja gave me also some excellent duck-shooting. At Tulwara Jeel we shot five hundred and fifty duck to eight guns in the afternoon, and at Gujner, close to the Palace, we shot six hundred duck in one afternoon to six guns, the two young Maharaj Kumars making the biggest contributions to the bag.

Demoiselle cranes (kunj) and smaller bustard (houbara) also provided sport, but these could hardly be compared to the duck and sand-grouse as affording the best form of sport.

Of course there could hardly be better ground for pig-sticking than in Bikaner, with its stretches of scrub and sandy desert, and the country is full of pig. I have referred to this in my chapter on pig-sticking.

I have never had a better time than those most enjoyable days of sport with the Maharaja, who is the very best of sportsmen and a marvellous shot, and the many hours we spent together in long expeditions were certainly very helpful to me in many ways, and I hope also of some advantage to His Highness.

During the five and a half years that I spent in India nothing ever ruffled our relations of mutual respect and affection, and I look back upon our friendship, which began twenty-three years ago, as one of the priceless incidents of my life.

Duck-Shooting in Bhurtpore

It was at the close of a long journey commencing in October 1912, during which I had visited several States, that I arrived at Bhurtpore on the 20th December, on the invitation of H. H. The Maharani Regent of the State of Bhurtpore, during the minority of her son the Maharaja. The object of the journey was to pay an official visit to the Maharani and State and to take part in the annual duck-shoot for which Bhurtpore was famous all over India. I was the guest of Mr and Mrs Holland at the Residency, where I received the warmest hospitality in surroundings of the utmost comfort.

Bhurtpore is a small State of Rajputana with a population of rather more than half a million and of small revenue, but it was admirably administered by the late Maharani during her son's minority. The Maharani was a charming lady, speaking English well, as is almost invariably the case with well-born Indian ladies. The Viceroy, being regarded as sexless owing to his high position, is permitted in many cases, as in Bhurtpore, to interview Indian ladies without the usual purdah restrictions, There were none in the case of this Maharani, and on the three occasions that I visited Bhurtpore I was permitted to have long conversations with Her

Highness, whom I found to be a very intelligent and capable woman. My official functions on this occasion were very few, but I inspected the Bhurtpore Imperial Service troops, a corps of transport which rendered good service with the Indian Expeditionary Force during the Great War.

On the morning of the 21st December I proceeded with my Staff to the lake (jhil) where I found a large company of sportsmen already assembled, fifty in all. It looked a formidable gathering with all the loaders, servants, etc.

The jhil was a shallow lake, probably about eight miles in length and about half a mile broad at the centre. Here at the centre there was a 'bund' a few yards broad stretching out to the middle of the lake with high grass on each side. This, from long custom, was named 'The Viceroy's bund.' I was asked whether I would like to shoot from a position where the duck flew high, or whether I would prefer to obtain a larger bag where the duck would be flying low and much easier to shoot. I naturally replied at once that I would like to stand somewhere where the duck would be numerous and would fly high. I was therefore placed at the extremity of the Viceroy's bund stretching out into the lake.

The system adopted for driving the duck was that at each end of the lake there was a line of elephants which slowly but steadily walked through the shallow expanse of water towards the central bund on which I was standing, and which was equidistant from the two ends of the jhil. As it took the two lines of elephants two hours to meet I concluded that the total length of the jhil was about eight miles from end to end.

I was taken to my stand, where I found everything arranged with great comfort, including drinks and cigarettes, etc. There was an Indian whose task it was to count the birds I shot, and in addition I saw some fine young Indians, almost naked, sitting in the long grass. I asked what their functions might be, and was told that they were troopers of the Imperial Service Troops and were my retrievers!

When I took my place duck were already flying around in every direction, but it had been agreed that not a shot was to be fired until the signal was given by a bugle at ten o'clock exactly from the tents near by. When given, a fusillade started on every side which continued without cessation for two hours until noon, when a further signal was given by bugle to cease fire. At that time the elephants were close up to the central bund on each side. During those two hours the duck were flying on every side and in every direction and as time went on flew higher and higher, but the stream of duck never seemed to cease. They gave difficult but very sporting shots.

At noon there was an interval of two and a half hours for luncheon and rest.

As my tally showed that I had shot more than a hundred duck I was disappointed to find that my four retrievers had only brought about sixty to land. I mentioned this to a friend, and he told me that it was customary to offer a reward to the retrievers based on what they retrieved. I need hardly say that I followed this advice, which certainly, judging by results, appeared to be very sound, for on resumption of shooting in the afternoon I offered my retrievers one anna (about a penny farthing) for every duck that they retrieved, and I really believe that afterwards I hardly lost a bird, and if they could not find one of mine I have a suspicion that they pinched one shot by one of my neighbours. I have never seen such keenness in retrievers, either male or canine.

After a very excellent luncheon and rest till 2.30 p.m. shooting was resumed till 5 p.m. The elephants marched in line up the jhil, but this time they had not, I imagine, gone to the extreme limits of the water, since there would not have been time for them to do so. As a consequence the duck were not as numerous in the afternoon. Still there were plenty and more than enough to shoot, and I personally was glad to finish before 5 p.m. owing to my shoulder being black and blue and my face swollen on one side!

At the close of the shooting all the game was laid out and the scores given on the respective sportsmen. The grand total was three thousand five hundred and eleven duck, amongst which I counted no less than thirteen distinct varieties, many of which were quite unknown to me.

The biggest scores were Mr Cruickshank two hundred and fifty, the Maharaj Rana of Dholpur one hundred and fifty, Colonel Impey one hundred and fifty-seven, Major Munn one hundred and fifty-one. There were thirteen centuries in all. My own score was one hundred and ten.

On the following day I left Bhurtpore and arrived at Delhi in the morning of the 23rd December for the State Entry into the new Capital.

I had two more duck-shoots at Bhurtpore on the 3rd December 1914 and the 13th November 1915.

On the 3rd December 1914 the duck were far more numerous than the previous year, and our total bag to forty-nine guns was four thousand one hundred and twelve ducks. The proceedings carried out were exactly similar to those of the preceding year. There were three scores of over two hundred, the highest being Mr Rowan's two hundred and sixty, closely followed by Mr Cruickshank's two hundred and fifty-six. There were in addition eleven centuries. My own score, thanks to my retrievers, was one hundred and eighty-three.

My last duck-shoot at Bhurtpore was on the 13th November 1915. Owing to the war it was carried out on a much smaller scale. There were only thirty guns, and the total was one thousand seven hundred and sixteen. The best scores were Captain G. Herbert one hundred and fifty-five, and Mr Gamble one hundred and five. There were no other centuries. My own score was seventy-nine.

When in India two years ago I inquired whether the duck-shooting at Bhurtpore was as good as ever, and I was told that owing to various reasons the water level of the jhil had become lower and that consequently the area of the lake had diminished, and that the duck came there in very reduced numbers. In fact, I gathered that the bags obtained there now are comparatively small.

Excerpted from On Hill and Plain *(London: John Murray 1933).*

GUY FLEETWOOD WILSON

A 'Criminal' Lion in Kathiawar

By the turn of the century, anyone who wanted to 'bag' a lion had to go down to Kathiawar. But most lions were in the confines of the state of Junagadh, whose Nawabs rarely gave consent to have one shot. High officials, princes, governors and viceroys could, however, get an invitation. Fleetwood Wilson was granted permission to shoot a lion that was a danger to people—a rare occurrence in the area. He was also among the few to have shot a lion in India the same year in which he shot a tiger in the Sundarbans, Bengal.

No matter how many or how deadly be the sins committed by George Sydenham Clarke, otherwise His Excellency the Governor of Bombay, I feel confident that the Recording Angel will, for his superbenevolent treatment of me, wipe them all out and accord him the benefit of a clean slate.

To his kindness I owed the exceptional privilege of securing a very, very rare trophy—to wit, a maneless lion of India and an exceptionally fine one.

I am assured that not one had been shot for over thirty years, and on the last occasion the lion killed one of the then Governor's Staff. The fact is there are very few of them. Indeed, at one time only sixteen were believed to exist. They are to be found in only one spot in the whole continent, in the forest of Gir, and they are strictly preserved. It is only at rare intervals, generally when drought has killed off all the cattle, that one takes to man-killing, and then he has to be tried, condemned, and executed.

Of the lions of Gir an early Italian writer makes the quaint remark that they are like honest men and truthful women—'Esistono, ma rari.'

Gir Forest is an extensive, sunbaked district covered with bushes that resemble dried-up mimosa, very plentifully furnished with sharp thorns. Such trees as exist are few and far between.

It is owing to the lions having for centuries forced their way through this thorn-covered bush that their mane (incessantly torn off) has ceased to grow. I understand that there is a district in Africa where the lions carry little or no mane from the same cause.

It fell to my lot to be in India when a 'criminal lion' came in evidence, and it was my good fortune that my sojourn in the East synchronized with that of so sympathetic and generous-minded a Governor of Bombay.

I had always longed to complete my list of the big game of India, but I had given up all hope of filling up the lion gap, when Sir George one day said to me: 'I can and will give you what no one else in the world can give you—a Kathiawar lion.' And he kept his word.

On April 15th I started for Kathiawar State, and I shall not easily forget the journey there and back.

It is not a season when sane people travel across the plains of India, and I had two Indian servants down with heat apoplexy on the return journey.

On the 16th I reached Viramgam and proceeded by narrow-gauge railway in the most terrific heat, passing Rajkot and through a badly famine-stricken district. The cattle seemed to have all died off and I believe the human mortality had been somewhat heavy.

Kathiawar, which once formed part of the old kingdom of Guzerat, is a Native State in the Bombay Presidency on the Arabian Sea. Gir or Gier Forest is in the south-eastern extremity of the Peninsula of Kathiawar on the Bay of Cambay, opposite Diu Islands, which belong to Portugal, and from whence it is said a nice little smuggling trade is carried on.

When I increased the import duties it was remarked that a constant stream of women flowed up-country. They came under especial notice because they appeared to be in pain and seemed to walk with extreme difficulty. Investigation showed that their legs were absolutely covered with very thick rings of smuggled silver which in this fashion they carried into the interior. *Se non è vero, è ben trovato.*

On Wednesday, the 17th, I reached Verawal, where Robertson, the Resident, met me and drove me twenty-two miles to the camp at Kokhra. From first to last he devoted himself to making my expedition a success, and I shall always retain a vivid recollection of his unselfish kindness to me. We rode sixteen and walked two miles. Midway we were met by a deputation of the inhabitants. They gave me quite the most exquisite coffee I have ever tasted and a cordial welcome. They appeared to me to be as nearly as possible pure Arabs, and they had all the attributes of the Arab race: a fine physique, great dignity, marked courtesy, and much reserve. They are a fine, manly, independent people.

In overwhelming heat we had a lion drive. I saw no lion, but I was assured that one had been 'moved.'

This moving is a curious process.

A 'kill' is watched, and when the gorged lion has eaten and drunk his fill his lying-up place is located and sufficient time is allowed to elapse to ensure his going off into a sound sleep.

In due course a number of men very silently creep as near as they dare to the lion's sleeping chamber, which they almost surround, leaving unguarded the line of retreat which it is considered the lion will probably take.

At a signal from the head shikari all the men give utterance in unison to an ear-splitting, demoniac yell which wakes up the lion under conditions calculated to make him about as ill-tempered as a retired Colonel of the old school with a severe attack of gout.

The men are careful not to show themselves, and are especially careful not to make the slightest noise. The head shikari allows about a minute to elapse and then gives the signal for yell number two.

The plan has never been known to fail. The lion thinks the whole performance uncanny and trots off or bounds away according to his temperament in the direction whence no sound has reached him, and if all goes well passes the spot where the sportsman has been posted—sometimes in a tree, sometimes behind one.

In this instance my post was beside a rock about the size and shape of a grand piano. But on this occasion no lion passed me. On the 18th and 19th, there being no news of lions, we stood at ease—ease which I very greatly needed. We were cheered by a visit from Mrs Robertson, to whom I am greatly indebted for much trouble taken on my behalf, and who possesses the admirable qualities which characterize the Mufassal Englishwomen in India.

On the 20th we got *khabar* of a lion—a well-known, unusually big, old and grey lion—and started very early after him on foot. I think we were some ten hours following him. He at last lay up in a clump of bushes in a very small wood, almost the only wood I saw in Gir Forest. Gir Forest, like a Scotch forest, is treeless and equally bears its name on the *lucus a non lucendo* principle.

I was posted on rising ground behind a bulky tree-stump and the head shikari, a splendid specimen of the class, stood beside me. The 'move' succeeded, but the lion did not wait for the second yell. Immediately he heard the first he bounded out and came past me at a hand-canter. He was inclining towards Robertson's post, when that kind, unselfish young man fired two shots over him, pushing him towards me. I hit the lion rather too high and too much behind to kill him straight away. He went a short distance and then lay up on a bare spot exactly in the attitude of the Nelson Monument lions, and we, the head shikari and myself, had to walk up to and finish him.

I fully expected him to charge us and confess to feeling at the moment in a 'blue funk,' but my shot had injured the spine and he could not rise, much less charge.

He was a grand beast, and when I looked at him lying dead, I confess that I would have given all I have in the world to have been able to restore him to life.

He looked superb when he passed me, taking great leaps rather than galloping; he looked very formidable when sitting up wounded, and he looked noble in death. Of course, I am beyond measure delighted at getting a Gir lion, but the day is not far distant when I shall hate to kill any of God's creatures.

On the 21st I returned to Viramgam, and next day reached Delhi, a veritable wreck from intense heat and exhaustion. My damaged knee resembles a large and overripe pumpkin, and from my devoted personal assistant, Mr Baker, I shall receive a well-merited rebuke.

Excerpted from Letters to Nobody *(London: John Murray, 1921).*

J. W. BEST

Visitors and Big Game

In order to 'bag a tiger', the great and mighty travelled by train away from the densely populated coast or the valley of the Ganga into the woodlands in central India or the Himalayan foothills. But for the forest officer whose work routinely kept him in the jungle, such visits were a challenge. J. W. Best tells of how he laid out the red carpet and used all his skills to ensure his visitors went off with a tiger-skin in their bags. The hunt had by now been reduced to a ritual, but the tiger was no easy quarry, and could become shy and secretive. Perhaps these traits helped it survive the high noon of the imperial hunt.

I have, I think, already shown that the forest officer's lot in the old days before cars were the fashion, and railways developed as they are now, was a lonely one. Made roads were scarce, some districts boasting only one or two such luxuries. Long journeys in the districts meant long rides, or uncomfortable mileage cramped within the limited space of native bullock carts. Those quick dashes out meant organization—*bandobast*—ponies had to be sent on ahead so that one could move quickly, with changes of mounts every ten or twelve miles; and there is nothing so restful in a long ride as a change of mount; or, if one travelled by bullock cart, the trotting bullocks had to be replaced by fresh animals at shorter intervals.

There was a fixed tariff prescribed for the hire of the bullocks according to the mileage, and with each stage there was a new driver, who took his animals home again on relief. Riding was, of course, quicker, but even the bullocks could do their six miles an hour or even more in the case of a well-bred pair. Some were obstinate, specially at the *nalas*. There would be a quick run down the steep bank to the ford, then strained toiling through the heavy sand, but a willing pull to the welcome water where there would be a great splash and a sudden stop while the bullocks drank deeply. Then away again with the cool water splashing their flanks, followed by the hiss of sand as the wheels came on to the dry again cutting deep into the soft bed of the *nala*. The driver, knowing the way of his beasts, would tickle them under the tails with his toes, and if that were not enough encouragement twist their already knotted tails, leaning well forward as they heaved and strained up the steep bank where the ruts cut deep into the hard dry mud.

Sometimes they were obstinate; they would lie down in the road simulating death as though utterly weary of this world's toils. Then they would be prodded with goads and submitted to all manner of cruelties, including the ordeal of fire, till the humane man cried halt to the brutalities. Once started again, however, the bullocks knowing that their driver was indeed the master, would go on willingly enough. The driver had called their bluff.

But the horse for me. The bullock may be sacred to the Hindus, although it is hard to believe it considering the cruelties to which he has to submit at times; but if any animal can be called sacred to an Englishman surely it is the horse. Indeed, when London, after our civilization is gone in a few thousand years, is excavated by the antiquarians, and learned men are trying to reconstruct our national life, it seems probable that they will decide that the noble animal, the horse, was the principal object of worship in our times. No other creature receives such flattering attention in statuary or the arts.

Every forest officer in those days had to pass a test in riding before he could come to India. He was expected to keep at least one horse and found difficulty in having his travelling allowances passed unless he did.

It seems a pity that the noble animal should be replaced by the car. I can remember the time when on news coming in of a sounder of pig near the station every officer turned out for the hunt, and the cultivators would join in a riotously joyful day followed by a feast of pork. Alas! I was to see the day too, when I was the only Englishman in a moderately-sized station to turn out with his horse. I have noticed that the man well mounted always excites respect. In an Indian bazaar he gets his salaam and the bold glance from the ladies; in rural England the touch of the hat and the welcome upward twist of the head that means 'we be of one blood, you and I.'

And talking of horses, there is a story told of a well-known forest officer in charge of a division who scored over an unpopular conservator whose duty it was to pass his travelling allowance bills. The rule was, that if an officer came into headquarters, he could charge to government the cost during the period of his stay of the maintenance of his camping equipment, such as the hire of carts or camels. This officer kept two ponies and when opportunity offered—which was not often—he used them for playing polo. He needed his ponies for his work, however, and felt justified in charging for their keep during a short stay in headquarters during the camping season. So he put down their feed and keep during the time. This was disallowed by a hard-hearted conservator, because the ponies, so he said, were kept for the officer's own pleasure—a ruling which the officer noted.

A month or two later he was in camp some twenty-five miles away from the nearest railway station, a dull spot consisting of a few sheds set down alongside the line in a vast treeless and dusty plain. The conservator wanted to see the officer urgently and quickly. A wire was sent to him in camp, which he received, telling him to come to '———' railway station at once to meet the conservator who was waiting there for him. The officer started; it was a long walk and took two days. He met a very angry conservator.

'Why the devil could you not come more quickly?' he was asked by the enraged man.

'Twenty-five miles is a long way to walk, sir,' was the answer.

'But you have two ponies, why not ride?'

'You have told me, sir, that I keep my ponies only for my pleasure. This is not pleasure,' he answered as he glanced round the bare arid plain.

Communications in those days were not easy. We were expected to spend the best part of the year in camp, and we did. A stay in headquarters during the camping season gave occasion for an explanation. A forest officer's place was in his forests and among the people entrusted to his care. If he met people of his own race when on tour he was lucky; sometimes a missionary, more often a soldier on a shooting trip; perhaps a district official on his tours. But any such meetings were rare. Once or twice in my career I managed to meet the forest officer of the neighbouring division to settle small affairs of our mutual frontiers, but as a service we saw little of one another, and we were so keen on our jobs that we resented being called into headquarters to discuss affairs with our chiefs. Often such summonses were sent purely out of kindness to break the spell of the lonely jungle. It was thought that too much of our own company was not good for us. Our chiefs had been through it and knew its influences, and when I was in the lofty position of being conservator myself I was guilty of the same offence. It is not good for a man to go wholly '*jungly,*' although that was my nickname with some—of which I am proud. One got into a rut and formed the habit of making plans of tours long ahead, resenting any interference with them. The annual volunteer camp in those days was not very serious soldiering, it was more of a yearly reunion and ten days under canvas for those whose lot lay in offices. Few forest officers attended it. They were held by the shade of their own forests and the claims of the people that lived in them.

But visitors were nearly always welcome. They were of all sorts, and, except for the visiting tour of an inspecting conservator or that rare bird the Inspector-General of Forests, came with one object—to shoot.

The most welcome were those from England—home—as we call it in the East. I always found it a great joy showing off my jungles to these people—they appreciated it. They knew nothing, and were keen to learn. More, they knew that they knew nothing, and did not try to interfere in the arrangements made for their amusement or to criticize after events. I know nothing so irritating after a hard day, when, as the result of careful plans a tiger beat has been held, to have a man who has taken no part in those plans and knows nothing of the terrain, to start criticizing; trying to show how, in fact, if he had been running the show he would have done it much better. Then there is another sort of interference, when having successfully laid plans to get a tiger into a certain place for a beat, and even had the *machans* placed a week beforehand so as to take the beat in a certain direction, one is desired to alter arrangements at the last moment, and by one with no real first-hand experience of *shikar* but who wishes to pose as an expert.

The Anglo-Indian brass hat often comes into this category. He has probably killed a few tigers in his time, but they are what are popularly known as 'Viceroy's tigers'; beasts that have been driven over time-honoured ground to a place where they are certain to come out for the brass hat to massacre. The wretched man is cursed with a knowledge of the language and so by questioning the *shikaris* is in the position to find out what the plans are; having done so he tries to show his superior knowledge by criticizing.

I remember one in particular. He had shot a good many tigers in his time, yet I doubt if he had ever run a beat in his life. We were in new country where beats had not been held before. For that reason it was all the more interesting to me. I put him where the tiger was expected to go and the beast walked straight beneath the tree where he sat. He had a very easy shot at a few yards' range and killed his tiger in one shot. Another added to his total! That man never saw the kill, he did nothing about helping to plan the beat (I did not want him to, of course), he did not help place the stops or do a single thing to help in the general arrangements. Yet after the beat was over he had the gall to tell me that the *machan* was badly placed. I don't quite know what he wanted. A tower perhaps such as the Rajahs produce, with a lift and iced champagne ready to hand.

The next beat was a very scratch affair over a natural kill; a piece of luck that one receives with thanks as a gift from the gods; the tiger again went under the man's tree and he missed it. Once more he pointed out how he could have run the beat so much better. That was a hurried beat that we had to do before the limited hours of daylight gave out. He was lucky, with such hasty plans, to see the tiger at all. The trouble is that that sort of brass hat is usually very highly placed and cannot be put in his proper jungle position, which is a low one in my view. I did answer one of them back on one occasion, scoring heavily.

I was asked to run some small-game beats at a difficult place before I had time to cut rides or to bait the ground to collect the game. Moreover, it was the wrong time of year; the jungle was too thick. I said so. The person's reply was that he did not expect more than a day in the country, which is about all that he had. Our bag was three jungle fowl and two peafowl to four guns.

A year later I met the same brass hat in headquarters after I had left that division and another was in charge. There were a number of people present at the club bar and the brass hat amused himself at my expense. He asked my successor whether he could arrange some shooting for him in those forests at Christmas. 'You know,' he said to his listeners, 'a year ago I asked Best to arrange a shoot for me in that same place. Best thinks that he knows a lot about *shikar*, he has written a book about it. He arranged beats for me; we worked all day and what do you think we got?'

The crowd waited for him to answer his own question in respectful silence.

'Three peafowl and two jungle fowl!'

There was laughter in court.

'No, sir,' I said, 'you are mistaken. One more must be added to the bag.'

'Oh,' he said, 'what?'

'One beater,' I replied joyfully.

'Who shot him?' he asked.

'You did,' I said.

And it was true. The man had been only slightly peppered, of course, but it was enough for my retort. I am not usually quick on the uptake, my best answers coming after I have gone to bed.

It is right to add that the brass hat to whom I refer was a man for whom I had, in other respects, considerable regard and sympathy. He had the gift of humour and bore no grudge.

Perhaps one loses one's own sense of humour to a certain extent in the jungle. The monotony of the life plays on a man's nerves and the loneliness, although one does not realize it at the time, makes him irritable over small things that most men laugh at. India herself is a sad country where it is rare to see a smile on a native's face, and from close association with the people the white man is apt to react to the mood. It is natural that one's greatest sympathy should be for the young, and I must confess that I took far more pleasure in helping the keen young subaltern to get his shooting, than I did in the case of more senior officers.

In nearly all instances the type of sportsman who goes to the jungle plays the game. He knows that the shooting rules are made mainly in the interests of sport, and it was very rare indeed to have an infringement of the rules reported. There was such a thing as the black list, but it was rarely that a name was reported for entry. Once on it, and the name circulated to the divisional officers, no permit could be issued to the offender.

A block permit was once issued to a certain lady—I did not at the time know she was a lady, but it would have made no difference. I discovered her sex when she complained to me that my forest ranger had been rude to her. The good lady had wounded a tiger and wanted it finished off. Finishing off wounded tigers is dangerous work and most sportsmen do this job themselves. Not this one though.

She wrote complaining to me that the ranger refused to order the little jungle men to finish off her wounded tiger with their bows, arrows, and axes. Would I punish the ranger? My answer can be imagined. It was then placed on record that I was a very rude man—no gentleman, in fact. She received no more permits to shoot in the districts where I served.

Very different was a young subaltern from a battalion stationed in Bengal. He was given a block that bordered a native state where tigers were preserved more carefully than any home-bred pheasant. And there were man-eaters waiting for the rajah to shoot. Our young soldier tied up his kills within his own block in British territory, and being keen, he did the job himself. Over the boundary line and close at hand was the rajah's magnificent camp.

A tiger took the subaltern's kill. Beaters were out of the question, the rajah's party had seen to that. So our subaltern decided to sit up over the kill and wait for the tiger's return. He received a visit from the prime minister saying that His Highness wished to shoot the tiger. Naturally the request was declined. The soldier tied his own *machan* and sat all night, which was made hideous by the noise of drums, firearms, and other alarms. It would seem that no normal tiger would venture near such noise; but this one was the exception; it returned and was shot. Next day a polite note was sent to His Highness asking him to come over and see the trophy!

One of my most welcome guests was an officer in the Indian army. He knew a lot about *shikar* and was as keen as mustard. Whatever the result of the beat he wanted to go over it afterwards and try to learn something from it. He was very different to the brass hat type: you don't find *them* trying to learn anything, they know it all and give one to understand that that is so. All they come out for is to add yet one more tiger to their record of which they can boast in the clubs. Any fool can sit in a tree and murder a tiger as it walks under his feet.

There was a freshness about visitors from home; a breath of England that did one good. Perhaps the most amusing of all were a couple of naval officers ashore for the Christmas holidays. Here, indeed, was a change for them and for me. A difference in our respective outlooks that can be summed up in an excuse one of them made for not shooting as well as he might have.

'It is all right for you,' he said, 'shooting beasts hidden in this tangled jungle. Your job is to pry about into its depths. Ours is to shoot at targets on the sea with no trees and branches to obstruct the view.'

We were not, of course, always in the jungles. During the monsoon period from July to September the roads were no longer serviceable. They were mud, and all wheeled transport ceased. Some touring was possible in the hilly forests by means of coolie transport and living in rest-houses, but it was wet and unhealthy, although more interesting than at other times, because one could study the trees during their short season of astonishingly rapid growth. There were plants and flowers to be seen during the rainy season that were not there, or were inconspicuous, during the dry months.

In the stations we lived the normal life of the Anglo-Indian. Office work in the day, following a ride in the morning with tennis or golf in the afternoons. We played much bridge and snooker. The usual small station consisted of the Deputy Commissioner, Police Officer, Doctor, Forest Officer and possibly one or two assistants. All of these would be Europeans and we generally got on wonderfully well together. Club life was the centre of social affairs, and the clubs were kept almost exclusively for Europeans. Europeans started and built them. We played cricket and hockey as the weather permitted and often took on the next station, which was usually some hundred or more miles away, sometimes considerably more. Our teams were mixed and were as good an example as could be found anywhere of the friendly co-operation of European and Asiatic.

Then at Christmas we would meet again in camp—those of us who wished to enjoy more of one another's company. In the ten days' holiday I would be responsible for the amusements. We always tried to fix camp in some place that was near good small-game shooting, whether water fowl or true game birds, such as peafowl or jungle fowl. That would keep guests amused by day and sufficiently tired in the evening not to wish to play bridge, which in my view is almost sacrilege in those marvellous forest nights when one should be listening to the many voices of the wild. If we did not bag a tiger or two we were disappointed. Sometimes we killed a deer.

After the Christmas camp we were unlikely to see one another again until the rains broke in June. In that period the forest officer would be fighting fires, the policeman might be hunting dacoits, the doctor waging his chronic war against insanitary conditions, cholera, malaria, plague perhaps, and the thousands of other ills that afflict unguarded man; and over all would be the watchful eye of the Deputy Commissioner supervising and tactfully oiling the machinery of each department in his district, lest there be a squeak in an important bearing indicating trouble that might affect the efficiency of all. He was the Mussolini of the district; farming, engineering, mining, forestry, police administration, medicine, irrigation, what ever was being done came under his eye. He was the chief magistrate of the district as well. Nowhere in the world can there be such wonderful personal rule as there is in India under the I.C.S.

In running tiger shoots it is my experience, and sometimes that of others, that if the organizer takes a *machan* in a tiger beat the tiger is sure to come his way. It is naturally one's wish that the tiger should be shot by a guest. I have heard it said that if the beat is left entirely to the *shikaris* and to native subordinates, they do their best to drive the tiger to the *machan* of the most important person out. I do not think that this is the case, as I know how difficult that would be. It is true that the stops on either flank might, by making excessive noise at a certain time, drive the tiger to one particular flank where the important person is. But those who run beats are not stops in the flank, they are with the beaters, and in any case if the stops do make an excessive noise and an unnecessary tumult on a flank then the guns cannot fail to hear it even in a noisy beat. Whatever the reason, I found that if I took a *machan* the tiger generally came my way. So after some experience, when I had guests I went with the beaters. It had the advantage of giving them confidence, and was far more interesting than sitting in a *machan* for the tiger to be driven up. One avoided the long wait while wondering why the men delayed so long and whether the beat would be bungled.

Sometimes in the case of a tiger that was well known as a sticky customer I went with the beaters unarmed, as they were, save for their axes, which in an encounter with a tiger would be of little use. Then one had to pretend to be really brave tackling thickets that others carefully avoided. There were thrills in plenty and the work was intensely interesting. First, placing the guns in their *machans* and clearing the ground round them so that they could have as good a view as possible, then putting the stops up on either flank, one would work back as quietly and quickly as possible to where the beaters would be waiting.

Every sort and type of man would be there. Cultivators, cartmen, camp followers and squat little black men from the jungles, each with his axe slung over his shoulder as he sat on his heels waiting for the word to line up. With them would be uniformed forest subordinates, men of authority to keep them together and ensure the essential silence. Generally they would be waiting on a road defined by two deep cart ruts and the grey dust on the leaves at its side. Now it would be deserted of traffic. Everyone knows when there is a tiger beat on, and the road is deserted except for those who come on tiger affairs, and they are many; the Indian countryman is one of the best sportsmen in the world.

A few brief words to the beaters. 'Don't bunch together. Let no one get ahead of his brethren, for there danger lies for all. When the word is given shout your loudest and continue shouting till you hear a shot, then be silent. If you hear me blow my hunting horn climb trees till told to come down again. Then, when word is given that all is safe, drive on again.'

So we line out in silence. I take the middle, the elephant one flank. And as we spread along the road facing the jungle where we hope the tiger lies up after a full meal, we see a cloud of vultures spiralling to a point in the air and we are heartened. He has driven them off his kill.

Two shots start the beat, there is a mighty shout and we move forward, plunging into the undergrowth. The jungle is pretty thick here, as it should be if a tiger is to be enticed to lie within its shelter; on either side of the strip of heavier growth, which marks the sides of a watercourse, there is open country, park-like with a few trees standing over high grass. From where the kill was taken we find a well-marked trail along which, in his efforts to drag the carcass away from the vicinity of the man-haunted road, the tiger has broken down shrubbery and lower bushes as he pulled his clumsy burden behind him; the ground is scarred as though in places someone has dragged a heavy log; earth, leaves, sticks and stones are scraped away and piled up in little banks and heaps. There are patches of ground with no trace of dragging, where the tiger has lifted his prey over a stretch of particularly awkward ground, then the tell-tale marks are found again. We press on, looking anxiously ahead; it cannot be far off now; a crow calls in a tree to our left; the trail leads to the steep bank of a *nala* shrouded by creepers, the lip of the *nala* is scarred and stones have recently been displaced on its side, and there, screened by leaves the beast has left his kill.

A half-eaten calf, the buttocks gone, the entrails torn away by vultures, the whole carcass grey and fouled by bald-headed creatures unworthy the name of bird. Round about, the grass has been trampled by them and the branches of trees soiled where they waited for a safe moment to descend to the feast. They knew well enough whose meat they were stealing, and one caught napping by the returning tiger lies broken, dead, an untidy mess of dirty, dishevelled feathers hiding a carrion bald head.

But we push on. Here we find the tiger's track, fresh, where he hastened over the sandy bed of a dry stream. There he drank in a shallow pool. Freshly disturbed earth shows where he climbed the bank of the *nala*, rolling rocks behind him. He has hurried. We pass on, keeping our place in the line so far as the thickets and the tall waving grasses allow. The cries of the men have merged into a monotonous chorus. We find where the tiger has left the *nala*, seeking better covert for his escape through the tangled undergrowth. We lose the trail to find it again where he crosses the *nala* for the other side. There is a sudden crescendo from the beaters. Why? One wonders. A panic-stricken deer rushes back through the line, then a sounder of pig, and a man is knocked over but is none the worse, though regretting the loss of pork by an error in aiming his axe. Good sportsmen these fellows. They amuse themselves by throwing things at terror-stricken monkeys that dash here and there, lost in their panic and not knowing whither to flee.

We reach a thicket of thorns impenetrable to man and elephant. We line up in close formation behind it, then shout together with sudden fury, hoping to shift the tiger. A man shouts that he has seen him go ahead. He is probably a liar. Again we move on. We are more than half-way through the beat, and the shouting diminishes a little. Will there never be the sound of a shot?

Then in a pause of the shouting, while men as it were took breath for the final effort—it was tiring work—we had our message from in front. The full deep-throated roar of the tiger bellowing in baffled rage.

It began with a sudden nerve-shattering cough so loud that it seemed to shake the leaves of the forest, and it could be heard by all in spite of the confusion of human voices. At the sound men stopped in their tracks or took a pace or two back; many made for the nearest tree. Then there was a bellow repeated three or four times in quick succession, ending in a shattering roar followed by muffled grumbles like the echoes of receding thunder. The beaters are quiet. Most of them are half-way up trees, for the tiger is close—somewhere on the right flank where he has tried to break through the stops. He has clearly been turned, but must not be allowed to come back. So we shout our loudest, hoping that he is being driven on to the line of guns. There is a tendency for the beaters to group, a natural hesitation about going on. Stout-hearted men patrol the line and the main body is led on again, but more slowly and with cautious glances at trees.

Nothing happens, and one fears that the tiger has got away between the stops. The beaters regain confidence, growing less careful. They forget those terrifying bellows in the comfort of their own company and nothing happening. I become anxious. Why has there been no shot? Has all this trouble and scheming been in vain? One thinks of and hopes for nothing but the sound of a possible shot.

There are no animals left in the beat now. No monkeys tearing out through the beaters in their panic. The deer have been gone a long time and so have the pigs. One grey monkey, half grown, has been left behind by its terror-stricken mamma; it sits at the top of a bare tree, its little body clasping the grey trunk against which it tries to hide, while the howling horde of men passes beneath and beyond. A large red squirrel flies before the line, leaping from tree to tree till it reaches the last in a row and there is a gap before another can be reached. Will it jump? Scuttling up the highest branches in a corkscrew course it stops, looking fearfully below. The beat passes on.

At last! A shot. The dull roar of a heavy rifle, muffled in its echo by the leaves and branches. It is followed by another shot. Again there is a halt of the beaters while men listen. Two shots may be a bad sign in a small-game beat, but in a tiger drive it is a cheerful message. It probably means that the tiger has been bowled over with the first, and that a careful gun gives him a second to make sure that he will keep dead. There is little need for a warning, again most of the beaters are in the sanctuary of the trees; and they chatter. It is essential to know what has happened. Has the tiger gone on? Perhaps it is dead? But it may have been headed, and wounded is now approaching the line. No risks may be run.

I blow my horn and the few men remaining on the ground race up the trees there to watch, wondering perhaps what the Sahib will do if the tiger comes.

There is a stop on my right. He shouts to the next man and so word is passed to the guns asking what has happened. Why can't the beaters stop their chatter in the trees? It is difficult to hear what the stops say. Contradictory messages come back. At last we get reliable news. The tiger was knocked over, took a second shot and then went back towards the beat.

There is uproar of conversation among the beaters now crowding the branches of the higher trees. They hope to see something. Rather a vain hope, because a tiger that does not want to be seen is almost invisible. Each man gives his view of what has happened, and since the noise is great he must shout louder than his fellows to be heard. But none listen. One knows that shouting at the men is hopeless. So the horn is blown again. That is something new, different to the other noises. The shouting stops for a moment, and before they have time to reopen their babel they are ordered to stay where they are.

Then, with the two trackers, we carefully feel our way along the line of stops towards the guns. Overhead we hear a branch shake. It holds the Babu. A sportsman, but a very fearful man, keen to be able to say that he has seen a tiger shot and to help in the shoot.

'Hallo, Babu. Have you seen the tiger?'

'Sir, it is there.' We get behind a tree ready for a charge.

'Is it dead?'

'No, sir, it is not deaded yet.'

We avoid the place and collect the other guns. The Babu stays where he is. He is not likely to move voluntarily. The guns come down from their *machans*. Now we get direct news of what happened.

'That is where I first saw him. There I took the shot and there he fell over the creeper. As he went off into those bushes I fired again and I am sure I hit him a second time.'

Now is the danger time, when even the most experienced men do things that they know are foolish. There is a reaction after all the excitement, and an impatience to get the business over. There is, too, I often think, the risk of a nervous man not quite sure of himself taking chances that wiser men avoid. A natural course for most people. I always had nerves while I waited that half-hour before following up a wounded tiger. There is an urge to get along with the job and have done with it; many have fallen to it, and alas, many have paid for it. Practically every death from tiger wounds has come from that 'follow up.' Men who have killed their scores of tigers have in so many cases made this mistake and paid the price.

The beaters grow impatient. Some come down from their trees and are driven up again.

We start to follow up, all of us fully armed. The trackers in the middle, the guns covering them on either side. Behind us we put men up trees to search the ground ahead. We are too intent on our business to have nerves now. The trackers smile. This is their job. Another smile as one of them points to a red scar in the bark of a sapling, showing where the 'sure hit' went—not into the tiger. But the first shot is another story. There is a big splash of blood on the ground over which ants scramble, quickly sensing a meal, an overturned stone, freshly moved earth. A splash of blood on green leaves and a horrid smell of wounded tiger. More blood. We throw stones ahead into the bushes, more men are put up the higher trees as we move on.

'There he is!' shouts a man.

We close ranks, waiting.

'Is he dead?'

'I can't see,' is the answer.

Another man goes up beside him. There is much low talk as the first man tries to point him out to his fellow and then to us. We cannot possibly see from our position, but, with luck we may get in another shot before the charge comes.

The man says he can see the tiger plainly. In that case one of us can shoot it from the same place.

The rest get into positions of safety and I climb up with my rifle. There is one brief anxious moment as I part with my gun to climb the easier and a great feeling of comfort as I handle it again. The man, almost too excited to speak, points with his axe to a dark tangled mass of creepers. There, after much searching, I see a yellow patch darkened in the shade. With my left arm round a limb of a tree to keep my position, I fire, then try to recover my hold after the recoil for a second shot when the tiger comes out. There is a slight movement in the shadows and I fire again, this time without definite aim. A low moan and the tiger is dead. But not so dead that we approach without first throwing stones to make sure.

And that was the end of the beat. A typical beat and actually one from my own knowledge. Can anyone ask for more interesting experience of the ways of a wary beast, the panic terror of the lesser creatures of the forest, and the reactions of men to changing emotions, in pursuit of a beast whose name has been used throughout the world to denote the extremes of ferocity, cunning, and strength?

Tiger shooting need not consist of merely sitting in a tree and shooting the beast as he passes beneath. The real fun lies in outwitting him. First a place has to be found where tigers dwell (and they are not too common), then there has to be found a place with water, covert, and a sure line of retreat for the beast, and finally the tiger must be induced to take a kill in such a place that he will lie up within the ground selected for a beat. That is all the preliminary work. A beat I have described.

One of the last beats that I ran was shortly after I had recovered from two attacks of enteric and was not over-strong. I had two boys out from Home with me. Neither had been in India before and they came straight from Bombay to me. The most delightful form of guest imaginable. And they were as keen as boys can be. For nearly ten days the tigers refused to play up, and we passed our time with stalking and shooting small game, or rather they did; I had my work to do. Then at last, while I was away from camp, news came in that a tiger had killed. Time is precious on such occasions and they started out, one with his copy of *Shikar Notes*, the other with *Forbes' Hindustani*. And they proceeded to work the thing out. I reached them in time to arrange the *machans* to cover the line of the tiger's retreat. I was not fit enough to go into the beat on this occasion and took the right-hand *machan*. J was in the middle and W on the left.

> A tiger came out in front of J. He fired.
> 'Have you killed him?' I asked.
> 'Yes. It is dying in the jungle behind.'
> 'All right. There may be another.'

This was optimistic on my part, but there is always the chance of a second tiger in a beat. For which reason the order that guns should fire at tiger and tiger only in a beat, holds good even after the first tiger has been shot. I hoped that if there was a second then it would have been sufficiently frightened by the first shot to go across to W, who was on the left flank. There was little chance of one coming my way, the ground was not suitable.

There were two more shots from J. I felt annoyed. The wretched boy was shooting at deer or pig.

'What are you firing at now?' I asked, rather peeved.

'I have killed two more tigers,' was the reply.

It seemed absurd to expect a fourth tiger, but, in case there should be one, I fired at and killed a peacock that was scuttling towards my position. I hoped that the noise of my rifle-shot after those of J might turn towards W any other that might by great good fortune be about. And it did so, W killing the fourth in the beat. We had little trouble in collecting them.

That was a great day and gave enormous pleasure to our visitors. And although the bag was never so large, there were other days with friends that were as amusing. And not only in tiger beats. Imagine a Chief Commissioner (Provincial Governor) with a reputation for fierce temper but a heart of gold, standing in a small-game beat armed with nothing larger than number six shot, face to face with a tiger that said things in a loud voice that no human would have dared to mention to so august a personage. A fat commissioner in a peafowl beat in a like case; but it was a harmless hyæna which he mistook for a tiger, both being similarly striped. He climbed the nearest tree, which was a thin sapling, too thin for his weight, for it bent like a fishing-rod dangling a tempting bait, while rude people laughed.

Excerpted from Forest Life in India *(London: John Murray, 1935).*

PART IV

Trappers and Trackers

Junagarh Tragedy: Shikari fatally attacked, wounded by lion

A tragedy occurred recently in the Gir mountains, Junagarh state when Sindhi Tamachi, the head of the shikar party of His Highness the Nawab of Junagarh, was fatally attacked by a lion.

It is understood that the shikar party shot a lion which lay unconscious for some time. Thinking the lion to be dead, Mr Sindhi Tamachi approached it when the lion suddenly jumped at him. The members of the party at once shot the lion dead. Mr Tamachi was taken to hospital where he died.

The Statesman, Calcutta, 28 May 1937

G. P. SANDERSON

Elephant-Catching in Mysore

G. P. Sanderson set up and oversaw elephant-catching operations in Mysore, southern India and the Chittagong hills, Bengal. The 'kedah', the technique of enticing or driving elephants into stockades was only of the many ways in which to catch these animals. But Sanderson was more than an 'elephants-only' man and was a fine naturalist. It is, of course, for his refinement of the tactics of trapping elephants that he is widely read to this day.

It was in September 1873 that I arrived at Chamraj-Nuggar—a large village ten miles from the foot of the Billiga-rungun hills—commissioned to try and capture some of the herds of elephants which frequently left the hills and trespassed into the cultivated lands adjoining. I knew nothing of elephant-catching at the time, nor had I any men at command who did; but I knew where there were plenty of elephants, and I was well acquainted with their habits. Some of the Maharajah's mahouts who were amongst my following had been accustomed to catch single elephants with trained females, and in pitfalls, but they had never heard of any one attempting the capture of a whole herd. It was said that Hyder had made a trial, a century before, in the Kákankoté jungles, but had failed, and had recorded his opinion that no one would ever succeed, and his curse upon any one that attempted to do so, on a stone still standing near the scene of his endeavours. Consequently all the true Mussulmans who were with me regarded the enterprise as hopeless—though they judiciously kept that opinion to themselves.

It was owing to this general inexperience that the Chief Commissioner of Mysore had been reluctant to sanction the expenditure required for the attempt. The proposals originated entirely with me. I had been soliciting permission to make a trial for the past eight months, and it was only granted when the season for finding elephants in ground where it would be practicable to catch them— June to December—was far advanced. However, when I did get permission, I commenced the work with the hearty support of an officer of high influence in the province, a keen and experienced sportsman, and who warmly assisted my scheme. The Amildár, or head native official of the district, was an able and energetic person, and obtained for me the willing co-operation of the people required for carrying out the works I decided upon.

My first step at Chamraj-Nuggar was to send for my old sporting friends, the Morlayites, whom I questioned about the number of elephants in the jungles, their principal haunts and routes, and other particulars. I had not met these men for more than two years when we used to hunt together; and though they were not very clean, I could almost have hugged them with pleasure at getting back to them and my old hunting-grounds; whilst, as I had always behaved well to them, they were delighted, and prostrated themselves in a body, declaring I was their father and mother, and that they had been as children bereft since I left them! I put them in good spirits by asking about such little grievances as Indian villagers generally imagine they have, regarding their lands, taxes, and so forth, and promised them that the Amildár would pay particular attention to anything that they had to represent if they rendered effective assistance in elephant-catching.

Next day I moved camp to Morlay, and occupied the hours between sunrise and sunset in tramping the jungles and examining places that seemed likely to afford facilities for circumventing elephants. I knew the whole neighbourhood well, so was able to decide upon a certain ford, A, on the Honhollay river, at which to make an attempt. The river was here about twenty yards wide, but ordinarily with only a narrow and shallow stream flowing over its clean gravelly bed. In the rainy months heavy but short-lived floods sometimes rose twenty feet in a few hours. Wild elephants crossing from its east to its west bank used this and two other fords X, X (the banks were not practicable except at these places). They also retreated by the same routes. When on the west side of the river it was their custom to seek shelter in covers D or E, and we calculated that by stopping the two fords X, X we could drive a herd out of D or E across by ford A, which was indeed their favourite route.

Upon these considerations I marked out a kheddah at A, on the east bank of the river, consisting of a horse-shoe-shaped piece of ground surrounded by a trench. The trench was about five hundred yards round, and the entrance to the enclosed space was by the ford. The elephants would enter by the heel of the shoe, as it were, and would have to go some two hundred yards before they came to the farthest point, the boundary trench. The trench was eight feet wide at top, six at bottom, and eight feet deep (this I subsequently found was a greater section than is necessary to confine wild elephants), and in a few days it was finished, except where rock was met and had to be blasted. There were eight hundred men at work, whose wages were about three pence each per diem. They removed about one cubic yard per man per diem.

It was nearly a month before all was in readiness, as the removal of the rock was laborious work. The personal labour I spent on the enclosure, severe though it was, was not greater than the anxiety I had to endure. Some Job's comforters suggested that if one elephant fell into the trench the others would make a bridge of him and hie them back to the hills; others, that the gate which I had devised for closing the entrance, and which was hauled up on a single rope, to be cut away in the joyful moment when the stern of the last elephant cleared it, would be carried away like chaff before the wind by their backward rush! whilst a few did not hesitate to say that no elephants would approach a place bearing traces of new earth-work and the recent presence of so many work-people. I lived under canvas at Morlay, three miles distant, as the jungle was too unhealthy to admit of my camping at the work, and I frequently got drenched by the heavy September rains, which was not conducive to either comfort or health. I remained at the kheddah daily till late in the evening, and then rode to camp as fast as my pony could carry me, unattended, though there was the notorious man-eating tigress of Iyenpoor afoot, and many others of her race which I stood a chance of falling in with. They would not in all probability have interfered with me, but still it was exciting to my pony, who quite understood jungle-life, if not to myself. I was determined to make the scheme succeed if possible, not only from my love of adventure, and the necessity for executing what I had suggested to Government and undertaken to carry out, but from the desire to prove to several officials who considered the scheme to be the vision of a lunatic, that their croakings were rather the utterances of Bedlamites. Pleasantries appeared in the Bangalore papers regarding the probable effect the kheddah operations would have on the price of salt, which it was represented was being laid in by me in large quantities for application to the caudal appendages of any elephants I happened to meet with!

At last all my plans were completed. Fortunately the elephants had been absent from the neighbourhood up to this time—there were three herds which commonly frequented it—but on the 5th of November the trackers came in early to say there were about thirty elephants in cover D! Immediately messengers started to all the villages near, where orders had been given to the people to hold themselves in readiness to help in the great Government elephant-catching scheme. Still it was twelve o'clock before they collected. I fumed and chafed at the delay, and I am afraid some of the last to arrive did not find me in the best of humours. However, shortly after twelve I started with about five hundred of them—far too many, as I afterwards found—and when we approached the temple I ordered one body to the left, to station themselves along the north-east bank of the river; a second to the right, to cut the elephants off from communication with cover E; and a third, composed of the best men, chiefly Morlayites, to drive the elephants out of cover D. They were to begin to beat at the temple, and we hoped that the elephants would be kept straight for ford A by the guiding-lines of stops. I took my own station near the ford on the west side of the river, with the object of giving the elephants a final impetus forward as they approached it, and to guard the gate with my rifles when they had entered.

After the usual delay, inseparable from anything natives have to do, I heard the beat begin, half a mile distant, and presently five elephants approached the crossing of the river, but kept themselves concealed in the thick jungle between it and the Honglewaddy channel. I observed that they were looking back wistfully as if for their fellows, nor did the beaters follow them up as quickly as they should have done. After some time the five went back, whilst the shouts and shots of the beaters continued near the spot from which the elephants had been originally started, I did not like to leave my post at the ford; but at last, as no news came, nor was there any sign of more elephants approaching, I stationed a man, in whom

I thought I might repose confidence, at the gate, and went with my rifles to see what was the matter. I found that the main body of the elephants had not left cover D, chiefly on account of numbers of the men forming the guiding-line on the south having left their places, and so confused the elephants by joining the beaters, and shouting in all directions, that they did not know which way to flee. They had therefore ensconced themselves in an extensive and almost impenetrable thicket of thorns, whilst the fiends in human shape who had spoilt all my plans were mobbing them in every direction, at a respectful distance, yelling at the top of their voices, and apparently quite oblivious of what the object to be attained was. I gesticulated to them to clear the side towards which we wished to make the elephants break, shaking my fist at them in a fury. The villains redoubled their cries, beating their sticks with heavy thuds on the ground; they thought I was angry at their not exerting themselves sufficiently! Talking was useless; a trombone could hardly have been heard in that din; so arming my gun-bearers with rattans, I sent them amongst the rascals, whom they quickly dispersed, and most of them bolted, and, happily, did not appear again.

I now made the best re-disposition I could of the Morlayites, and we managed at last to start a number of the elephants on the right road. Some of the best men and I pursued them, determined to catch even a small number rather than fail altogether, and they were going fast and straight for the crossing, when, just as they reached it, we at their tails, a sudden shot in front saluted them. A momentary halt and crush ensued; the leading elephants turned, the others followed, and back they came, heads down, tails twisted, going their best, and evidently oblivious of us and everything in their path. The river-bank was close at hand on our left, the channel on our right, whilst the herd almost filled the intervening space. I was maddened by the ill-luck and failure of our measures, and I determined if the elephants got back now it should be over my body; so, shouting to the men not to give way, I fired at and floored one elephant in the front rank. The beaters with me behaved very pluckily, some even throwing the blankets which they carried rolled up on their backs into the elephants' faces before making off. The fall of the leading elephant acted as a momentary check on the others, but they were resolved to be back to the thick cover they had left; so, swerving to their left, they bustled across the channel in mad baste, and with a prodigious amount of splashing, struggling, and roaring, gained the far side, and continued their flight, the wounded elephant amongst them.

The fatal shot that had turned the elephants, in the moment when success was all but grasped, had been fired by my trusty friend at the gate, who must have become frightened at their rapid advance. But the exact circumstances of the case are involved in mystery, as, when I went to have a little conversation with him, I found he had left his gun against a tree and had bolted, and I have never seen his face from that day to this!

The Morlayites now lost their heads, as every one else appeared to do on that memorable occasion. They pursued the retreating elephants with shouts and brandishing of clubs, and as the huge beasts again shuffled across the Honglewaddy channel to regain the cover, some of the boldest actually struck at them from the bank with their long bamboos, the blows sounding loudly on their broad croups. The elephants might have turned and rent them many times during the hunt, but they seemed to have been deserted by the intelligence and sagacity with which they are popularly accredited in as great a degree as the men were by common-sense, and to have no ideas beyond using their legs.

It was now evening. I was drenched with perspiration, bruised, scratched and hardly able to speak for hoarseness. I threw myself down on an elephant-pad under a tree, lighted a cheroot, and applied myself to a review of the day's proceedings, as it was worse than useless to continue the hunt. This, then, was the result of my plans and pains. Things could not have looked more promising at the commencement of the action, yet in four hours the elephants had been terrified beyond hope of their returning to our side of the river for months, and my men demoralized by our failure. However, in the midst of discouragement there was something to be thankful for. No one had been killed, as might well have happened, and the attempt had clearly demonstrated the impossibility of succeeding with such untrained though willing material. This was something gained; and as I conceived that greater eventual success might be evolved from our present failure, did not feel greatly discouraged on a consideration of all the circumstances. I had had too many reverses in my sporting experience to be surprised at this one. The Morlayites had shown great pluck, and I believed if they were disciplined they would act more judiciously on another occasion.

They also had seen how frightened the elephants were at them, and their confidence would rise in proportion. I had made the mistake of having too many men engaged. Elephants must, as the butcher says of beef-steaks in *Martin Chuzzlewit*, when Tom Pinch is trying to cram his purchases into his pocket, 'be humoured, not drove.' The collapse of my immediate hopes was certainly rather depressing, but reflecting that I probably felt it more at that moment than I should in a few hours, I mounted my elephant and rode home, followed by my chop-fallen heroes.

I had a long and earnest consultation with my right-hand men over the day's events round the camp-fire, when dinner and the soothing pipe combined to enable us to review them with some calmness; and long after I turned in I heard the trackers considering what we should do on the next occasion. Some of the Morlayites were again quite confident, and were agreed that if such and such things had happened that did not, and others had not that did, they would have been keeping a joyful watch over impounded elephants at that moment instead of looking wistfully towards the dark and distant hills in which they had doubtless already found safe shelter. 'Yes,' said Marah, a cautious old hunter, 'and if your aunts had had mustachious they would have been your uncles!'

During the next few days I hit upon a plan for the future which had the great advantage that few men would be required to execute it, and even undisciplined ones could hardly spoil it. This was to fortify cover D, so as to prevent the egress of elephants after they had once entered it, and to catch them in it, instead of trusting to a drive in open country.

The elephant season in the low country—June to December—was now over, and the herds had betaken themselves to the hills, but I commenced in January 1874 to put the cover in readiness, during the dry weather, for the coming rainy season. I employed a European overseer, Jones, to help me, and it was fortunate I had such assistance, as I was frequently prostrated during the hot weather by attacks of ague and fever, the result of the exposure I had unavoidably been subjected to for the past few months. I found leisure to superintend the building of a rough bungalow instead of living in a tent, and I also amused myself by shooting a few of the tigers in the neighbourhood. Amongst these was the Iyenpoor man-eater.

The Honglewaddy channel approaches the river to within 30 yards at a point, B. It then runs inland, owing to the levelness of the country, but again approaches to within 90 yards of the river at C, near a temple. The space (cover D) bounded by the channel between B and C, and the river, is about 50 acres in extent, and consists of a jungle of large trees, forest creeping-plants, and several strong thickets. In this retreat it had been the immemorial habit of herds of elephants to take shelter at certain seasons, and to issue forth at nights into the adjacent cultivated country. The north bank of the river was so steep hat they could not cross at any point between fords X, X; whilst there were only five places where they could cross the channel on the west, as it was deep and had perpendicular banks. I, however, had the banks cut to a uniform vertical height of 10 feet, except at the crossings, to make sure of the elephants not getting out of the cover when once in. To barricade the channel crossings, each about 10 yards in width, cocoanut trees, which are exceedingly strong and light when dry, were kept in readiness; and to prevent the elephants escaping by passing up or down the river (past B and C), the bed was spanned at those points by barriers composed of five rows of heavy chains. As soon as elephants entered the cover (of their own accord), it would only be necessary to connect the channel and river at B and C by cross trenches to make the surround complete.

All was in readiness by May. After a few showers the early rains set in in good earnest, and on May 5th a large herd of elephants came down the hills into the low-country jungles. On the 19th five of them visited enclosure D during the night, and after feeding about returned to the herd, which was three miles distant. From this time till the 9th of June small parties visited the cover occasionally, but always returned to the headquarters of the herd. This was very tantalising. We were kept constantly on the stretch; and each morning, until the trackers returned to camp, the villagers of Morlay who were to help were detained at home so as to be mustered at a moment's notice if required, whilst a man was stationed on the wall of the Hurdenhully fort to fire a small cannon I had mounted there, as a signal to other villagers to collect at Morlay in case we wanted more men. Tools for digging the trenches at B and C, baskets for carrying earth, ropes for securing the barricades, and provisions and cooking-pots for the multitude, were stored in the temple buildings. Special services were held daily by the Poojaree and trackers at that celebrated shrine, and the promises of gifts held out to Koombappah for success were sufficient to have moved the heart of even as stony a deity as himself.

On the 9th of June I was at a hill some six miles west of Morlay looking after a bear. The trackers had brought in their usual morning report before I left my bungalow, to the effect that the elephants were still at the foot of the hills, five miles from cover D; so, not expecting them to make a move during the day, I had sent the trackers back to their duty of surveillance, and with a number of men from Oomchwaddy was busy in the pursuit of the said bear, a female with a cub. It was afternoon, and I was seated on the top of the rocky hill, which rose some five hundred feet from the plain, amused by the chase of the bear by my men along the hillside below. The bear had broken wide of me when she was roused from a thicket, and I had not had a shot; but being encumbered by her cub, which was riding on its mother's shoulders after the manner of young bears, the old female could not get along so fast as to keep much ahead of my men, who terrified mother and cub so much by their hot pursuit that the cub fell off; and before it could follow its mother—being very young—a blanket was thrown over it and it was secured, whilst its mother held on for a cave close at hand, into which she fled.

This scene was enacting when I heard the distant boom of my old cannon on the fort-wall of Hurdenhully. I waited to hear it repeated. Yes, another shot! No mistake this time. There goes the third! Hurrah! That is the signal that the elephants are on our side of the river! The smoke of a fire lighted on the highest ground near Morlay—the sign that I was required at camp—now attracted my attention and that of the men with me, so down the hill we went pell-mell, thinking no more of the bear; and making the men fall in, I mounted my elephant and we started for Morlay. We passed through Oomchwaddy and Hurdenhully, where the people were hastily collecting, and soon reached camp.

Here the lately despondent but now rejoicing trackers met us with the gratifying intelligence that the whole herd had made an unexpected move after mid-day, and had marched straight to the river, which they had crossed after bathing and drinking, and were now revelling in the succulent rushes and grass growing along the channel. Anxious though we were to begin, we agreed that it was too late to do anything that day, as the herd must be already scattered for the night's grazing, whilst the proper time to deal with them was when they were collected during the day.

I accordingly gave orders that no one should leave camp, but that all should be entertained 'by Government,' whose guests they were to consider themselves as having the inexpressible honour of being. Most of them were Oopligas and Torreas, both meat-eating castes (except as to beef); so I ordered my flock of sheep to be driven up immediately, and as I named the headmen who were to choose for their people, they made a dash amongst them and dragged out the sheep they preferred, amidst great amusement and comments upon their respective notions of mutton. These were speedily carried off and slaughtered, whilst another man of each group received cooking-pots, ragi-flour, curry-stuffs, and tobacco, at the stores, where Jones presided.

What a night of pleasant anticipations and merriment it was! Everybody was happy, and we occasionally heard the trumpet of the elephants, fully three miles distant, as they fed and disported themselves about the river. I visited the various knots gathered round the fires dotted about the cleared plain before my bungalow, and said a few words to them about their conduct on the morrow. Agreeable fellows the rustics of Mysore are to entertain. They do not drink, and where the greatest dissipation is smoking or snuffing, there is no likelihood of quarrels or too noisy mirth. In this respect my Oopligas were a great contrast to the tame-elephant attendants, chiefly Mussulmans, with a sprinkling of Pariahs, or low-caste Hindoos. When it was necessary to treat these for any special services, the only thing was to give them a few sheep and bottles of spirits—without which it would have been no treat—and to order them not to approach the camp till next morning. Their revels seldom concluded without a fight, though when the effects of the *bhang* they smoked, and their potations, passed away, they resumed the natural quiet demeanor of Asiatics.

Every one was astir betimes on the eventful 10th of June. I have caught a good many elephants since, and have witnessed many exciting scenes in the work, but I shall never forget the pleasurable anticipations I experienced on this occasion. Every contingency that could be foreseen had been carefully considered; nothing had been left to chance. The men had had their respective posts allotted to them weeks beforehand, and we had even had a rehearsal, or review day, on which my tame elephants, under the direction of their mahouts, led by a Morlayite experienced in the ways of the wild ones, had represented a herd, whilst we took steps to meet their various moves. I had also practised the men in deep-hunts, &c., when I gave prizes in the shape of coloured handkerchiefs for turbans as well as rupees, to those who distinguished themselves. I certainly felt that I now had a very different following to the undisciplined band that frustrated the first attempt. I had imbued them with some notions of obedience in executing instructions, whatever they might be; of working together; and of silence. The difficulty of getting natives to do anything without noise can only be fully understood by those who have had to deal with them. I considered it a triumph that I could march three hundred of them on an exciting expedition, without a whisper being heard. Despite all this I experienced a good deal of anxiety, now that the time for testing our arrangements had come; but I daresay this added to the pleasure of the occasion, as had the result been beyond doubt, where would the excitement have been?

At 9 A.M. we started for the temple. Early in the morning I had been joined by Major G., Deputy Inspector-General of Police for Mysore, and a keen sportsman, who happened to be encamped at Chamraj-Nuggar, and to whom I had sent word overnight. As Gaindcully, the elephant we were riding, swung along, followed by the long serpentine line of beaters in single file (the jungle-path being narrow), I felt proud of the comments my friend bestowed on my men, as he was in a position to appreciate the state to which they had been brought, having to drill and reduce natives to order for the ranks of the police.

When we reached the temple, the trackers, who had preceded us, informed us that all the elephants were not in cover D; some were scattered feeding on the upper side of the channel, and would have to be driven to join the main body. This was quietly effected by a handful of men, though a female with a young calf, an albino, gave us some trouble, threatening to charge. Had the men acted as of yore there would doubtless have been a scene, but by giving her time to retire safely with her charge we got her pounded into D with the others. Having ascertained that all the elephants were now in, all hands were engaged in barricading the crossings and cutting the trenches between the channel and river at B and C. To render this latter work easy I had previously had the trenches dug and filled in again, a small drain covered with flat stones being left at the bottom of each. Water was now admitted to these from the channel, whilst the end near the river was kept closed, and as the water had a head of some ten feet, it speedily blew up the superincumbent earth and scoured out the trenches to the depth and width required. It was past mid-day before we got all the elephants into the cover, and not a moment's rest did any of us get till 11 P.M. Captain C., of the Revenue Survey, came over from his camp at Surgoor, and Major G. and he helped to superintend the people. At one point the supply of tools was insufficient, and Captain C. was superintending and encouraging a body of men who were digging with sharpened sticks, and even their bare fingers! The elephants were very noisy in the cover, but did not show themselves. At every twenty yards three or four men were stationed to keep up large fires. These were reflected in the water of the channel and river, which increased their effect. We all had a most exaggerated idea of what the elephants might attempt, and the strength of our defences was in proportion, and greater than they need have been. I was kept on the move almost all night by alarms at different points, fortunately groundless ones. One tusker showed himself on the bank of the channel, but met with such a reception from firebrands and stones that he retreated in haste. The river was an advantage, as the elephants had easy access to water. The lurid glare of the fires, the gaunt figures of the lightly clad watchers, their wild gesticulations on the bank with waving torches, the background of dense jungle resonant with the trumpeting of the giants of the forest,—formed a scene which words are feeble to depict, and that cannot fade from the memories of those who witnessed it.

By 11 P.M. the defences were thoroughly secured, and I had leisure as I stood by a log-fire with nothing but my trousers on (my flannel shirt and coat were drenched with perspiration, and were being dried before the blaze), a piece of bread in one hand, and a bottle of claret and water in the other, to reflect on our complete success so far. That the elephants could not now escape was certain, unless indeed they carried some of our barricades, which were, however, so strong as to be almost beyond their power. The men differed as to their number. I had seen about twenty; some declared there were fifty, but I could not believe this at the time. The number, however, was fifty-four, as we subsequently found. I tried in vain to rest. The excitement of the scene was irresistible, so I betook myself to walking round the enclosure at intervals throughout the night, followed by a man carrying a basket of cheroots, which I distributed to the people. The rest of the time I lay upon my cot, which my servant had been thoughtful enough to bring from Morlay with his cooking paraphernalia, enjoying the wildness of the sounds and scenes around, and soothed by cheroots and coffee. When the elephants approached the place where I was the guards thrust long bamboos into the fires, which sent showers of sparks up to the tops of the trees overhead, and they also threw joints of a bamboo-like reed into the flames, where they exploded with a sound as loud as pistol-shots.

The first crow of the jungle-cock was the most grateful sound I think I ever heard, as it showed our anxious vigil was drawing to a close. We knew that during the day the elephants would give us less trouble. My headmen now joined me from the points where they had been stationed during the night, and we set about considering the next step to be taken—viz., making a small enclosure or pound off cover D, into which to get the elephants confined. Of course this would take some time to carry out.

If driven from the east we knew the animals would pass between the temple and channel, at the west end of the cover, with a view to crossing the river below the temple, and regaining their native hills, which, however, they were fated never to see again. I therefore laid out a pound (F) of 100 yards in diameter, surrounded by a ditch 9 feet wide at top, 3 at bottom, and 9 feet deep. This was connected with the large cover by two guiding-trenches which converged to the gate. It was completed in four days by the personal exertions of the Amildar with a body of labourers, who worked with a will, as their crops had frequently suffered from the incursions of elephants, and they appreciated the idea of reducing their numbers. The last thing completed was the entrance-gate, which consisted of three transverse trunks of trees slung by chains between two trees that formed gate-posts. This barrier was hauled up and suspended on a single rope, so as to be cut away after the elephants had passed. The news of the intended drive attracted several visitors from Mysore. Tents were pitched in an open glade close to the river, and we soon had a pleasant party of several ladies, the cheery Deputy Commissioner of the district and his Assistant, two officers (Captains P. and B.) of Her Majesty's 48th Regiment, M. of the Forests, and Captain C. and Major G., who had remained from the first day. The evening before the drive all assembled within view of the point where the elephants were in the habit of drinking at sunset, and were gratified with an admirable view of thirty-five of the huge creatures disporting themselves timidly in the water.

On the morning of the 17th, everything being in readiness for the drive, Captains P., B., and I proceeded with some picked hands to drive the herd from its stronghold towards the pound. We succeeded in moving them through the thick parts of the cover with rockets, and soon got them near to its entrance. A screened platform had been erected for the ladies at a point near the gate, where they could see the final drive into the enclosure from a place of safety.

The elephants, however, when near the entrance, made a stand, and refused to proceed; and finally, headed by a determined female, turned upon the beaters and threatened to break back down an open glade. P. and I intercepted them, and most of them hesitated; but the leading female, the mother of the albino calf, which had been evilly disposed from the beginning, rushed down upon me, as I happened to be directly in her path, with shrill screams, followed by four or five others, which, however, advanced less boldly. When within five yards I floored her with my 8-bore Greener and 10 drams; but though the heavy ball hit the right spot between the eyes, the shot was not fatal, as the head was carried in a peculiar position, and the bullet passed under the brain. The elephant fell to the shot, almost upon me, when P. fired, and I gave her my second barrel, which in the smoke missed her head, but took effect in her chest, and must have penetrated to the region of the heart, as a heavy jet of blood spounted forth when she rose. Probably one of the large arteries was cut by this shot. The poor beast moved off a few paces and halted, a stream of blood issuing in a parabolic curve from her chest, and making a loud gushing sound as a pool was formed in front of her. For some moments she swayed from side to side, and then fell over with a deep groan, to rise no more. This was a painful scene; the elephant had only acted in defence of her young; but shooting her was unavoidable, as our lives, as well as those of the beaters, were in jeopardy.

The next scene partook of the ridiculous. The herd had dispersed and regained its original position. The little albino calf, seeing P., screamed wildly, and with ears extended and tail aloft chased him. He, wishing to save it, darted round the trees, but was near coming to grief, as he tripped and fell. The result might have been disastrous had I not given the pertinacious youngster a telling butt in the head with my 8-bore. His attention was next turned to a native, who took to his heels when he found that three smart blows with a club on the head had little effect. After some severe struggles, in which a few natives were floored, the calf was at last secured to a tree by a native's waistcloth and a jungle-creeper.

While all this took place the beat became thoroughly disorganised. When the elephant had charged P. and me, our men had given way, and the herd regained its original position at the extreme east end of the cover. After a short delay we beat it up again to the spot near the gate from which it had broken back. The elephants here formed a dense mob, and began moving round and round in a circle, hesitating to cross the newly-filled-in trench which had reached from the channel to the river, but which was now refilled to allow them to pass on into the kheddah. At length they were forced to proceed by the shots fired, and by firebrands carried through the paths in the thicket. The bright eyes of the fair watchers near the gate were at length gratified by seeing one great elephant after another pass the Rubicon. After a short pause, owing to a stand being made by some of the most refractory, the last of the herd passed in with a rush, closely followed into the inner enclosure by a frantic beater waving a firebrand. P. and I came up third, in time to save any accident from the fall of the barrier. C., who was perched on a high branch of the gate-tree, cut the rope, and amidst the cheers of all, the valuable prize of fifty-three elephants was secured to the Mysore Government.

I often think of the rapture of that moment! How warmly we 'Sahibs' shook hands! How my trackers hugged my legs, and prostrated themselves to P. and B. An hour of such varied and high excitement as elephant-catching is surely worth a lifetime of uneventful routine in towns! Sore disappointment had been undergone by myself and men. Many tedious days and nights had we laboured against discouraging incidents and hardships. But all was forgotten in the success of that moment. We lost no time, however, amidst our self-gratulations, in thoroughly securing our prize. Guards were immediately posted round the kheddah, and my own tent pitched outside the gate; but the elephants gave no further trouble. The jungle inside was dense, and they kept so quiet that, large number though there was, we could scarcely see anything of them from the outside for some hours, until they began to move, when they soon trampled down much of the jungle. They never attempted to cross the trench. The most noisy animal of the herd was the little albino calf, which had broken its bonds during the second drive and made its way with the others into the kheddah, and which continued to roar lustily for its mother, and in pain at the kicks which were freely administered to it by the other elephants when it endeavoured to push its way amongst them. If the writers who have stated that female elephants suckle and tend each other's calves indiscriminately were but subjected to half the pummeling the unfortunate orphan underwent the first day and night in the enclosure, they would have but a poor opinion of indiscriminate suckling, I imagine.

On the day after the drive we commenced the work of securing the wild ones. Out of seventeen tame elephants belonging to the Maharajah and Commissariat Department which I had in camp, ten of the most steady and courageous males and females were told off for work in the enclosure, and the rest to bring fodder for the captives. Water was supplied to them through bamboos across the trench, emptying into an improvised trough. As none of the mahouts had seen elephants caught before, except single ones, they were rather nervous about entering with but ten among so many wild ones. P. rode one pad-elephant in advance, and I another, to encourage the men. The wild ones all mobbed together when we entered and showed great interest in our elephants. After some little time we separated a few from the herd, and a mahout slipped off under cover of our tame elephants and secured a noose round a young tusker's hind-leg. The tame elephants then dragged and pushed him backwards nearly to the gate of the kheddah, where we secured him between two trees. We afterwards found, however, that it was much easier to hobble each elephant's hind-legs, and then to let it fatigue itself by dragging them after it for some time before we finally secured it, than to proceed as we did at first.

In ten days, during which time the visitors remained, and we had a merry camp, we secured all the elephants. Calves were allowed to go loose with their mothers. The captives were led out of the enclosure by our elephants as fast as they were secured, across the river, and were picketed in the forest. Water-troughs were made for them of hollowed lengths of date-trees. These were pushed within their reach by a bamboo, and withdrawn with a rope to be again filled. Two men were appointed to each large elephant, and one to each small one. They made themselves shelters of boughs and mats just beyond the reach of their charges, and by constantly moving about them, singing to, and feeding them, many could handle their elephants in a few days. The elephants at first kicked or rushed at their captors (they very seldom struck with their trunks); but as soon as they found nothing was done to hurt them they gained confidence, and their natural timidity then made them submit without further resistance. There was a great variety of temperament observable amongst them. The small elephants, about a third grown (particularly females), gave the most trouble. The head jemadar ascribed it to their sex and time of life. 'Wasn't it so with human beings?' he said. 'How troublesome women were compared to men, who were always quiet!' He was a Mussulman, and had several ladies in his establishment, so, as I was an inexperienced bachelor, I did not presume to question his dictum. One young elephant lost the sole of one foot with three toes attached after it had become loosened from her violence in continually kicking up the ground, and died soon afterwards. A mahout and I mounted a full-grown female on the sixth day after she was removed from the enclosure, without the presence of a tame elephant, which shows how soon elephants may be subjugated by kind treatment.

The ropes were changed from one leg to another every day, otherwise the wounds made by them would have been very serious. Whilst this was being done it was necessary for a tame elephant to stand near the wild one, as it became alarmed on seeing men on foot near. We were much troubled by maggots in the wounds of the new elephants. In a few hours after they were dressed they would swarm again. The animals kicked up sand and blew it upon their sores to keep off the flies; this sopped up the oil and dressings we applied, and the chafing of the ropes was much more severe when sand got under them. The mahouts used various substances, as lime, tobacco, the juice of certain plants, &c., to kill the maggots; but they were unfortunately all agents of an irritating nature, and though fatal to the maggots, were far from conducive to the healing of the wounds. I have since found camphorated turpentine a valuable remedy. On the present occasion, with a bucket of margosa oil (called also *neem* oil, most offensive in smell, and deterrent to flies) at hand, and a mop for applying it, the men managed in about a month to heal their elephants' wounds.

During the tying-up process in the kheddah several amusing incidents occurred. Active fellows would constantly cross it on foot with ropes or other things that were required, and at first they were pertinaciously chased by the wild ones. The men made for the protection of the tame elephants, and it was considered creditable to do this with as little hurry as circumstances would admit. The arena formed a centre of attraction to the onlookers, as the theatre of a Spanish bull-fight may do, and the men who showed the greatest coolness were loudly applauded. The elephants, however, soon gave up pursuing when they became accustomed to seeing people. The wild ones did not attempt to interfere with the men when they gained the shelter of the tame elephants. On one occasion a friend in the Forest Department, who was riding one of our elephants, was swept off, as well as the mahout, by an overhanging creeper, when their elephant was dragging a captive across the kheddah. Having but a confused idea of the points of the compass when they gained their legs, they rushed toward the nearest elephant for protection. It was a very fine animal, but unfortunately a wild one, which they mistook for a friend! The elephant was rather startled and did not take so prompt an advantage of their mistake as it might have done. They meanwhile made some remarkably good time towards the gate of the enclosure, which they reached in safety.

The largest tusker amongst the captives began to be troublesome a day or two after the herd was impounded. He would approach our elephants as if to measure his strength with theirs. A prod with a long spear in the head kept him off at first, but the novelty of that treatment wore away, so I told the riders of our tuskers to set their elephants at him if he gave more trouble. Amongst them was one called Jairam, not taller than the wild elephant, and with the disadvantage of having blunt tusks; but he was of a most warlike temperament amongst his own kind, though remarkably gentle and good-tempered to his keepers and strangers. It had been necessary to restrain him hitherto from attacking the wild tusker, but I now gave his rider permission to gratify Jairam if the wild elephant required chastisement. Whilst we were at work that day in the kheddah I heard the clash of meeting tusks, and a tremendous scuffling behind me. I turned and beheld the valorous Jairam with the wild tusker's head jammed between his tusks, whilst he ran him rapidly backwards towards the trench, urged on by his delighted rider. The scuffling of even a pair of bullocks makes a considerable noise; that created by struggling elephants may be imagined. The tusker having got his head into chancery could do nothing but run back to clear himself. He fortunately managed to do this when just on the brink of the trench, and made his escape, pursued round the enclosure for some minutes by the gallant Jairam, who, amidst the plaudits of all, added to the tusker's discomfiture by administering some nasty prods behind whenever he could catch him. I sent for some money and rewarded the mahout before the spectators, as his position had been a highly dangerous one during the tilting-match. Mahouts are always pleased when their elephants deserve commendation, and Jairam had a double allowance of grain and a large bundle of sugar-cane that evening as a mark of his master's approbation. The wild tusker was thoroughly cowed by this encounter; and it was amusing to see the riders of the elephants told off to guard whilst the others were engaged in tying the captives, jockeying the late combatant round the enclosure when he did anything which afforded them an excuse for administering correction.

One great piece of excitement was the capture of a single male elephant in the elephant-lines. Unfortunately I was the only spectator amongst our party. I was just getting up at dawn one morning when a mahout rushed into my tent saying, 'Wild elephant, wild elephant!' and away he went again. The word he used for elephant might mean one or any number, and imagining a herd must have come, and was threatening interference with our captives, I ran down to the elephant-lines just as I was, in my flannel sleeping-suit. I found the men unshackling three of our best females, and seizing spare ropes, and they now told me that a single male elephant was amongst the new ones picketed across the river. I jumped on to Dowlutpeary behind the mahout. We only had girth-ropes on her, no pads, and not even dark-coloured blankets to cover ourselves. Crossing the river we saw some mahouts in a tree, who pointed to the jungle on the left, where we found the elephant, a fine tusker, but with the right-hand tusk missing. He was a young elephant, and would be a prize indeed. We all lay flat on our elephants' necks. Presently the tusker approached us, and my elephant's mahout turned Dowlutpeary round with her stern towards him, that he might be less likely to see us. He put his trunk along her back, almost to where I sat. I took the goad from the mahout, so as to job his trunk if he came too near me, but he seemed satisfied. Bheemruttee and Pounpeary, the other two elephants, now made advances to him under the direction of their mahouts, and he soon resigned himself unsuspiciously to our company.

He now led us through the lines, interviewing several of the captured elephants, whose position he did not seem to be able to understand, and then retired to a shady tree, as the sun had risen. I signed to the hiding mahouts to get the other tame elephants quietly across the river, but to keep them out of sight; and as soon as the elephant stood perfectly still, my mahout and Bheemruttee's slipped off, whilst Pounpeary's rider and I kept the three elephants close against the wild one to prevent his seeing the men. They had been at work tying his hind-legs for a considerable time, when he attempted to move and found himself hobbled! The critical knot had just been tied when he shifted his position! He was on the alert in an instant. Our elephants sheered off with great celerity, as he might have prodded them with his sharp tusk. The mahouts each threw a handful of dust into his face in derision before they retired, and now the fun began. Men came running from all directions with ropes, to the dismay of the tusker, who trumpeted shrilly and made off at an astonishing pace, scuffling along with his hind-legs, which were not very closely tied to each other, and which he could use to some extent. He rushed away through the low jungle, the whole of our elephants and men in hot pursuit. He was red with a peculiar earth with which he had been dusting himself, and formed a great contrast to the black tame elephants. Our tuskers were all slow (their pace might have been improved by an application of the Assam elephant-hunter's spiked mallet), and we did not gain on the elephant for nearly half a mile. The men on foot were running in a crowd alongside him to his intense terror. At last he turned into a thicket and halted, and we quickly surrounded him. Dowlutpeary and Bheemruttee again went in, and he was secured and marched back between four elephants in triumph. I sold him subsequently (for Government) for £175; had he had both tusks he would have brought double that sum. I gave the three mahouts who secured him £5 each—a small fortune to them—the moment the elephant was made fast, and said a few complimentary words upon their activity. I have always found that, in rewarding natives for any service, the value of a present is greatly enhanced by its being given on the spot in presence of their fellows: and the Canarese proverb, 'Though the hand be full of money, there should be sweet words in the mouth,' should not be forgotten; a few pleasant words go well with rupees.

The captured herd consisted of sixteen male elephants of different sizes, of which three were large tuskers—the highest was 8 feet 5 inches at the shoulder—and three *mucknas*, or tuskless males; thirty females,* full or half grown; and nine calves. Of the largest elephants nine were allotted, after careful selection, for the Maharajah's stud, ten to the Madras Commissariat Department, nine died, chiefly young ones, and twenty-five of the least valuable were sold by public auction at Chamraj-Nuggar three months after capture, when most of them were tame enough to be ridden away. These latter brought an average price of £83, 8s. each, or an aggregate of £2085; and the total realized for the fifty-four (deducting deaths) was £3754, which, after deducting £1556, the total expenditure from the commencement of operations in 1873, left a surplus to Government of £2199. The elephants drafted into the Maharajah's and Commissariat establishments were the most valuable animals, but were only credited to the Kheddah Department (by the orders of the Chief Commissioner) at the same price as the second and third rate animals sold for at auction—viz., £83, 8s. each. At least £100 per head more might have been added, when the surplus receipts would have been £4099.

The Chief Commissioner complimented me on the performance of my task in an order on the subject as follows: 'The success that has attended Mr Sanderson's skilful and energetic arrangements in this matter is in the highest degree creditable to that gentleman, and the Chief Commissioner cordially congratulates him thereon, and will have much pleasure in bringing his excellent services in organising and carrying the same out to the favourable notice of the Government of India.' The experiment having succeeded so well, the scheme was sanctioned for a further extended term, and the officiating Under-Secretary to the Government of India addressed the Chief Commissioner of Mysore as follows: 'I am directed to state that his Excellency the Viceroy and Governor-General in Council is pleased to sanction the grant to Mr Sanderson of a bonus of £200, in acknowledgment of the skill and personal daring displayed by him.'

Not long after this, I was deputed to Bengal on temporary duty for elephant-catching, leaving the work in Mysore in abeyance for some time, though my trackers and best men were allowed half-pay

---

* This includes the female shot in the enclosure on the day of the drive.

until my return. An account of the expedition which I undertook after elephants into the wilds of the Chittagong hill-tracts will be given in the next chapter. I have not caught elephants in Mysore since my return from Bengal in 1876, owing to the disastrous famine prevalent in Southern India, the cause of which, lack of rain, affected the fodder upon which we are dependent for maintaining newly-caught elephants. But everything is kept in readiness at Morlay for the continuation of operations as soon as affairs improve, and it will be strange if, with our extended experience, my Morlayites and I are not able to do even better than in 1874. The herds in Mysore are large and numerous. I calculate that there are at least 800 elephants in the jungles where catching operations can be carried on.

A few remarks on the breaking of newly-caught elephants may not inaptly close this chapter. As soon as a wild elephant is secured, two keepers are appointed to it, who commence, one on each side, to fan it with long branches, keeping out of its reach. At first the elephant is furious from fear, and attempts to strike or kick them. They keep up a wild chant, addressing their charge by any extravagant title they can think of, such as 'King of a thousand elephants,' 'Lord of the jungles on the summit of mighty hills,' &c. The elephant is well fed from the beginning, and it is a remarkable circumstance that they eat from the first. They do not seem to be able to break through their habit of constantly feeding—a wild elephant grazes or browses almost incessantly—and if an elephant refuses its food it is generally something more serious than alarm that ails it. A not uncommon idea that elephants are starved into submission is quite unfounded. In a day or two the elephant pays little attention to the men—being engaged on the choice fodder with which it is supplied when they are at work at it. They gradually approach till they can clap its sides, its legs being secured for fear of a kick, which might kill them on the spot. The elephant soon learns to take sugar-cane, fruit, &c., from the hand, and allows them to be put into its mouth, which all elephants prefer to taking food in their trunks. I found a small allowance of rice for each elephant useful, as a pinch can be wrapped up in grass, with a little sugar, and the constant feeding with such morsels forms a bond between the animal and its attendants. Girth-ropes are soon tied round its body, and under the tail as a crupper, and the men climb on to it by these. When an elephant once gives up striking at its attendants (which it generally does in a few days), it is very seldom that it subsequently does anything intended to injure them, unless terrified by haste or excitement in their movements. Nor are there any elephants which cannot be easily subjugated, whatever their size or age. The largest elephants are frequently the most easily tamed, as they are less apprehensive than younger ones.

The elephant should not be taught to kneel, nor be subjected to other unnecessary restraints, until well over the immediate effects of capture, say in four or five months. It may then be taken into the water, and the downward pressure of a pointed stick behind the shoulder near the spine will soon make it kneel to avoid the pin.

Elephants are taught to trumpet by the extremity of their trunks being tightly grasped between the hands, when they are obliged to breathe through the mouth, in doing which they make a loud sonorous sound. They are rewarded and made much of for this, and so learn to 'speak,' as it is termed, on an indication of what is required. In Dacca the Government elephants are particularly well trained, much more so than in the south of India. They are taught to collect their own fodder where it is plentiful, and hand it up to the coolie on their backs, who packs it,—and many other useful services.

Excerpted from Thirteen Years Among the Wild Beasts of India; Their Haunts and Habits from Personal Observations; With an Account of the Modes of Capturing and Taming Elephants *(London: W. H. Allen & Co., 1878).*

A. MERVYN SMITH

My Shikaree Friends

The jungles of Chotanagapur housed an array of wild animals—the wolf and the now extinct Indian cheetah, the tiger and the elephant. The hills and valleys were also the home of various communities of people that had and still have an intimate knowledge of its fauna and flora. In much of the shikar literature, they only make a guest appearance as guides or assistants. Mervyn Smith, an officer stationed in Chotanagpur was exceptional: his book includes pen-potraits of a 'Bagh-Maree', a community of expert tiger-slayers and a Dom, one of the 'untouchable' castes.

BEEMA, THE *BAGH-MAREE* (TIGER-SLAYER)

'I wonder how you can tolerate such fellows. I would whip them out of the place. They are the curse of all *shikar*, with their cowardly method of killing tigers. There will soon be no tigers left to shoot if you encourage these rascals.' So said a planter friend who was staying with me, to whom I had introduced 'my friend the *bagh-maree*' (tiger-slayer). Now, I do not in any way share these sentiments, but on the contrary I regard the *bagh-maree* as a most useful member of society. Beema, the *bagh-maree*, has to my knowledge accounted for two panthers, two cheetahs (hunting leopards), one tiger and one bear. Their skins now adorn the verandah of my bungalow. These animals were all killed within a mile of the village of Somij (Chota Nagpore district), and in about two months' time. When it is remembered how vast is the number of cattle and goats annually destroyed by tigers and panthers in India, it will be seen that my friend Beema is anything but deserving of the hard words used by my European *shikaree* friend. In the last twelve months twenty-six head of cattle had been killed within two miles of my bungalow. This of itself is a very serious loss to the villagers, whose chief means of subsistence is agriculture, for which cattle are essential; and if Beema had not thus come to their assistance several families of *ryots* (cultivators) would probably have been ruined.

So much to establish the claims to gentle treatment of my friend the tiger-slayer; and now for his story:—

'I was not always a *bagh-maree*, Huzoor (Sir). I am a *tantee* (weaver) by caste. But a *chota bagh* (panther) did it. It made me a *bagh-maree*. It was not a tiger at all; it was a witch that had entered the body of a tiger, to do me an injury. I paid Gagee the Gond two rupees for a charm, and after three years I killed it. *Huzoor*, you know that I killed it behind your *bawarchee-khana* (kitchen) the day after it had killed and eaten Madho's son last year. I know the animal to be the witch by the piece out of its ear which I cut off with my *bulloova* (battle-axe) three years ago at Bara, my village. Gagee the Gond made the charm out of that piece of its ear, and I have the charm yet. Here it is, and there (pointing to the skin on my verandah wall) is the devil. I became a *bagh-maree* out of revenge. You see my head; it is nearly bald, and the girls laugh at me and say I am old, pointing to my baldness, but the witch did that three years ago.

'I made a cloth for an old woman of our village, and charged her one rupee and a half for it. It cost me one rupee two annas worth of cotton thread, so my gain was only six annas. She only paid me Rs 1-2, and when I pressed for the balance, six annas, she refused to pay, and cursed me, saying a tiger would eat me.

'A few nights after this a *chota bagh* (panther) got into my goat pen and killed two goats and was carrying off a third when I aimed a blow at its head with my axe, but only cut off its ear. It clawed me on the head and the wound caused all the hair to drop off. I vowed revenge and learned how to set the *thair* (spring bow) and poisoned arrows, from Maun Sing the Kowtia. I have been a *bagh-maree* for three years and I have killed two tigers, ten panthers, two hunting leopards and five bears. The Sirkar (Government) gives me Rs 25 for each tiger and Rs 5 for each leopard or panther. They don't pay for bears. The villagers also give me four seers (= 9 lbs.) of paddy each, whenever I kill a tiger or panther that has carried off any of their cattle. I also get fed when I am staying at any village. I do all the killing within ten miles of Bara. There are other *bagh-marees* elsewhere.

'We *bagh-marees* chiefly use *dakara* (aconite) for poisoning our arrows. *Dakara* is a root about a span long, and as thick as my wrist. We buy it at Chyebassa from the native medicine shops at four annas a tola. We grind it up with a little boiled rice to make a paste. This paste we rub over a rag, and wind the rag round the back of the arrow-head just behind the barb. The head fits loosely into the shaft of the arrow, so that when the animal is hit the poisoned rag enters the wound with the arrow-head and the shaft drops off. The animal dies within a few hours, and we easily trace it by the blood and broken twigs. Bears are the most difficult to kill. They will sometimes live a whole day with the poisoned arrow inside them. Tigers die very soon. We sometimes use cobra poison, but it is difficult to get. I keep two cobras from which I take the poison once a month. If I take the poison oftener it is of no use. I cannot take the poison while the cobra is changing his skin, which he does once every two months or so. He has no poison then, and won't bite the plantain. How do I get the cobra's poison? Why, I take a ripe plantain and tie it to the end of a stick, and with this I irritate the cobra until he bites the plantain. If he turns his head when he bites, I know the poison has come. He sometimes bites without giving a twist of his head, and then no poison comes. We rub the plantain over the rag, just as we do the *dakara*. A plantain with two bites in it is enough for a large tiger. Cobra poison is the best, as it never spoils; *dakara* gets weaker the longer you keep it. *Dakara* does not grow here; it comes from Calcutta. How do we know where to set our spring bows? *Huzoor*, you know that a tiger never crashes through the brushwood. That would alarm the game. He always takes paths through the jungle. He will not take a narrow path. He sticks his whiskers out straight, and with these he feels the brushwood and knows if there is room for him to pass. He also crouches low when walking. In the dry season there are many paths in the jungle, and as we know not which the tiger will take, we don't usually set our traps in the dry weather. During the rains, when the underwood has grown, we know that the tigers must take the beaten paths, and we set our traps accordingly. The bow is set on V-shaped twigs about eighteen inches from the ground. The bow is placed on one side of the path and a string connected with the trigger stretches across the path, about eighteen inches about the ground, and is tied fast to a twig on the opposite side. If a tiger or panther attempts to follow the path he must breast the string and the strain sets free the poisoned arrows (we generally use two to each bow), which enter his side, and he dies in a few hours within a few hundred yards of the trap. In case men or cattle should stray on to the path, two other strings are attached to the trigger and tied to twigs three and a half feet off the ground and three or four yards away from the trap. This greater height allows a tiger to pass underneath, but should a bullock or a man come that way, he brushes against the higher string, which sets free the arrows before he comes up to them, and they pass harmlessly into the brushwood.

'There is no danger in following up a tiger wounded with poisoned arrows, for even if he is not dead he is so weakened by the potency of the poison that we easily despatch him with our battle-axes. I have never been hurt by a tiger since I was wounded by the witch-panther I told you about. But that was before I became a *bagh-maree*. When we go to a new village we generally make *poojah* (worship), sacrificing a white cock... If we don't make sacrifice we lose our tigers.'

This briefly is the story of Beema the *bagh-maree*, elicited from him by a series of questions. The trap used by these native *bagh-marees* is most ingenious and seldom fails; and the danger from them is nil;. The weird story of the witch-panther of my friend the *bagh-maree* must be reserved for another occasion.

PURDASEE, THE DOM

My first acquaintance with Purdasee was under circumstances of so terrible a character that I can never forget them. It was during the dreadful famine of 1877-8, when upwards of five million persons died from starvation and disease engendered from a scarcity of food in South India. Out of a population of five and a quarter millions, Mysore lost a million and a quarter, while the Bellary district suffered even more severely. In Madras mountains of grain in bags were stacked all along the sea-shore, brought in by ships from Calcutta, Burmah, Gopalpore and elsewhere, but transport into the interior—to the districts most affected by the famine—was utterly inadequate. The Madras Railway in those days terminated at Bangalore on the south-west and at Bellary on the north-west, while the whole of the large stretch of country between these two towns was entirely without railways. Cattle had suffered even more severely than human beings during these two seasons of drought, so that even transport in bullock-carts was sadly crippled. In parts of the Bellary district and in North Mysore grain there was none, and whole villages were depopulated, the inhabitants literally dying of starvation. I was through the worst parts of these two districts during this terrible time, and the awful sights of mute human suffering that met my gaze I have no wish to recall.

To come to my story. I was riding along the high road between Chitaldroog and Bellary a few days before Christmas, and was anxious to make the latter town in order to get in to Bangalore by Christmas day. I had ridden across country some twenty miles and had just struck the high-road and hoped to fall in with my camp, which I had sent on a few days in advance, to await me at a large village I had named. The country through which I had ridden was extremely desolate. At that time of the year the fields should have been laden with *cholum* (millet), which thrives wonderfully on the black cotton soil of Bellary; but the failure of the North-east monsoon had resulted in a very scanty crop, which was plucked and eaten by the starving population before it had even had time to ripen. I had gone about a couple of miles when I noticed a few huts a hundred yards off the road, and as I was anxious to hear of my baggage I rode over to see if the villagers could tell me whether my carts had gone on. I shouted when I came to the huts, but no one answered. Some of the huts were closed, others open, but there did not seem to be a soul about. I was just about to ride off when I heard some low moans near a thicket of milk-hedge (euphorbia). On going to the spot I was witness of a most horrible sight. A couple of village *pariah* dogs were tugging at the legs of a man, trying to drag him out of a small hut of millet stalks. The poor wretch was so emaciated and weak from starvation that he had not the strength to beat them off, but was clinging convulsively to the sides of the hut moaning faintly now and again. A shower of blows with my whip failed to drive off the dogs, which had grown ferocious by feeding on human corpses, so that I had to draw my revolver and shoot one of them before the other took to flight. My terrible experiences of the previous few months had taught me that the village dogs, grown savage with hunger, had taken to feeding on the bodies of the dead and dying villagers, and had I not opportunely arrived when I did, they would have made short work of the poor wretch in the hut. My syce now came up, and I sent him on to hunt up my camp and to bring some villagers with a *charpoy* (village bedstead) on which to carry the poor fellow to my tent. A few drops of brandy from my flask soon revived him, and he greedily devoured a biscuit moistened in brandy. I could see that hunger was his chief ailment, but I would not for the present give him more than a second biscuit, as I knew that in extreme cases such as his, food must be administered with caution. After a little time he was able to sit up, and he then told me he did not belong to the village, but was one of a party of Doms or Pahariahs (hill-men), who were on their way to the wooded tracts to the South-east. They had been without food for days, and on the bare plain in this part of the country there were no birds to snare. They found the village deserted, and as he was too weak to follow his people they had left him in the hut to die. He had lain there all the previous day, and not a soul had come near the village. At night the dogs had smelt him out and attempted to attack him, but he had beaten them off. They renewed their attack in the morning and he again kept them off, but only for a time, as they had recommenced their attack and would certainly have killed him had I not come up and saved him from being eaten by them. 'Ough!' the poor wretch quivered and fainted off.

In a little time my *chuprassee* (messenger) came up with several villagers and a *charpoy*, and the poor fellow was carried to my camp and taken care of by my servants.

On my return to camp after the Christmas vacation I found Purdasee much better. Purdasee was not his real name, but on being asked who he was, he said he was a 'Purdasee' (literally a man from foreign parts, but used colloquially to designate anyone extremely poor); so the name stuck to him. He belonged to the great clan of wanderers and outcastes found all over India and known under various names, such as Doms, Ghassias, Bhujs, Kooravers, etc. They are wanderers all of them, having no settled habitation, but with a few donkeys to carry their household pots and baggage may be seen on the outskirts of most Indian villages. The women weave mats and tell fortunes; while the men snare birds and lift hen-roosts. They are notorious thieves, and not a fowl, kid, or cat is safe for miles round their encampment. Of the flesh of the cat these people are particularly fond, and when later on I occasionally took Purdasee with me into Bangalore many a fine Tabby mysteriously disappeared to the surprise and grief of its owner. It was no use expostulating with Purdasee. His sense of *meum* and *tuum* was dead as regards tabbies. His gratitude to me for saving his life was heartfelt, but I really believe that even that would have counted for little if weighed in the balance against his love for cat's flesh.

It took several months before Purdasee thoroughly recovered his strength and was able to accompany my camp. He then attached himself to my tent, would assist in pitching it, and would hang around all day for some word or notice from me. In the wooded districts he proved a great aid to my commissariat, as never a day passed but he brought in quail, partridge, pea-fowl, jungle cock, etc. He was expert at all kinds of snares, traps, nets, nooses and devices for trapping birds and small animals. He was also a most perfect mimic and would imitate the cry of the jackal, partridge, quail and jungle fowl. I have several times been present when he has decoyed birds, hare and jackal into his traps, and I could not for the life of me tell the difference between his call and that of the animals he imitated. Was it jackal he wanted—he would partly shade his mouth with his left hand and a series of yells would break forth, as if all the jackals had assembled to join chorus. He could make the notes sound distant or near by merely opening and closing his hand. If he were after a quail, the 'ronk' of the male bird was heard to perfection. From the thigh bones of a cat he shaped a whistle from which the strangest sounds would issue at will. Far off larks would come down in flight, or crow-pheasants and pea-fowl would answer the harsh scream. Sometimes he would be absent for days, and then he would return with a low flat basket filled with partridges and quail on his head, a long rod slung across his shoulder and a peacock perched on either side. To prevent the pea-fowl taking to flight he would sew their eyelids together with a small feather, so that they could not see, and in that condition they remained perfectly quiet on their perch and could be handled.

Purdasee was delighted when work took me to the wooded districts. There he was in his element, snaring game. On the plains he could only exercise his ingenuity on the village roosters, and when he found that I compelled him to take them back to their owners, and that he was in disfavour for the remainder of the day, he brought me no more village fowl; but I felt sure that the thieving went on all the same, as he would find ready receivers in my camp servants, who were not so scrupulous as to how the fowls were come by. I asked him one day how he managed to catch the fowls without noise. He said he drugged them with rice which he kept in his mouth all night. It fermented there and became very intoxicating. The fowls eating this rice became drunk and stupid, and were easily caught.

Poor Purdasee! He fell a victim to his devotion to me. I was out fowling one day and had shot a couple of duck in a tank (pond), but found that they were too far out and the water too deep to recover them by wading. I sent Purdasee to see if he could find a villager who could swim, but all of them said they could not. I turned away disappointed, and was wending my way home when a villager ran up and said that Purdasee had gone into the tank after the duck and had not returned. We instituted a most careful search, and constructing a couple of rafts, I had the tank carefully dragged in the direction where he was last seen, but without result. His body was not recovered till two days later. Poor Purdasee!

Excerpted from Sport and Adventure in the Indian Jungle *(London: Hurst and Blackett, Ltd., 1904).*

DIVYABHANUSINH

How to Trap and Train a Cheetah

Cheetahs, the fastest four-footed animals on earth have been captured and used to run down gazelle and antelope since the twelfth century in India. Akbar, the great Mughal emperor kept as many as one thousand of these elegant, swift and gentle cats. A highly sensitive animal, the cheetah is maintained in good health and bred in captivity in only a handful of modern zoos.

Until recently, little was known about the keeping and tending of cheetahs by Indian princes. Only the persistence of a self-confessed cheetah addict, Divyabhanusinh—a hotelier by profession and a diligent researcher by choice—has uncovered their secrets. In his book, The End of a Trail, *the author catalogues the history of the Asiatic cheetahs in India and sheds light on how they were caught and kept in captivity.*

From the mighty elephant, to the African grey parrots trained to fire ceremonial golden and silver cannons at Baroda, Indians have tamed and trained an amazing variety of animals and birds through the ages. Two members of the family *Felidae* were the subject of considerable attention, namely the cheetah and the caracal, both of which were used for hunting in India. The elephant and the horse were tamed for warfare, the red and grey jungle fowl for food, and the cow and the buffalo for food production and allied uses. What was the need to train the cheetah? It can be palpably argued that in Arabia the cheetah, falcons and hawks fulfilled a necessity to obtain food from the inhospitable desert sands. This certainly could not have been the case in the Punjab, the Indo-Gangetic plain and the Deccan plateau. Hunting with cheetahs appears to have been developed in India primarily as a sport of royalty and the nobility.

In ancient India, dappled cats took part in royal processions and Indians brought them as gifts to their kings as has been noted earlier. While these cats were probably cheetahs, handling them had become a specialized technique by the 12th century and the *Manasollasa* passage details this royal activity at some length. However, through the ages the techniques were refined further and cheetahs found mention in several court histories and compendia of animals prepared for hunting. The most celebrated of them were the accounts of the *A'in-i-Akbari*. Before examining the knowledge on the subject of hunting with cheetahs as was extant in the last phase of the animal's existence in India, it would be appropriate to take a close look at some of the sources, their nature, and content, as they are either unpublished or are extremely rare.

The *Baznama* of Muhibb 'Ali Khan Khass Mohalli is a treatise on falconry as the name suggests and consists of 61 chapters. The last chapter is devoted to cheetahs and is a source of very valuable information.

The Muslim state of Tonk in Rajasthan, though established as recently as the early 19th century, soon became a centre of Islamic culture. There is a Persian manuscript from Tonk which is of specific interest. Only about 150 pages of it survive and the pages at the beginning and end of the manuscript are missing. The authorship and date of the work are not known. Dr Muzaffar Alam examined the manuscript and its text, and opined that the style of the script would indicate that it was written around the middle of the 19th century. It is in all probability a copy of an earlier work. The first 130 pages of the manuscript are on falconry, that is care and treatment of falcons and hawks and related matters. Pages 131 to 139 are devoted to the cheetahs—their trapping, training and treatment of their ailments. The rest is devoted to arms and armaments.

The small Bundela Rajput state of Ajaigarh in central India had a remarkable ruler in the person of Sawai Maharaja Sir Ranjore Singh who came to the *gaddi* in 1868 and ruled the state till his death in 1919. An eclectic person, he enquired into several activities of court life. He authored or commissioned 30 books on various subjects, including a two volume, one thousand page, work on culinary practices of the time, the *Ranjor Pak Ratnakara*, and wrote a book entitled *Yudhabodha Mrigayabinoda*—the name reminiscent of *Manasollasa's* chapters on hunting—which was an exposition of the art of warfare and *shikar*. Coursing with cheetahs was one of the favourite sports of the Maharaja. It appears that the cheetah trainers of his state had become restive and did not train the royal cheetahs with care and diligence. The fate of the errant keepers is not recorded, but consequently Kanwar Balwant Singh Panwar, the Maharaja's *risaldar* (master of the stable), took on the task of training cheetahs and chronicled his labours in a work entitled *Sa'idnamah-i Nigarin*, consisting of 52 pages divided into six chapters on the catching of cheetahs, determining their age and swiftness, taming and training them, hunting with them and treating their ailments. In addition the work has 46 illustrations most of which show the various activities of cheetah handling.

Kolhapur state in the Deccan remained at the forefront of the sport of hunting with cheetahs and in fact was the last among Indian states whose princes coursed with these animals in the subcontinent. To commemorate the visit of the Prince of Wales (later the Duke of Windsor) in 1921, the Maharaja, Sir Shahu Chattrapati commissioned his court photographer, A. R. Pathan, to publish a Shikar Album. This album contains 57 photographs in all of which 29 are photographs of cheetahs, recording the various stages of trapping, training and coursing. The rest of the photographs are of other tame animals and birds including tigers, caracal, hounds and hawks.

In Baroda State, coursing with cheetahs was a well-established practice too, and the animals were left in the care of noblemen who were close to the royal family. However, from 1917–18 onwards the cheetahs were maintained at the expense of the state.

During the rule of Maharaja Sir Sayaji Rao Gaekwad, the procedure was streamlined and four cheetahs were regularly maintained at a cost of Rs 10,176 in 1924. Possibly, the high cost of maintaining cheetahs caused the administration of the state to record all activities connected with it and thus a detailed account was compiled by none other than Major R. S. Parab, the *Khangi Karbhari* (Private Secretary), of the Maharaja with the help of R. Annasaheb Bhavsaheb Jadhav, Shikar Superintendent of the state. The work was entitled *Chityasambandhi Samanya Mahiti*— Common Information regarding Cheetahs. It was probably meant for restricted circulation within the administration of the state. It contains 38 pages divided into five parts, one each on general information; capture, taming, training techniques and on buying cheetahs; on keeping of cheetahs; on hunting with cheetahs; and on the food and other requirements for cheetahs in captivity.

With the demand for cheetahs developed professionals for catching these animals. Among the rich variety of tribals in India are the *bavariyas*, who are to be found to this day in Rajasthan and the surrounding areas. They were classed as a criminal tribe but in fact they are people who are primarily hunters, with highly developed skills for trapping and their knowledge of wildlife is consequently second to none. According to M. Azizuddin, the son of the last cheetah trainer of Jaipur state, cheetahs were caught by these tribals and subsequently sold to the *darbar*. Around Baroda and the Deccan, the hunter/trapper tribe of *pardhis* caught these animals.

Over the years the *pardhis* came to be divided into several groups identified by the name or nature of their activities or dress. Among these were *shikari* or *bheel pardhis*, ones who used fire arms to hunt; *phanse* or *phase pardhis* (also known as *haranshikaris*—hunters of antelopes—in Karnataka), ones who used traps and snares and *telwale pardhis*, ones who sold crocodile oil. The smallest group of them were called *cheetawala pardhis*, ones who made a living by trapping cheetahs and selling them to princes, hence they were sometimes called the *raj pardhis*. This group also used the cheetahs for coursing in their own right. The *pardhis* to this day make a meagre living by hunting and trapping animals and birds, often finding themselves on the wrong side of the law. The late Professor D. D. Kosambi, who studied tribal peoples of the Deccan, stated that *pardhis* who were experts at making delicate traps for snaring hare, partridges and the like, were employed at the ordnance factory at Kirkee near Poona in their bomb disposal squads during World War II.

According to some English authorities the cheetah catchers used to call the animal by its Persian name *yuz* and one may conclude that some of them could have been Muslims. The cheetah trainers were a different set of people, as the skills required were specialized. In Jaipur, Bhavnagar and Kishangarh they were Muslims. Information from Baroda and Hyderabad is unavailable, though photographs of the keepers at Kolhapur show them wearing the Maratha state's court dress including the turban. Hence it is more likely that they were Marathas, though as per the practice of the time, Muslim retainers, if any, would have worn the same dress.

Two photographs, however, show a bearded person, obviously a Muslim, wearing a fur cap and a long coat with the bearing of a head trainer. At Ajaigarh on the other hand, the Rajpur *risaldar* of the state undertook the task of cheetah-keeping himself, but the religious persuasion of the state's intransigent trainers is not recorded. Interestingly, the book written by the *risaldar* is in Urdu which was the court language of many states in north and central India.

The blackbuck was the cheetah's most preferred prey as is evident from Mughal records, paintings and from the observations of British naturalist-sportsmen. The *chinkara* often occupied undulating terrain which is not the best ground for the cheetah's burst of speed, and was second to the blackbuck in preference. Two races of *Antelope cervicapra* are now recognized. The north-west blackbuck, *A.c. rajputanae,* extended over the plains of Jammu and Kashmir, Punjab, Haryana, Delhi, westernmost Uttar Pradesh, Rajasthan, Sind and the Gwalior region of Madhya Pradesh. The other race, *A.c.cervicapra,* of the southern and eastern region extends eastwards from Kheri in Uttar Pradesh, Bihar, Bengal, Orissa, Madhya Pradesh from Bhopal eastwards, Maharashtra, Andhra Pradesh, Karnataka and Tamil Nadu. It may be noted that of the 24 largest horned specimens mentioned in Rowland Ward's Records of Big Game that were obtained from known localities, 21 came from Punjab, Haryana and the adjacent tracts of northern Rajasthan. The blackbuck of the north-west were larger. As a result, the cheetahs of the north-western region would have been larger than the cheetahs from the Deccan. It is possible that the blackbuck of the Gujarat area were larger during the Mughal times than the animals subsequently recorded or studied in this century. The larger cheetahs from Gujarat and the north-western regions could have coursed more successfully, and hence the repeated mention in these records of their efficiency.

In the final phase of the cheetah's existence in India they were being trapped in Berar and North India according to one source and if *bavariyas* and *pardhis* caught them, as has been noted, the statement stands corroborated. The *Baroda Manual* categorically laments that the cheetah population

has diminished along with diminishing jungle cover in parts of south India, especially in the western parts of Southern Hyderabad, Ballari [Bellary in Karnataka], the Anandpur [Anantpur now in Andhra Pradesh] district of Madras [Presidency], around Mysore, Konkan, Karad [Maharashtra], Agra, Ajmer, Multan [Sind, Pakistan], and Gujarat.

Both these sources appear to have recorded more from history and traditions of the past than from the current status as the cheetah had virtually disappeared from India by the turn of the century. The *Sa'idnama*, compiled at about the same time, also gives similar locations, namely, Pali in Rajasthan, Isagarh near Guna, Rajangarh near Bhopal, Mundyawah in Ajaigarh State and the Kathiawar (Saurashtra) peninsula in Gujarat.

Since the object of catching the cheetah was to course with it, it became necessary to know the finer points of the animal and these were meticulously recorded. The *Tonk Manuscript* states:

Amongst these [cheetahs from different parts of India] the best for coursing with are those who have broad face and eyes, medium-sized head and neck, wide chest, slim belly, large back, small tail.

These are characteristics of a racing animal like a greyhound. The *Sa 'idnama* on the other hand gives detailed instructions to find the age and agility of the animal. Firstly, in a young cheetah, the yellow on the tip at the back of the ear is of 'two fingers' breadth' which decreases in an older animal and turns blacker with age. In a leopard the same feature has been noticed, but with age the colour fades. In a cheetah, however, when the backs of the ears become completely black, it has attained its full life span. Secondly, the tip of the tail of a young cheetah is thin and sharp and it becomes less sharp and more fluffy with age. This is a characteristic peculiar to the cheetah and not to other cats. Thirdly, the pads of a young cheetah's feet are soft which harden with age. Fourthly, a young cheetah's teeth are small and white which yellow with age as in all large cats.

On the methods of trapping itself, the oldest recorded one appears to be that of making pits called *odi* in which the cheetahs would fall,

but they [the cheetahs] often broke their feet or legs or managed by jumping to get out again. Nor could you catch more than one in each pit...[Emperor Akbar] therefore invented a new method, which has astonished the most experienced hunters. He made a pit only two or three *gaz* [1.83 metres to 2.75 metres] deep, and constructed a peculiar trapdoor, which closes when a leopard falls into the hole. The animal never gets hurt. Sometimes more than one got into the trap.

Apparently, the method of catching cheetahs in a pit was used in Afghanistan and may have come to India with the Mughals. The

Baznama records that the Sultan Husain Mirza of Herat who came to the throne in 1468 was very fond of hunting with cheetahs as well as capturing them.

Generally they used to dig a pit to catch them, to a depth of nine *gaz* [8.23 metres]. After a huntsman had covered over his pit and gone home, he was unable to visit it for a week. On the first day a female cheetah with her six cubs happened to fall into the pit. For the whole week she brought her offspring food each day, leapt the nine *gaz* down into the pit, fed her cubs, leapt out and departed. When the cheetah-keeper became aware of this a snare was set at the edge of the pit. The end of the snare was fixed to a stone (of some kind) which stood beside the pit at a little distance(?). When the female cheetah, according to its daily habit, was coming to feed its cubs and reached the edge of the pit its foot was caught in the snare. The stone was dislodged(?) and she fell into the pit. They caught the female cheetah in this way and brought her and her cubs before the Mirza. After that they dug pits to a depth of eleven *gaz* [10.06 metres], but at this depth most cheetahs ended with broken tails. In the end the original nine gaz was settled upon.

The record of the incident appears to be exaggerated. A female with six cubs would be a rare occurrence to start with. It is also doubtful whether a cheetah could clear a height of eight metres without a run up. And yet the experiment was made to make the pit deeper, though it was unsuccessful. Akbar's innovative shallower pits with a trap door were a definite advance on the extant practice. Akbar is also believed to have caught cheetahs 'by tiring them out which is very interesting to look at'. It is not clear what is meant, but it could suggest chasing the cheetahs on horseback until they were tired and became immobile due to exhaustion.

Apart from this method there is one particular one which is generally accepted by all the sources examined—snaring the animals around a favoured tree.

According to the *Sa'idnama* a cheetah becomes sexually active at the age of two and a half to three years when it comes to play around a tree trunk after dark where it remains for 10 to 15 minutes before moving on. The particular trees in the jungles, to which the cheetahs thronged, were well known. Such 'playing grounds' are called *akhur* [or *akhori*, 'home territory']. The use of a particular tree or 'playing ground' by the cheetahs is indicative of their territory. In the Serengeti plains the cheetahs are not territorial and they follow the movement of Thompson's gazelle, their preferred prey, and contact with other groups of cheetahs is avoided by means of scent and smell. According to another study, the cheetahs employ 'a time-plan spacing system which allows individuals to occupy the same home range concomitantly, without disruption of hunting efforts in the Nairobi National Park'. If the cheetahs came across fresh markings of another group they moved on and thus avoided meeting them. Another study from Africa shows that a female cheetah leaves a scent trail when it comes into season through the discharge of sex hormones in its urine which the males follow. In India the cheetah's prey did not move vast distances as in Africa. It would appear that individuals or groups of cheetahs visited the same area repeatedly as a result of a time-lapse spacing system. The reference to sexual maturity of the cheetahs in the text may be incidental to adult or sub-adult males hunting in their own right and using the same territory. Why they should visit the tree at night is unclear, as cheetahs are diurnal. Around the tree, a trap made of 'deer gut' rope was laid. It was fixed to the ground with wooden nails with nooses (*phande*) of the rope standing upright. The arrival of the cheetah would be heralded by a rustling of the leaves or by the bell tied to one end of the rope. The cheetah approaching the tree would be entrapped in the nooses and that was when the cheetah catchers hiding in a leaf covered pit nearby swung into action. The vigil had to be maintained for days on end as there was no certainty as to when the cheetah would arrive.

According to the *Sa'idnama*, the moment the cheetah was trapped the cheetah catchers approached the animal, waving before it a leafy branch. The cheetah would bite the branch in anger and in the process its eyes would get covered by the leaves. At this moment a hood (*topi*) was put over its eyes and a cot quickly placed on the animal to restrain it. A cloth would bite put over the animal thereafter, and a collar (*galida*) and waist band (*kamarkach*) would be tied around the neck and waist of the animal to make a harness. The feet were also tied together with a knot (*ant*). The cheetah was pulled up unto a cot; the process was called *santra*.

According to the *Sa'idnama* the next stage of the training of the cheetah was to get it to recognize the keeper, take instructions from him and in the process become attached to him. It gives a detailed account showing that the cheetah had to be made to stand with three ropes from the *bhanwarkali* (swivel at the neck) held fast, similarly two ropes from the *kamarkach* (waist band) again held with a cloth screen at the back of the cheetah. The animal was made to stand in this fashion daily. It was irritated deliberately by an attendant carrying a *chorkathi*, and who harassed the animal with such sticks and taunting calls. The *Baznama* differs here insofar as according to it the purpose of the *chorkathi* is to cover the eyes of the cheetah with the cloth lest it attack the trainer. The keeper, who remained aloof, came to the cheetah only when the *chorkathis* were taken away and slowly made it take food from a *hankana* (long handled wooden ladle). As the process went on, the cheetah was made to eat the food first with the *topi* on, then without it. Then the number of ropes were reduced, and ultimately only one was left. A cloth was kept at the back of the cheetah the purpose of which was to ensure that the animal was not distracted during training from anything at the back. If it did get distracted during training, it would become stubborn (*kara*). The animal was fed boneless pieces of meat offered in a *hankana*. The *Baznama* states that the wooden ladle had a ring fixed to it which would make a sound when moved. Once the cheetah got used to it, it would associate the sound with food. According to the *Sa'idnama*, the amount of food to be given was three to four *seers* (one *seer* = 933.12 grams) a day. The animal was also made to eat/lick butter, of which it became fond, and was also made to drink milk and eat specially prepared cottage cheese. While butter was given in summer, *khoya*, dried fresh whole milk, or *pedas*, a sweetmeat made of milk, were given in winter. The amount was not to exceed 125 grams. The last stage of intimacy was attained when the cheetah licked butter smeared on the hand of its trainer.

To start with, the animal was fed only in its enclosure or while standing. Once this stage was completed it was taken to the jungle and fed. The reason for this was to get it used to accepting food in the jungle, so that it could be trained to give up its kill and come back to the keeper to be fed. Feeding the cheetah at night made it get used to its keeper faster.

Through feeding, the animal was taught to climb a tonga; the command for climbing up was *di*, and for climbing down, *parti*. The object of the training was to create a bond between the animal and its keeper. Terms of endearment to be spoken to the animal were strongly recommended as well as sounds such as '*Aa-ha-ha-ha-ha-Aai*' during feeding. The animal was to be called by words such as '*aao beta, wah beta* (come on son, well done son), '*chale chalo bahadur'* (march on, brave one) and so on. The animal was also taken for walks in the bazaar to get it used to crowds. This ritual was called *'pherna pheri par'*.

Taming a male cheetah generally took four to five months, whereas a female needed seven months. This difference in the training period was not explained. The total number of persons required to train a cheetah could be as high as twelve. The *Baroda Manual* gives a far shorter description of the process, though it does not differ in the essentials from the *Sa'idnama*, except that according to it the total time for taming the animal was not more than three months and the number of persons required was six. However, according to M. Nirmalkumarsinh it took one year to fully train a cheetah at Bhavnagar. The Kolhapur Album has photographs of the entire process and the *Sa'idnama* has drawings which illustrate the same.

After taming the cheetah, the next stage was to train it to hunt. The *Sa'idnama* states that before this stage commenced, the trainers had to ensure that the animal in question was completely tamed and confident. A sure sign of this was when the animal's tail stuck out when it was being fed. If it kept its tail between its legs, it meant that it was still nervous.

A blackbuck which had been incapacitated by a deliberately inflicted injury was let loose in the presence of the cheetah. The latter was released after removing the *topi* but along with the *satka* (a long rope) tied to the *kamarkach*. The moment the cheetah had caught the antelope, the keeper ran after it following the dragmarks of the *satka*, and the cheetah was caught with its help. The cheetah was made to release its grip on the blackbuck by luring it to the *hankana*, and a *tamancheki topi* (box-like cap) was fastened over its eyes.

If an antelope was not available, a goat could be substituted for the purpose. Care had to be taken to ensure that training was given on flat land, clear of bushes. A cheetah with a *satka* tied to it should be released only in a ploughed field or field of wheat or rice. Since the quarry was already incapacitated to make it slower, the only purpose of releasing a cheetah in a ploughed field would be to make it slower so that it could be retrieved easily should it decide to make a bid for freedom. However, the *Sa'idnama* does not give any reasons. The process was to be repeated often enough until the cheetah was fully trained to hunt. The *galida* and the *kamarkach* were always retained during a hunt.

The cheetah was taken to a herd of antelope on a bullock cart. The cart was driven in an uneven and circular motion called *dāv karna*, meaning to perform a strategy, in order to get close to the herd. It was suggested that the cart be driven upwind, and wherever possible, the cheetah should be released in front of the herd. The cheetah was trained to only hunt a male, which was done by showing it a male when unhooded. It was sure to hunt the blackbuck down successfully if the male antelope was following a herd, or when two males were sparring, as the quarry was obviously distracted in such situations.

Once the cheetah had caught its prey, the keeper would go to the animal, give the same call that was given during feeding at the time of its training and show it the *hankana*. The moment the cheetah released the antelope, the *tamancheki topi* was put back over its eyes. The antelope's throat was slit and its blood filled in the *hankana*, to which the cheetah was attracted. It would leave the kill and not bother about the antelope. The hind leg of the antelope was to be cut and the meat from the leg chopped into small pieces and fed to the cheetah so that it felt that it had eaten its prey. However, if the cheetah was to take the field again on the same day, it should not be allowed to feed to its satisfaction. Also a young cheetah recently trained should not be released on *chinkara* as they were believed to be very clever and would lose the cheetah easily in ravines.

A cheetah with three to four years' hunting experience could also be used on moonlit nights. Care had to be taken to ensure that the hunting ground was open, else it would become difficult to locate the cheetah. The procedure was to drive (*hankna*) the antelope towards the stationary cheetah or to get them to move towards the cheetah pushed forward by men moving slowly behind them (*talna*).

In addition, other methods of hunting with cheetahs were (a) *agot*, that is by stationing a cheetah near the path of animals; (b) *dwari* (*dwar*, gate), hunting animals as they would come out of the only exit of a forest area; (c) *pani* (water), hunting animals on a water hole or watering place; and (d) *cheroe* (*churai*, grazing), hunting animals at their grazing grounds. In such methods, the cheetah would be made to sit in a depression (*gadhdha*) along with its keeper. The place would be surrounded by foliage and the cheetah made to keep quiet by feeding it butter and *pedas*. Once the cheetah was released on the quarry, the procedure was the same as when it was released from a bullock cart.

The keepers simulated the *shikar* environment. A man ran with a blanket on his back towards and past the keeper with the cheetah in tow. If he reached the keeper, the latter would command the cheetah to stop. If not, the runner would lie down on the ground and the cheetah would run on towards the keeper without attacking the runner. The entire emphasis was on the cheetah obeying the commands and stopping when commanded to do so by its keeper. Thus from 5 a.m. to 10 a.m. daily the cheetah was allowed to roam the jungles and from 10 a.m. to noon it was made to go through the command drill. In the evening, the cheetah was taken out into the town until about 9 p.m., when it was given a call command, fed and returned to its cage.

Though neither the *Sa'idnama* nor the *Baroda Manual* dwell at length on the speed of the cheetah, their authors recognise this specific quality of this feline in running down its quarry. The *Baznama* on the other hand considers the speed of the cheetah to be the one attribute which sets it apart from other animals. Chapter 61 begins by stating that 'Although a description of the cheetah is not demanded here [as the work was concerned with falconry], nevertheless, because of its ability to soar (*parandagi*, literally 'to fly') and its swiftness it is classified together with flying birds of prey'.

Excerpted from The End of a Trail: The Cheetah in India *(Delhi: Banyan Books, 1995).*

SUYDAM CUTTING

Cheetah in the Kolhapur Deccan

By the time the intrepid travel writer and photographer Suydam Cutting visited the princely state of Kolhapur, the ruler was importing his hunting cheetahs from Africa. India had hardly any left; and the cheetah's main prey, the blackbuck, was also on the retreat. Cutting's work includes memoirs of his trips to various exotic places—the mountains of Tibet and the islands of the Galapagos. But the recounting of a cheetah hunt is significant, because the art was already dying out. It was expensive to import the animals from Africa. The Maratha rulers of Kolhapur had one of the largest collections of cheetahs: it was also among the last of its kind. Cheetahs still survive in parts of Africa and in Iran but nowhere in India do they roam free. Cutting's is an account of a unique association of the only big cats trained to kill antelope for a prince's table.

A whole literature has grown up around the fox and the hound, but of cheetah hunting, a more ancient sport and a much faster one, very little has ever been written.

The cheetah is one of the most curious and interesting animals in existence, and the sport in which he becomes the leading actor deserves special attention in a book on travel and exploration.

Although cheetah hunting has dwindled today in India to such an extent that there are but few places where it may be witnessed, it was once quite common, and we learn from miniatures that it was popular with the Mogul conquerors of North India many centuries ago.

One of the places where it survives with its old-time vigour is the native state of Kolhapur, situated in the Deccan of South India. The Maharajah, an ardent sportsman and famous pigsticker, is enthusiastic about the cheetah.

While my wife and I were staying with the Maharajah he arranged a hunt for us, although it was not the proper season. I regarded it as a completely new experience—outside the usual hunting tradition. I write about it now as a novelty unknown to most sportsmen and hunters.

It was an impressive sight to walk into the long, high buildings in which the Maharajah housed his cheetahs and see the animals sitting on their individual *charpoys* or native beds along the wall. Each cheetah wore a hood, a black hood fitting snugly around the head. Two personal attendants were watching over the animals as if they were royal infants.

There were thirty-five of them lining the walls, the youngest being around three years old, the oldest eight or ten. Since the animal is becoming scarcer and the African breed not only grows bigger and stronger but acquires greater speed, the Maharajah imported his from Kenya. Brought to India, they had to be domesticated and trained. Each animal was worth something between £200 and £250.

It was obvious that the cheetahs took kindly to their keepers and mode of training. They responded much better than leopards, on which the ruler had conducted some abortive experiments.

Having trained his imported cheetahs in admirable fashion and revived an ancient sport, the Maharajah sold some of his surplus animals to other princes of India, including the late Gaekwa of Baroda, his uncle by marriage.

The cheetah, a member of the cat family, has many qualities of the dog. The height of the shoulders is about the same as those of an adult greyhound. The torso and legs also resemble a greyhound's. The cheetah lacks the back leg knee-bend common to the lion, leopard, and tiger. Again the feet are those of a hound with heavy, nonretractable claws. In size, however, the feet are much larger.

Yet, for all these resemblances to a dog, one glance classifies the cheetah as a cat: his markings are spots, his skull, eyes, and teeth are feline, his habits are purely carnivorous. Lastly the cheetah has a definite purr.

Like falcons, cheetahs wear hoods at all times except when they are being fed and exercised and at the moment when they take part in the chase. The hood on the cheetahs serves the same purpose as it does on falcons. It keeps them quiet and tractable. Contrary to what one might expect, it does *not* cause their eyesight to become any less keen.

One of the first things I learned about the Maharajah's cheetahs was that fondling them even when they were in their *charpoys*, properly hooded, was a risky pastime.

In Kolhapur the quarry was always black buck. The cheetah was strictly trained to kill none but adult males. Because these are readily distinguishable in any herd by their dark colour, a slow and curious process of familiarizing the cheetahs with this shade was carried out. The men who fed them dressed exclusively in black, while the regular keepers wore white. Having killed, the cheetah was invariably allowed to feed providing he had dispatched a buck. But if he killed a female he was haltered and pulled away: the punishment soon taught its lesson.

The cheetah's daily exercise varies. Usually he is led up a road on a halter by one keeper in white and then encouraged to run back to another keeper in black, who holds a piece of meat in his hand. Unlike whippets, on the leash the cheetah does not strain, and released he carries out his act without enthusiasm. Another exercise is a real black-buck hunt to keep him in good trim for big occasions.

The speed of the cheetah is amazing. He is well aware that no other animal can rival him in this regard. The tiger, lion, and leopard rarely rush their quarries more than a hundred yards. After that they desist, knowing they cannot catch up with game that has attained its maximum speed.

Trained cheetahs proceed slowly and methodically, choosing their quarry out of the herd. Only when the quarry has attained its greatest speed does the cheetah run him down with that tremendous burst that has no parallel in the entire animal kingdom. The cheetah's *lasting* powers lie somewhere between a cat's and a dog's, but nearer a cat's.

This ancient sport of India does not affect the abundance of game, for no great number of buck is destroyed and the herds are plentiful. Furthermore, the slain buck are eaten.

The day of the hunt a car called for my wife and me at six in the morning. An hour later we arrived at the plains where the hunt was to be held. Here we were greeted by the Maharajah, the Maharani, the Maharajah's sister, his niece, and two men guests, both Indians.

The genial Maharajah, who tips the scales at 300 pounds, is one of India's greatest sportsmen. His cheetah hunts are famous. He is an expert pigsticker, a daring cross-country rider. His stables hold 300 horses of Indian, English, and Australian breeding, his kennels bird and hunting dogs of every kind, 275 in all, and his fields fine herds of Brahminy cattle. Even his pets are unusual. Near his house were to be seen at that time a young lion and lioness and a pair of two-year-old tigers gambolling around on thirty-five-foot leads. Formidable sloth bears, that most dangerous of animals that attacks without provocation and has been known to tear a man to pieces for no reason at all, were sometimes seen walking down the village streets with their keeper.

The Maharajah's sister also has a unique place in the Indian sports world. She is the only woman pigsticker in all India. White women are not allowed to join the clubs, and native women do not participate. But at the age of forty, the Maharajah's sister has made a great reputation in a difficult sport. An active rider and hunter, she is a handsome woman, but belongs to a type that may be found among sportswomen of England and America, never in India. She is keen, trained down, alert.

The Maharajah usually followed the cheetah hunt from a brake or light wagon, specially built according to his own designs. It carried one eight feet above the ground and was drawn by four Australian walers. But although it was now early October, the beginning of the dry season, the grass, still high from the recent monsoon rains, covered many blind ditches that would have made riding in a brake impossible. So for this special, off-season hunt, the Maharajah elected specially built motor-cars with high shelves for the cheetahs. In addition there were two lorries, one for additional cheetahs and their attendants, the other to supply us with tea and sandwiches.

After a cup of tea we were off on a wild ride by seven o'clock, bumping and crashing along at thirty-five miles an hour. In one car went the Indian ladies and one of the guests. In another my wife rode in front with the driver, while I was sandwiched in between the Maharajah and his other guest.

A cheetah, still hooded, was lying on a platform built into the car at about the level of my knees. His keeper, crouched on the running-board, a precarious perch, had to keep his eyes on the animal and also get out at times and have a look at the trappy ground. On we careened, sometimes on lanes, but more often across open country where our thirty-five miles an hour was a dizzy speed. Everyone held on like mad.

Having found a herd, we manoeuvered for a proper position. Then began the real strategy. In confronting a large herd, an attempt was always made to detach the males—quarries for the cheetahs. They were kept in a continuous stampede. Since they did not run in a straight line, the car bucketing along was able to keep up with them.

I was trying to operate a camera, and because it required two hands I had to relinquish my grip. The car swerved violently. The cheetah and I were thrown forward, landing on my wife's neck. No more did we get settled than it happened again.

Then, with everything perfectly timed by the Maharajah, the car stopped. The cheetah was unloaded, unhooked, and hustled out on the grass. For a half minute he stood there sizing up the situation. At a gentle, slow lope he started off toward the herd. The black buck, about two hundred yards away, began to move off. The field was alive with galloping forms, their bounds increasing progressively in length. By now the cheetah had chosen his particular quarry. He rushed towards it with incredible speed.

The quarry, realizing too late that he could not match the cheetah's speed, attempted a downhill slope. The cheetahs prefer to run uphill: going down they are liable to a false aim and then a bad tumble.

Undaunted by this manoeuver, the cheetah soon overtook his buck. He sprang with front paws directed at the hindquarters of the quarry. The violence of this blow threw the animal. Then the cheetah caught him by the throat.

At this point we arrived on the scene. The cheetah lay full-length, with the buck's throat held tightly in its slightly curved canines. Gradually the victim ceased his violent attempts to tear himself loose. He was choked to death. The cheetah lay perfectly still in apparent ecstasy. Slowly he opened and closed his great greenish eyes, gently emitting a soft, rumbling purr.

The cheetah was allowed to feast on one hindquarter of the buck. Then he was gently and firmly led aside while one of the attendants disemboweled the victim. Some of the blood and the steaming viscera, placed in a long spoonlike bowl, was offered to the cheetah. This was his reward, and he seemed quite satisfied.

Another cheetah was brought from the lorry. The motor was started and off we went again. This time the car, overtaxed by the speed over such rough terrain, broke down. The Maharajah coolly beckoned for another car, of which there were several for such emergencies. It zoomed over and we continued the hunt.

There were dramatic variations in all the seven hunts staged that day. Once a cheetah, forced to carry the pursuit downhill, while going full speed, landed in a blind ditch and turned head-over-heels. He was given a rest and drink and the mud was cleared off his head, but he refused to run again.

Another cheetah killed a female, a regrettable incident. He was dragged off his feed so that he would never repeat the performance.

Once again a motor broke down. Sometimes the staff had difficulties in manoeuvering the herd. Sometimes we drew close to them but they succeeded in escaping. Then, perhaps, we would pick them up again. To add still more variety, the cheetahs were unloosed at various distances; so we had a good chance to watch all phases of the approach and attack.

By eleven o'clock we were through. Six cheetahs had overtaken animals. The cars and lorries went back to the bungalow rest-house. Then, with the Maharajah and his party, we made merry at breakfast, later viewing the kills or what remained of them, as they lay in line on the grass. Natives in a carnival spirit brought wreaths of flowers to hang around our necks. Made of plumaria and jasmine, they were larger versions of the leis of Honolulu.

One last word about the speed of the cheetahs. To begin with, one should remember that when they are apparently running fast, they are by no means running their fastest. Any one who has had experience with cats will realize that it is extremely difficult, owing to their peculiar temperament, to train them to perform any action out of keeping with their normal behavior. Cheetahs would probably never run at their maximum speed on a track. Furthermore, it is doubtful if they would sustain their greatest effort for any distance beyond two hundred yards. But in their habitat, when pursuing buck, their final charge is terrific. It must surely attain a speed of sixty-five or seventy-five miles an hour—for a hundred or more yards.

Excerpted from The Fire Ox and Other Years *(London: Collins, 1947).*

R. S. DHARMAKUMARSINH

Following the Lion's Trail: The Lion Trackers of Mytiala

Dharmakumarsinh was a younger son of the Maharaja of Bhavnagar, a princely state in the Saurashtra region in Gujarat. Like many princes or members of the landed gentry at the time, he was a keen falconer. The men who actually trained and cared for the falcons were experts at the art, custodians of a craft that had been passed down for generations. Similarly, the tracking of lions in the thorn forests of the Mytiala tract of the Gir forest could only be successful with the help of highly experienced pagis *or trackers. Unike many contemporaries, the author of the pieces that follow was a meticulous observer. He writes of the men who could read a lion's life-history in its pug-marks and of falcons that could strike at air-borne prey, a pigeon, a duck or a partridge. Many such skills survive among forest peoples but they are rarely drawn upon for protecting wildlife or for captive breeding schemes today.*

The Mytiala hills lie on the eastern extremity of the Gir highland system and extend over some seventeen square kilometres. They are an isolated group of low hills named after a village situated within their range. The river Dhantarvadi meanders through the area and despite its remote location, one can glimpse the shining waters of the Gulf of Cambay from Mammai, its highest peak, on a clear day.

Erratic rainfall and shallow stony soil makes the region arid and the only vegetation cover is a stunted dry deciduous thorn forest predominated by the ubiquitous salai, *Boswellia serrata*. Though there is no teak, bamo or other such varieties of flora found in the Gir, this is an excellent habitat for lions, leopards and their prey species, the nilgai, wild boar, and the ghuntavda or four horned antelope, hyena, porcupine and even the peafowl. The birds found here are the painted sandgrouse, painted partridge and nightjar.

In 1856 when the Asiatic lion disappeared from the thorn forest of Sihor in the present Bhavnagar district of Gujarat, some nomadic groups moved into the Mytiala hills. In 1931 the preservation of this small lion habitat began with a plan to afforest the entire hill range. Cutting trees and grazing were strictly forbidden and small dams were built to store water in a region of scarcity. This successful scheme extended the Asiatic lion's habitat and their numbers increased in consequence. It became a comfortable refuge for the animals to breed in and though the harsh summers often drove the lions back to the Gir, they returned during the monsoon and winter.

The Mytiala lion trackers were specially trained to track the many leopards that lived in these hills. They were hardy men, efficient and competent with a comprehensive knowledge of the behaviour patterns and habits of the leopard and its prey. With time their tracking skills were passed down from father to son and it was interesting to see the young being trained in shikar lore.

Bhavsinhji was a man of principle who judiciously adhered to a strict daily routine, dining and sleeping early to enable him to be up and out in the field before dawn. He disliked sitting over a kill after sunset and the shikaris tried their best to satisfy this whim. If a leopard failed to appear, Bhavsinhji would blow a whistle, climb down from the machan and try again another day. Most of his leopards, however, were shot on foot in a beat drive. The trackers had perfected the art of attracting the leopard to the bait and I remember that when Koli Kheema whispered in my ear, 'look the leopard is watching us from the top of the hill,' or 'be careful how you enter the machan,' the animal would indeed be there, his head and chest a tiny white speck on the horizon.

These expert shikaris were both Muslims and Hindus and the head *jemadar* was a Muslim Gaddhai called Naan Jemadar. He was the most experienced of the lot, short and sturdy with a flowing beard. His deputy, Kheema, was a Koli who was the complete opposite; tall, lanky, clean shaven and with perfect eyesight. They worked together as a team yet a good humoured competition existed between the two communities based on their individual hereditary skills and expertise. The Kolis had extensive knowledge of the wild boar whose meat they also ate while the Gaddhais knew more about the roz or nilgai. Both were remarkably silent workers though Naan often narrated wonderful animated stories of his experiences in the forest. Most of these men could use firearms and sometimes wielded swords and other such weapons.

It was a singular delight to watch these trackers hot on the trail of a lion or leopard. Completely sure of the territory they moved as on a chess board, aware of their quarry and its every movement. Their close acquaintance with the habits of the animals enabled them to predict their movements and locate their resting places. On the basis of both intuition and knowledge they would arrange beats with a minimum of beaters and noise. There was no cracking of firearms or shouting though occasionally an alarm would convey to the hunters that a quarry had been driven out. An average beat for a lion required six to twenty beaters. This number increased for the unpredictable leopard as a leopard drive required greater team work and an experienced shikari to plan and execute the venture.

Once a Mytiala lion tracker drank a bowl of blood from a slain lion's skull believing that this would give him the animal's strength and courage. The trackers would start work early in the morning before sunrise, following spoor along dusty paths or roads or cart tracks, walking briskly or jogging to return to camp by eleven in the morning with news of wildlife. They would climb the low hills with agile ease and had great stamina, running for miles carrying information. It was easier to track a lion because its pug marks were larger and it generally followed a straighter route sometimes revealing its presence with a roar before it settled down under the shade of a tree. The leopard, however, frequently deviated from his regular route, following a winding course, climbing steep hillslopes, hiding in holes or burrows or lying under thick cover. He was also more alert and elusive though he would often betray his lair by leaving signs which the tracker was able to identify.

Trackers kept watch from hill tops and other vantage points observing the animal's movements before they settled down for the day. A lion had been hit by a bullet and he had disappeared. Later, a *chadika*, as these watchers were called, saw it roll over dead and the mystery was solved. It was remarkable how able and skilled these trackers were to recognise and decipher the spoor of individual animals and thereby deduce their intentions. These trackers were a great asset to me in stalking as they not only knew the quarry's habits but their keen vision could discern the prey before it had seen us and they could judge whether it was a ghuntavda sitting under the shade of a tree or a leopard crouched under a rock or bush. Sleeping lions and wild boar could be spotted in similar fashion. However, a lion that had made a kill was easier to locate if the grass was not too tall.

If a lion took shelter in dense cover, the trackers would suggest sitting over the bait, depending on the terrain and season. Even during the monsoon these trackers could arrange a beat. One such beat was organised when Lt Col A.H.E. Mosse, the retired British political agent, shot a lion. The big cat was driven out but hidden by the long grass he was only visible when he was about ten to fifteen yards away from the shikari. As he raised his head and turned, Mosse lifted his Paradox gun and shot him in the neck, killing him instantly.

Lion trackers would never follow wounded animals. However, the shikaris were excellent marksmen and this was seldom necessary. When encountering a wounded lion in dense undergrowth, the Mytiala trackers would ask for buffaloes to help them locate the animals. My first Asiatic lion, when shot through the heart, escaped into a deep ravine where later we heard two loud, moaning death roars. Despite the dark night and heavy downpour the trackers followed its trail to retrieve it. They were also not deterred by the inherent dangers in tracking wounded leopards and had full confidence in their skill, taking every precaution. An easy familiarity with their forest environment included individual names for every hill or valley and from this they could identify the particular rock, tree or watercourse and thus judge the location of an animal. The trackers were trained not to give exaggerated or false accounts to the hunters and thereby raise expectations.

I recollect an instance when a close friend had come to Mytiala for a hunting expedition. News of a leopard's presence reached us late in the day, but without any evidence of a kill. The *jemadar*, nevertheless, suggested that a machan be set up instantly so that our guest could sit over a live bait. Our guest was sceptical about the chances of the animal appearing without a proper sighting. I reassured him saying that there was a ninety nine to seventy five per cent chance and that was exactly how it turned out. Such prophecies were not their custom and they dislike seeing the shikari leave the machan without firing a shot. If a leopard did not appear it was never the Mytiala trackers fault. Once some friends followed a leopard's tracks to a ravine where they waited for the animal with a live bait. I sat silently in the car on the main road some distance away. There was only the slightest chance of the leopard coming my way and it did. Leopards are whimsical and the trackers were more confident about the behaviour of lions.

The bungalow at Mytiala was situated on the spur of a hillside overlooking a valley through which the lions entered the forest. A cart track passed below and traversed the valley. Sometimes when I awoke at sunrise I could see lion pug marks near my bed out in the open. Lions are known to investigate human settlements in the dark hoping to steal the odd domestic animal.

Mytiala lion trackers measured the diameter of the front foot pad of a lion or leopard to judge its sex and size. A three and three fourth inch lion foot indicated an adult male and a two and quarter to two and a half inch foot pad indicated a large leopard. Three inches meant a very large animal. Females and lions smaller than the sizes mentioned were not shot at.

Camps were arranged once, twice or thrice a year for which the shikaries were amply rewarded and always received extra bonuses for a successful shoot. If a wild boar was killed the Kolis got most of the meat whereas if the animal was a roz, it was for the Gaddhais. However, if a big cat was reported in the vicinity no other game was shot even though there was abundant painted sandgrouse, partridge and rock bush quail.

Earlier shikars in certain areas were allowed in limited doses to check indiscriminate killing and the Mytiala lion reserve was one such place. My admiration for these shikaris and their knowledge of the wild does not mean that I support the practice of hunting. Often beats and baits were arranged to film the big cats. However, today there is a greater awareness and a deeper insight of wildlife behaviour and modern wardens can do more about preservation without actually destroying the animals and their habitats. Careful planning and encouragement is required to provide a strong incentive to study wildlife. Hunting is no longer relevant and is not the only way of learning about animals and the wild. This sport has been replaced by the camera. The tribals who live in close proximity to the wild realise the need to protect and preserve even though most of them kill to eat. But, all is not lost. National parks and sanctuaries have trained professional staff who track, observe and study wildlife ecology, and as a result people are learning more about the wild and our experience continues to grow.

Originally published in The India Magazine, *March, 1986.*

R. S. DHARMAKUMARSINH

Gulam Hussain Baazdaar: The Falconer

Kathiwar... a winter morning... a hazy horizon. Gulam Hussain is there with his hawks. Soon after sunrise the call of a Grey Partridge is heard from the scrubs. The Basha, an immature female Sparrow Hawk is ready. Bush beating along the dry river bed starts. Up flies the partridge a little higher. One follows the other. The hawk is making up before the quarry gets into the next bush. The hawk is now under it. And then, up she trusses, bringing it to the ground. The partridge is well held. The Basha is fed. The distance of the flight is about forty yards.

Many years in training hawks enabled Gulam Hussain to master the art of falconry from which experience he devised new methods of training birds of prey. A speciality was the style of swinging the lure which made his falcons stoop with greater zeal and accuracy. As an expert Austringer, one who trains short-winged hawks, he believed that if a hawk is in *yarak*, top form, the use of the *jungoli* or halsband worn on the hawk's neck, was not essential. The *jungoli* must be used when it is necessary to give a hawk a *batola*, a racing start by a forward swing of the arm. Also in the scrub the *jungoli* can get caught in the thorns and become a handicap.

Gulam Hussain, Baazdaar, the eldest son of the famous falconer Makekhan was born in 1901 in Bhavnagar, Kathiawar. He joined my father's service at the pay of just one rupee. Later he was called by Durbar Shri Kanthadwala of Bilkha who became his guru and taught him more about falconry. After his guru's death he returned to Bhavnagar and joined my brother, Krishnakumarsinhji's private department. Here he attained a high standard in falconry and was given independent charge of a falconry unit.

Amongst his outstanding achievements was the training of Peregrines, a species he liked best, at duck, Black Ibis, Grey Heron, Houbara, C. Kite, but most important the Eastern Common Crane, a bird almost the size of the Sarus Crane. Gulam Hussain knew how to train falcons to *wait on*, whereby a falcon circles above the falconer waiting for the quarry to be flushed, and qualified himself for duck hawking. He was also a maestro in training falcons for *ringed flights*, aerial dog flights, and considered the flight of the Saker at the Kite or Shorteared Owl the best. He also trained Sakers on hare. I encouraged him to train eagles which he did successfully. There was no bird of prey he could not manage and he was always prepared to break tradition to learn the intricacies of training new species. His proficiency reached the highest standards attained by falconers of the Mughal period and of Emperor Frederick II of Hohenstaufen, Italy. I considered him to be of the highest international standards. On hearing of his and his father's talents, two expert American falconers and conservationists, the Craighead twins, visited India in 1940. They were sent by the National Geographic Society to learn the art and practice of Indian falconry.

Gulam Hussain was a hereditary professional falconer, a shrewd judge of hawks and responsible for selecting the best birds from the famous Amritsar market in Punjab and trapping those that migrated to the plains of India. He knew how to get the most out of his birds and had no sentiments or love for them. He believed in following a rigorous course of training by using ancient drugs to condition them to fly better. His slogan was 'see my hawks perform and judge'. With his long experience he could without weighing his birds judge their form by the mere feel of his fingers and their weight on his hand. Gulam Hussain's forte was in training Peregrines and Shaheens, as well as Short-winged Hawks.

After Independence, Gulam Hussain specialised in Bashas, Sparrow Hawks, favourites of Emperor Akbar. He trained them on partridge and adopted *manning*, taming them by the *gaddi* method using a cloth brail instead of leather and by keeping the birds awake at night. However, in the late sixties falconry in India was on its decline and Gulam Hussain had little to do.

In falconry there is success and failure; the chance for prey to escape is as good as any. No bird goes wounded, though frequently falcons and hawks are lost in flight. A falconer's labour lost after many sleepless days and nights of training. King James of England once said that falconry would be the grandest sport in the world were it not for its disappointments. However, shedding blood for the sake of sport is not out of date. People like to see wildlife alive in nature. Yet, the history of sport is a most absorbing subject in a reading room.

Gulam Hussain stands aloof with the hawk on which is looped a *jungoli*. With difficulty the partridge is driven out and as it breaks cover, Gulam Hussain delivers a *batola* to his hawk, by catapulting it from his arm, giving it a flying start of about twenty feet which almost reaches the partridge and with a few quick flaps has caught hold of it. This is the advantage of a *batola*, an art which has to be mastered. In the next flight, a covey of Grey Partridges have been marked down in a bush and we surround it so that when flushed they rise high affording a better flight. The Jurra or adult male Goshawk, has now no *jungoli* on him as a starting handicap is not necessary and the hawk may choose his bird from a distance of about thirty yards. As we face the wind, an assistant flushes the birds which expode in all directions; the hawk selects the partridge closest to him which rises high and off goes the hawk flying low and now it is a fairly long chase in which the Jurra gets below the partridge, timing it like a cricket fielder taking a high running catch, and swooping upwards holds the bird as it descends before it can reach cover. 'Well caught Sir!' After this another Jurra is ready to be flown, a Choose Jurra, male Goshawk in immature plumage. More walking and beating of a farm hedge in which partridges have taken cover. Beating from inside the farm with Gulam Hussain with Jurra walking on the outside of the hedge, we follow slightly behind him and at his side. A partridge is flushed and off goes the hawk, and the partridge hard pressed, swerves into the hedge further up and the hawk crashes into it but in vain. This partridge was a clever bird, he passed through the hedge and out the other side to escape. Then another flight by the same Jurra but with a *jungoli* on. With a whirr of wings a Grey Partridge breaks cover. Gulam Hussain swings round for better positioning and gives the hawk a *batola*, and no sooner said than done, the hawk has taken the quarry within six feet or so. An accurate throw-in by Gulam Hussain indeed.

We now drive out in search of Black Ibis. We have an immature passage Peregrine falcon, Choose Bheri, which was snared on our seaside. Soon we find a flock of five Black Ibises feeding in an open field. The open car stops. Gulam Hussain stands up from the rear seat and the birds are flushed. The Peregrine is unhooded and after a few nods of her head she takes off. Seeing the falcon the Ibises are alarmed and rise almost vertically, simultaneously beating their wings rapidly and rising in the air almost in leaps. The Bheri closes in but the birds are well above her and she must circle up with rapid wing-beats. Higher and higher they all go spiralling into the blue sky, the Bheri chest up against the wind and gaining height towards the Ibises, she accelerates her speed as she aims towards them. As she closes in, an Ibis emitting a characteristic scream peels off from the group at a downward angle followed closely by the Bheri. The falcon dives on her quarry but misses and then swings up high above it, by that time the Ibis drops lower hoping to reach safety in a lucerne crop below, but before that happens, the Bheri has dived again and with an accurate swoop has caught the big bird and down they come tumbling to earth. The car races ahead to reach the spot and an assistant jumps off to hold the Ibis down for the *coup de grace*.

We are up early morning in search of small flocks of Common Crane. The Peregrine is trained on the Eastern Common Crane, a bird much larger than the Demoiselle Crane and slightly smaller than the endemic Sarus Crane. The Common Crane is known as the Kunj and is a winter migrant like the Demoiselle. Although found in large flocks it has a tendency to divide into small groups while foraging and one may then find it in fours, trios and even pairs scattered in fields of picked groundnut. Driving through the countryside we see many crane but not in small numbers. At last we find a pair in an open field. It is getting late and the feeding time of cranes is nearly over. The car stops about two hundred yards away to assess the situation. This is the chance to release the Bheri. Leaving the main road we move closer and we notice three Sarus Cranes standing a little on the side and further down the pair of Kunj. We close on to the Kunj to about a hundred yards when they suddenly take wing and Gulam Hussain not quite ready when I stop the car, hesitates, but not to lose precious time removes the hood and casts the Bheri off. As the falcon leaves the glove, the three Sarus Cranes also take wing which is unusual for such confiding birds. We follow the chase. Within a few seconds to our surprise we realise that the Bheri has set her eyes upon the full grown juvenile Sarus, which is of a ginger-grey colour and not unlike a juvenile Kunj, though larger. It is something we had never expected nor desired but it all happened in a few seconds. In a few minutes the falcon reached the young Sarus and has tackled it from above forcing it down to the ground. Our job is to reach both before the parents return and attack the falcon. As we charge forward I see a pair of Saruses turn to circle back but in the meantime I have stopped the car within fifty yards of the two birds jostling on the ground. An agile assistant leaps off the car and runs to hold the Sarus down. The Sarus is larger than the Kunj but it does not have the courage to defend itself. The whole thing was so sudden and bewildering not only for us but even for the Saruses which never would expect a falcon to attack them. Not even an Eagle would attack a Sarus Crane. Furthermore, it is a protected bird under moral sentiments and nobody kills a Sarus. In this case it happened accidentally. Perhaps the bird could have been saved but in all the excitement no one thought of it. We called it a day. Gulam Husain was overwhelmed at his falcon's success. How is one to blame him or the falcon for what occurred? It was unique. For some days we poignantly brooded over this amazing but rather sad event.

Originally published in The India Magazine, *March, 1983.*

K. M. KIRKPATRICK

Aboriginal Methods Employed in Killing and Capturing Game

Tribal techniques of kill and capture were often highly sophisticated, combining a deep knowledge of local terrain with an array of weapons and traps. Besides birds, reptiles and small mammals, large predators could also be killed by such methods, the former for food and the latter mostly for self-defence.

Before the introduction of the fire-arm, the aboriginals were hunting in the forests of India. Their whole life being devoted to the filling of the family stomachs, it is not surprising that they invented methods of killing and catching game to supplement the efforts made with their bows and arrows, axes, and—sometimes—spears. The aboriginal kills for two reasons. The primary reason is food, for as the Ho say 'merim he merim' (meat is meat), and they are not particular as to whether this meat is furred, feathered or scaled as long as it is edible. The other reason for killing is when animals prove themselves to be a menace to the security of the aboriginal. No aboriginal kills for the sake of killing, and they can see no object in destroying an animal for the sake of sport or for its hide, head or horns. They kill, primarily, for food and food alone, much as any other carnivore, and were, before the influx of outside hunters, an important feature in the scale of the balance of nature.

The aboriginals with whom I am familiar, having lived amongst them in their villages, are the Ho, the Munda, the Urang and the Bhuiya of south Bihar and the States of Gangpur and Bonai, the Kutia Khond and the semi-Hindu Oriya of Kalahandi and Jeypore Samasthanam, the Muria of Bastar and the wandering almost pigmy Bir-ho of south Bihar and northern Orissa. Their traps are described here as well as their methods of securing game. In order to attain some sequence, I am listing the animals and birds concerned, grouping several species where the trap is common, in order of classification.

MAMMALS.

The Common Langur (*Semnopithecus entellus*)

Amongst the forested hills of south Bihar and northern Orissa wander the small family parties of the Bir-ho ('Bir' meaning forest and 'Ho' meaning man), a jungle people of no fixed abode to whom the jungle is home and in which they live and die, erecting little shelters of branches or of grass as temporary shelters when the weather is wet or cold. They live on the forest life around them and their major source of sustenance is the Langur or Entellus monkey. In the tall tree forest, the langur very seldom comes to the ground but lives the major part of its life amongst the high branches. Up there in that leafy world, it has its special sleeping sites, its chosen food sites, and to and fro between these it has its well-marked travel routes, running along the same branch and jumping from the same springy take-off point to leap across and land on the same landing point. The Bir-ho is well acquainted with the fact that these monkey-gangs tread the same familiar path to and from their sleeping places and, on moving into a section of forest, they immediately scout out the Langurs' routes, the times at which they are used, and the frequency with which the gangs travel over them. Having gleaned this information, the route is reconnoitred and a spot, usually where the monkeys leap across some gap and are bunched together on some branch, is chosen.

One of the hunters climbs up the tree at a time when the monkeys are furthest away from the scene and very carefully cuts through the underside of this selected branch until it is literally hanging by a thread and will be disturbed by the slightest weight. Nets woven of the bauhinia creeper are slung loosely bag-like beneath the branch and the aboriginals settle down to await the monkey-gang's arrival.

As soon as the monkeys are near, one or two small boys create an uproar behind them, causing the monkeys to run for safety and this they do, leaping across the gap bunched closer together than normal on to the weakened branch, which gives beneath their weight so that some fall into the bag-nets, others lose their balance or jump wildly elsewhere on seeing the trap their leaders have fallen into, whilst the little Bir-ho shoot at them with their arrows; their shooting is fairly accurate and I have often seen an entellus shot in mid-air. The netted monkeys are not killed until later and it is usual to kill all the adults caught but to keep the young as pets—(for consumption at some later date when food is hard to find?)

The Tiger (*Panthera tigris*)

The tiger constitutes a grave threat to the aboriginal, especially in those areas where encroaching civilization has caused a depletion in the deer population of the forests. In such areas the tiger falls back on the killing of cattle, and sometimes human beings, and it is usually the forest dweller who suffers. Thus, during the breeding season, in Saranda, I have known a pair of tigers kill three cattle of Baliba village on one night, two cattle of Ponga village the next, and finally five cattle near Chota Nagra village on the third night—all these villages are roughly eight to ten miles from each other in a triangle. In such cases and where a tiger is making a nuisance of itself by preying on village cattle or on human beings, the villagers usually call for the services of a professional tiger-killer or 'Bagh-mari'. This gentleman contracts to dispose of the tiger for a consideration in cash or kind. His equipment consists of three large bows, capable of propelling arrows some five to six feet in length carrying barbed heads of iron some nine to ten inches long. These bows are set in a triangle around the kill, the lane of fire of each bow being cleared, and each bow is then set for firing with a trip-string. The bows are so aligned that the arrow, which travels with considerable velocity on release, will fly about twenty inches above the level of the ground. Some ten paces back from these trip-strings, the tiger-killer strings a 'Dharamsuta' (a life-saving thread) at chest height from the ground. This latter string serves to warn any human intruder that the trap is set. The tiger, of course, passes below this string on its way to the kill and usually fouls one of the trip-strings, whereupon the arrow is released and the tiger mortally wounded or killed outright. There have been instances where man-eaters, who have become gun-shy and who will not return to the kill after making the first meal from it, have been secured by the tiger-killer using himself as bait by sitting in the centre of his deadly triangle. Thus, in 1949, the Patharbasa man-eater in Saranda R. F. was disposed of, the fee being Rs 300.

Another trap, not known to the Ho, Munda, Urang (or Oraon), Kutia Khond or Bhuiya, is the 'Suri-phanda' used in Central India. This consists of tying the kill to a barked and slippery pole, which is lashed horizontally between two stout trees at some twelve feet above the ground. The ground below the kill slung on this pole is planted with upright spears, concealed as best possible in the grass or bushes. The tiger, on approach, sees the kill dangling out of reach and tries to secure it by climbing one of the support trees and then out along the pole, which, being both slippery and too narrow in diameter to afford a grip for the great paws, causes the tiger to miss its step, slip and fall on to the waiting spears below, its heavy weight usually inflicting a mortal wound.

Other methods, which I have not observed personally, but about which I have been told by the people of Erpund pargana, Bastar, are as follows:—

(*a*) A path frequently used by the tiger is carpeted with large Bauhinia leaves or Asan leaves, liberally smeared with bird-lime, made from pipal tree latex and mustard-oil. The tiger, walking on the path, puts his foot on a leaf, which promptly adheres to it; he struggles to shake it off and usually collects a few more leaves on his other feet and legs. He then tries to lick the leaves off and these adhere to his head until, finally, he succumbs to a terrible rage and rolls, roaring loudly, on the leaves. The waiting hunter, secure in some retreat, hears the roars, walks up and shoots the tiger with an arrow from a safe distance.

(*b*) Large, loosely hung nets constructed of Bauhinia creeper are slung on weakly planted bamboo poles across several paths used by the tiger around his favourite lying-up haunt. Two hunters are stationed near each net whilst beaters enter the area and make a tremendous uproar by beating drums, whacking bushes and shouting very loudly. The upshot of this is that the tiger decides to leave the area in a hurry and runs into one of the nets, the weak bamboos collapse and the net enmeshes the struggling tiger, which is then disposed of by the sentries. I have seen these nets, which are also used in securing deer, but I have never witnessed a hunt in action.

The Leopard, or Panther (*Panthera pardus*)

The panther probably constitutes the greatest menace or nuisance to the aboriginal. Virtually every village in or near the forest is haunted by one or more of these beasts, which prey on cattle, goats and dogs and sometimes become man-eaters. Their great cunning combined with their familiarity with man makes them all the more dangerous, and the aboriginal is only too glad to dispose of them but usually does not do so until they have become an extreme nuisance. The south Bihar-northern Orissa tribes usually employ a bow-trap which is fairly satisfactory. A bow is tied to two small 'Y' stakes facing the narrowly opened gate in a fence, through which the panther has been observed to move. Another 'Y' stake is planted in front of the bow and this supports a piece of bamboo, one end of which is rounded and holds the bow-string back, a small notch in this bamboo holds against the supporting 'Y' stake's fork when the pressure of the bow-string forces it forward. From the other end of the bamboo, a short string holds up one end of a small plank, which is placed directly in the open gate-way. The whole trap is set with great care and the bamboo is virtually a hair-trigger for the slightest pressure on the plank causes the bamboo to jump up and release the bow-string, propelling the arrow forward with great velocity. The bow is so set that the arrow flies diagonally upwards. The marauding panther usually tries to sneak through the gate, treads on the plank and releases the arrow, which strikes it in the chest or head inflicting a terrible wound or killing the animal completely. This trap is very common amongst the Urang.

The Muria of Bastar and some of the villages in Orissa use a regular trap with a trap-door. The trap is a two-compartment affair, built of stout stakes hammered well into the ground and roofed with heavy timbers. A partition of thinner stakes divides the two compartments. A heavy trap-door is poised above the entrance to the trap, held in place by a rope tied to a narrow stake against the partition. A pig is placed in the smaller of the two compartments. The panther, in an effort to secure the pig, enters the trap and its scratchings against the partition disturbs the key-stake holding the rope and the trap-door is released to trap the panther. It should be pointed out that the trap is so narrow as to make it impossible for the panther to turn around and attack the bulk of timber that serves as a door. The panther is then disposed of with an arrow through the gaps in the stakes.

Excerpted from 'Aboriginal Methods Employed in Killing and Capturing Game', Journal of the Bombay Natural History Society, 52, 1954.

S. R. DAVER
A Novel Method of Destroying Man-eaters and Cattle-lifters

In the early years after independence, there was still a continuing war against large predators, though attention was increasingly focused on those individual panthers or tigers that harmed people or livestock. Bounties were still paid for crocodiles in north India till 1970, and for tigers in parts of West Bengal till 1972. S. R. Daver shared the antipathy felt by many foresters to the carnivores, but he hoped to use the services of the tribe of Baigas in central India to destroy tigers. The Baigas were expert trappers and hunters who were now seen as a cost-effective means to kill off carnivorous animals. For over a century, the Baigas had been under intense pressure from foresters to curb shifting cultivation and hunting. Daver was a rare admirer of their trapping skills.

Foreword

In introducing this article I need only explain that Shri S. R. Daver, Deputy Conservator of Forests started his career in the Central Provinces as a Range Officer in 1916 and has been serving in these forests for 33 years, of which nine were spent as Chief Forest Officer, Bastar State and seven as D. F. O. Bilaspur. The article itself is, however, sufficient evidence of its genuineness. Whether tigers of other parts would show sufficient boldness to fall a victim to 'Soori Phanda' is a matter for debate but in many parts, where one has had experience of the extreme cautiousness of tigers and their disinclination to return to kills which have been disturbed, it would appear doubtful. Perhaps it depends on the relative ease with which a tiger or panther can obtain a second kill.

I commend the article as a slice of real life which will,.I am sure, provide plenty of discussion.

C. E. HEWETSON
I.F.S.

INTRODUCTION

During recent years man-eaters and cattle-lifters have increased in the Central Provinces and Berar at such an alarming rate that very few districts are free from these pests. The district in which the writer is serving is not only overrun with cattle-lifters, but a man-eating tigress played havoc for more than a year before she was brought to book. This fact led the writer to investigate the methods adopted by Baigas to destroy such carnivora. There is, no tribe or people in the world who could master the art of trapping animals better or outdo a Baiga in it. Someone has said that Nature never yields her secret lightly. A Baiga is Nature's child, and as such so secretive that one can get very little information out of him. But when the questions and information sought refer to trapping animals, one is treading on delicate ground, particularly when the questioner happens to be a Forest or Police Officer. In spite of all these difficulties the writer succeeded in winning the Baigas' confidence and they demonstrated their method of destroying cattle-lifters, when an actual 'kill' occurred during his tour in May 1949 in Mandla District. Unfortunately in this instance the tiger was not destroyed.

The writer does not claim originality in the device adopted for killing tigers and panthers. He has merely attempted to describe how tigers and panthers are destroyed by aboriginals. If by chance any reader succeeds in destroying cattle-lifters or man-eaters by this method he should thank and remember our simple friends—the Baigas.

MAN-EATERS AND CATTLE-LIFTERS

(1) Abnormal increase of Cattle-lifters and Man-eaters in Central Provinces and Berar:—

'With the departure of shikar-minded people from India, the cattle-lifters have increased,' said one of the senior ministers of this Province. To prove how true this observation is, one has only to turn over the pages of *C. P. and Berar Gazette*. The writer has analysed the notices in *C.P. and Berar Gazette*, dated 6-5-1949 in Part IV, pages 210–14 as detailed below:—

(a)
- (i) Total number of separate rewards offered — Rs 40
- (ii) Total amount of rewards offered — Rs 3,580
- (iii) Highest amount of single rewards offered — Rs 400
- (iv) Minimum reward offered — Rs 5 (for wolf)
- (v) Number of rewards offered for killing man-eating tigers — 17
- (vi) Number of rewards offered for killing man-eating panthers — 1
- (vii) Number of rewards offered for killing cattle-lifters — 21
- (viii) For wolf — 1
- Total — 40

(b) If the area infested by proscribed cattle-lifters and man-eaters is analysed district by district, we get the following result:—

(1) Bhandara, (2) Balaghat, (3) Sarguja, (4) Wardha, (5) Chanda, (6) Jubbulpore, (7) Hoshangabad, (8) Mandla, (9) Bastar, (10) Chhindwara, (11) Nimar, (12) Bilaspur, (13) Buldhana, (14) Raipur and (15) Amraoti.

Since the notices are published once every month, the list of districts infested is not exhaustive. In the writer's opinion, there could be no district free from cattle-lifters in the C. P. and Berar.

(2) Once a cattle-lifter, never a game-killer:—

When tigers take to cattle-lifting, it becomes increasingly difficult for them to catch and kill agile wild game. Firstly, the tiger finds it very easy to strike down a domestic animal. Secondly by eating bovine cattle, it develops into a heavy and weighty animal unable to chase and obtain wild game in the battle-field of jungle life.

(3) Heavy economic loss to cattle owners:—

(i) Man-eaters are mostly disabled animals. Either they are wounded by shikaris or by porcupine quills, or their canine fangs have decayed due to old age. Whenever possible they kill bovine cattle and sometimes human beings also. Hence a cattle-lifter in pursuit of its food requirements is also quite frequently a man-eater.

(ii) All sportsmen agree that a tiger or a panther must kill an animal at least once a week for food. Even on the very conservative estimate of only 3 kills per month, the number of cattle required by a single cattle-lifter works out to $3 \times 12 = 36$ a year.

Therefore, the 39 man-eaters and cattle-lifters notified in *C. P. Gazette* of 6–5–1949, if not destroyed for a period of 12 months, would require to kill $39 \times 12 = 468$ bovine animals. If the present value of a buffalo or a cow is taken only as Rs 100 then the economic loss entailed would be $468 \times 100 =$ Rs 46,800.

(4) Reasons for increase of carnivora:—

The increase of carnivora in the C. P. and Berar is due to the following reasons:—

(i) *Departure of shikar-minded people from India:—*
Every year civilians and army officers used to visit forest blocks and thin out tigers and panthers in almost every district of the C. P. But now-a-days civil officer's tours in forest lands have diminished considerably owing to pressure of heavy office work. On the other hand, few young officers of the Army, Navy and Air Force are keen to spend money and time on shikar. The writer has suggested that a copy of notices regarding offer of rewards, appearing in the *C. P. Gazette* should be supplied by the C. P. Government to Army, Navy and Air Force Headquarters and young officers should be made shikar-minded not for the sake of the rewards but to get training in jungle warfare. It is said that 'the jungle is the battle field of our play hours'.

(ii) *Withdrawal of ordinary rewards:—*
A few years back, rewards used to be paid in the C. P. for destroying carnivora as follows:

Tiger	Rs 15
Wild Dog	Rs 15
Panther	Rs 10

These rewards were no inducement to the class of sportsmen who are used to reserve shooting blocks in Government forests. Only local village shikaris used to sit up over 'kills' and shoot tigers or panthers, and draw rewards from the nearest treasury on production of a skin. Now-a-days, firearms as well as ammunition are very costly; and on top of this, Government, instead of increasing the rate of rewards, have withdrawn the ordinary rewards. No local shikari would care to waste his precious shot and gun powder in shooting a tiger or panther. The writer estimates that local shikaris used to destroy or thin out at least 50 tigers and panthers a year in the C. P. and draw the prescribed rewards from treasuries. In the *C. P. Gazette* of 6–5–1949, Government have announced an enhanced reward for destroying carnivora so that the total amount for these 40 cases now comes to Rs 3,580. If all these carnivora had been destroyed by local shikaris on payment of the ordinary reward, it would have cost Government $40 \times 15 = $ Rs 600 only. Is it not more economical for Government to keep down the number of cattle-lifters and man-eaters by reintroducing ordinary rewards?

(iii) Increase of Wild Dogs:—

Owing to withdrawal of rewards, wild dogs have increased by leaps and bounds. In one shooting block, a permit holder witnessed a pack of twenty in one day in the Bilaspur Division, in May 1949. Wild dogs destroy and wipe out game in forest, consequently tigers or panthers are forced to turn their attention to men or cattle for their food. The sooner the Government restore or even enhance the ordinary rewards for destruction of wild dogs, the better for the wild game and lesser chances of tiger and panther becoming man-eaters and cattle-lifters.

(iv) Increase of gun licences and decrease of herbivore:—

In many tracts, herbivora have been over shot due to the liberal grant of gun licences. In such tracts tigers and panthers have of necessity to depend on bovine cattle or even take to killing human beings. Gun licences should be restricted and strictly controlled, and Pardhis should not be allowed to trap game animals.

SUMMARY OF RECOMMENDATIONS

(a) Restoration of ordinary rewards for the destruction of tigers, wild dogs and panthers.
(b) Encouraging Army, Navy and Air Force officers to shoot in Government and private forests.
(c) Regulating gun licences and Pardhi's trapping licences.
(d) In the wildest inaccessible areas, where sportsmen cannot readily reach, or in areas where man-eaters or cattle-lifters are a menace, the 'Soori Phanda' device, described in the following pages should be encouraged.

'Soori Phanda'

The Baigas of Mandla District destroy cattle-lifters by a device known as 'Soori Phanda'. Being past masters in the art of trapping wild animals, they have used this device to destroy cattle-lifting panthers or tigers for generations.

In feudal ages, the game laws were so strict, that it was a grave offence for an ordinary man to shoot or kill a game animal. In India the tiger is considered the king of the forests, and only a few privileged persons like rajas, nawabs and high officials could shoot it. The shooting of tiger by a common person was viewed with great disfavour. Against such a background no wonder the Baigas consider it unwise and unsafe to advertise their 'Soori Phanda' to the world. Unofficially the writer came to know of the killing of one cattle-lifting tiger and two panthers within 3 years in one Range by means of the 'Soori Phanda' method.

'Soori' in Baiga language means spear-head. 'Phanda' means strap. The spear-head together with a wooden shaft is also called a 'Soori' but a detached shaft separated from spear-head is called 'ganj' or 'gauj'.

Equipment and Conditions Necessary for 'Soori Phanda'

The whole object of this device is to make the tiger or panther fall on a number of spears planted in the ground with spear-heads pointing skywards. To be successful, the following equipment and conditions are necessary:—

(1) There must be a 'kill' by a tiger or panther. The 'kill' may be a bovine carcass or a human body.

(2) There must be two young trees, forking at a height of 13 feet to 15 feet from ground level and standing 10 feet to 12 feet apart.

(3) A barked pole made from *kydia calycina* known as bargha or Baranga in Hindi.

(4) 12 to 16 iron spear-heads with an equal number of bamboo shafts.

(5) Rope made from the Mahul climber (*Bauhinia vahlii*).

(6) A pole for levering up and lifting a heavy carcass.

(7) Green branches and twigs for camouflage.

(8) A 'Khanita' or iron instrument (drill) for digging holes in the ground.

(9) A number of villagers to help in lifting the carcass and for arranging the 'Soori Phanda'.

I will take up the above nine items separately and explain in detail the precautions necessary.

1. The 'Kill'

If the 'kill' happens to be a small animal, 'Soori Phanda' is organised with ease. For this reason the number of panthers destroyed is much larger than that of tigers. Tiger 'kills' are usually large bovine animals, sometimes even full grown buffaloes. To sling and lift such heavy 'kills' 15 feet above the ground level by means of props requires about 30 to 40 men.

In Mandla District and elsewhere a cattle-lifter frequently kills more than one animal at a time. At Ramnagar, a village 4 miles from Karanjia, a cattle-lifter attacked 5 cattle and killed 3 outright. Of these, the tiger finished off one young cow during the night, and dragged and collected the remaining two in one spot—this was in May 1949. Hence before arranging 'Soori Phanda', it is essential to search thoroughly the jungle in which a 'kill' is discovered. If there are more than one 'kill', it is useless to expect a tiger to visit 'Soori Phanda'. To be successful there must be only a single 'kill', or if more than one the others must be removed or destroyed.

2. Selection of Two Standing Trees

To support a horizontal pole of 'Bargha' two forke trees should be selected. They must be 10 to 13 feet apart. The forking points should be 14 to 15 feet from ground level. Such ideal conditions do not often exist close to the 'kill'. If the 'kill' is heavy, it is difficult to move it a long distance, and moreover such removal may make the tiger suspicious. To get over these difficulties in the jungle, the Baigas sometimes select one or both the trees without a fork. In such cases, they cut one or two forked poles without disturbing the jungle, and these poles are strapped upright to the selected trees and the transverse pole of the Bargha rests on these props.

Size of the trees.—If the trees selected are too thick and large in girth, the tiger or panther finds it difficult to climb them. On the other hand, if the trees are too thin and small in girth, it is equally difficult for the tiger to climb. In making a selection the Baigas prefer trees of 2 feet to 3 feet in girth.

Choice of species.—If there be a choice, they always select Bija (*Pterocarpus marsupium*) trees. Tigers for no apparent reason always delight in climbing Bija trees. This is common knowledge among Baigas and forest officials in Mandla forests. One can see numerous Bija trees with the tell-tale marks of tiger climbing them. If a Bija tree is scratched with a sharp instrument, an astringent blood-coloured gum exudes. A tiger's claws are full of septic particles of the decomposed flesh of its victim, and sometimes the claws may have wounds. It is possible that the oozing astringent Bija gum may have a healing effect, or it may act as disinfectant. Research on the antiseptic properties of Bija gum may prove beneficial in medical treatment.

Other suitable tree species are Karra (*Cleistanthus collinus*), Amti (*Bauhinia racemosa*) and Sal (*Shorea robusta*). Baigas believe that tigers do not climb Dhaura (*Anogeissus latifolia*), Salai (*Boswellia serrata*) and Kulu (*Sterculia urens*).

As a general guide any tree of suitable size (2 feet to 3 feet girth) with rough bark may be selected. Two such standing trees are required.

3. A Barked Pole of Bargha or Baranga (Kydia calycina)

In the forks of the two vertical trees a horizontal pole of Baranga is fixed and secured with rope made of Mahul (*Bauhinia vahlii*) bark. The whole success of 'Soori Phanda' rests upon selecting only Bargha or Baranga pole 15 inches to 18 inches in girth. Before fixing this pole between the two uprights, it is barked and its cambium layer (living bark) is also removed. The smooth texture of Baranga wood and its sap make it so slippery and soapy that a tiger or panther cannot walk along the pole and reach its middle without falling off. This greasy property of the Bargha pole can be maintained and refreshed for a number of days, by sprinkling water on it in case the tiger does not visit the 'Soori Phanda' on the first or second day. No other species of tree is used for this purpose. The 'kill' is slung from the middle of this horizontal pole with 'Bakkal' (rope made from mahul bark) in such a way that the carcass dangles and swings freely. A twig of green leaves is slipped into the winding portion of the 'bakkal' for camouflage. It may be noted here that tigers or panthers do not reach the middle of the transverse pole, nor do they normally fall off on to the upright spears from this point. This slippery pole is merely a ruse to baffle a tiger or a panther, to rouse its anger and exhaust its patience. Only one out of a hundred tigers or panthers could possibly succeed in reaching the middle of the pole, and even then it would certainly fall on the spear-heads below and meet its death.

4. *Spear-Head and Shaft*

Spear-heads are forged by Agarias from iron ore smelted locally in Bilaspur and Mandla Districts. A Baiga pays 8 annas to 12 annas per spear-head according to its size.

Size of the spear-head.—The total length of a spear-head varies from 8 in. to 10 in. from tip to tip, of which the blade portion is 5 in. to 6 in. long and the hind-end about 3 in. or 4 in. The widest portion of the blade varies from 1½ in. to 2 in. The spear-head is thick in the middle portion of the blade tapering down to a pointed tip and to sharp edges. The thickest portion of the spear-head is situated at the junction of the tail-piece with the widest part of the blade so that the spear-head may not break or buckle up by the weight of a falling object. Baigas call a spear 'Sang' or 'Barchhi'.

Shape of the blade. The two edges of the blade of the spear-head are not symmetrical. One edge is prominently convex and the opposite one more or less straight in general outline. The Baigas believe that when an impaled animal struggles on a planted spear, the straight edge functions as a lever and the bulging edge cuts deeper into its flesh, and the spear-head thus penetrates deeper and deeper into the victim's body. But the writer has also seen many spear-heads with symmetrical blades being used in 'Soori Phanda'.

Margin of the blade.—The margin of the blade should not be 'entire' (to use a botanical term) but according to Baigas, the general outline of the edge should be slightly and irregularly crenulated or serrated. They consider that such a jagged margin hastens cutting action when the impaled animal struggles on the spear-head.

Spear-head in the shape of arrow-head more effective.—A spear-head made in the shape of an arrow-head with 2 barbs, and known as 'Bissar' by the Baigas, is more effective. A tiger killed by 'Soori Phanda' at Bhilki village, in Pandaria Zamindari, in December 1947 carried away three spears of which one was an arrow-headed spear. It extracted one spear and hurled it about 30 feet away from 'Soori Phanda'. The second was found lying just near the trap, but the third arrow-headed-spear remained in the tiger's body. The animal was found dead at a distance of 2½ furlongs from the 'Soori Phanda' on the following day. The Baigas had to cut open the body to retrieve their precious weapon. A tiger or panther cannot cast out an arrow-headed spear from its body nor can a Baiga afford to lose his arrow-headed spear. The writer suggests that if 'Soori Phanda' is conducted under official supervision, such barbed spear-heads should be used.

Shaft.—Shaft or handle of the spear is made preferably of well-seasoned bamboo 3 to 5 feet long. The thinner end of the bamboo is split longitudinally into four parts and only upto the first internode. This end receives the tail-piece of the spear-head. After inserting the tail-piece, the split end of the internode is tightly wound with 'bakkal' (rope made from 'mahul' climber).

It is essential that the widest part of the spear-head blade should invariably be larger than the diameter of the shaft, i.e. if the diameter of the shaft or bamboo is 1½ in., then the widest part of the spear-head should not be less than 2 in., so that the slit made in flesh of the tiger or panther may permit the shaft to penetrate deeply with ease. The split end of the shaft is tapered at its tip to help in this.

Shaft of unequal length.—The Baigas make two sets of 'Soori' or spears of different lengths. They erect one long and one short spear alternately. The longer set may project 3 to 3½ feet above ground level, the buried portion of the shaft being 1½ to 2 ft.. The shorter set may be 2½ to 3 feet above ground and 1½ to 2 feet deep below it. This arrangement is considered important because a falling tiger or panther received on fewer spear points has the probability of deeper penetration. The difference in two sets may be 9 in. to 12 in. It stands to reason that if a tiger or a panther drops on large number of spear points all in the same plane its weight will be distributed over a large bearing surface and consequently the penetration of spear-heads will be more superficial. The object is to impale the tiger on one or two spears only so that the penetration may be deep and prove fatal.

Bamboo spikes as substitute for iron spears.—In backward tracts where iron implements are valuable and scarce, the aborigines substitute them with some handy forest product easily obtainable. A typical example is the wooden bell for cattle which one can see in forest areas only.

At the time of a 'Soori Phanda' the villagers may not be able to collect the required number of 'Soori' (spears) normally 16. In such a case they select a dry and strong piece of bamboo, of which one end is pointed like the blade of a spear-head and the other firmly planted into the ground.

The writer suggests that a few of these bamboo spikes should always be used along with the iron spears as they harmonise so well with the surrounding objects and would not arouse the tiger's suspicion. Green bamboos should never be used for making these spikes as they are ineffective. When the man-eater of Talaidabra dropped from a height of 15 feet 2 inches on one of these green bamboo spikes, the point of the spike bent over and the man-eater got only a superficial wound in its neck, hardly skin deep. Where bamboos are not available, shafts for spear-heads and spikes are made from 'Khirsari' (*Nyctanthes arbortristis*) poles.

5. Rope Made from Mahul (Bauhinia vahlii)

Fibre from the mahul climber (*Bauhinia vahlii*) also called 'Mohlain' in Mandla district makes very strong ropes and this is freely used by Baigas in 'Soori Phanda' for the following purposes:—

(i) Fastening the two ends of the Bargha pole firmly in the forks of the two selected trees.

(ii) Strapping upright forked supports to selected trees, where natural forked trees of the right kind are not available, and securing the horizontal bar on to these props.

(iii) Tying the four legs of the carcass together for hanging it from the middle of horizontal Bargha or Baranga pole.

6. A Slinging Pole for Lifting Heavy 'Kill'

When the kill is a heavy carcass, it is lifted in the following manner:—

The three or four limbs of the carcass are drawn together and tied up with mahul rope. A tiger usually eats the hind leg of its 'kill' first, hence a carcass may have only three legs.

A pole of any species, strong enough to carry the weight of the animal is prepared and inserted between the limbs of the 'kill'. Forked poles of different lengths 8 feet, 10 feet, 12 feet and 15 feet are prepared in sets of two, and with the help of these the carcass is lifted in stages, first say 8 feet, then 10 feet and so on till the desired height of 14 to 15 feet from ground level is gained.

Then one or two men climb up to the transverse Bargha pole by means of a crude ladder propped against it and suspend the 'carcass' from about the middle of it. When this operation is completed, the props and the slinging poles are withdrawn.

A Novel Method of Destroying Man-eaters

Note of warning.—Until the above operation is completed and the men working on the Bargha pole have come down, never set the 'Soori Phanda' on the ground. Otherwise the men working on Bhargha pole may get killed, should they accidentally slip and fall.

7. Green Branches or Twigs for Camouflage

In order to conceal the rows of spears, the inter-spaces between the upright spear-heads or 'Soori' are planted with branches of green leaves. However, if there be tall grass of the proper height, (3 feet to 4 feet) or bushes in the place where 'Soori Phanda' is set, then the artificial camouflage is not necessary. But a green leaf is stuck on the tip of each spear-head so that this unfamiliar object may not be visible from above.

A branch or two of green leaves is also stuck in the middle of the Bargha pole where the 'kill' is tied with rope. This is done to hide the strong rope with which the 'kill' is attached to the pole.

The kill should swing freely but its position must be such that the head and tail ends always point towards the two supporting trees.

8. 'Khanita' (also called 'Sabar' by Baigas) or an Iron Crow-bar

To dig holes in the ground for planting about 16 spears upright, an iron crow-bar locally known as 'Khanita' or 'Sabar' is necessary. The holes are 1¼ to 2 feet deep according to the nature of the top-soil.

9. Number of Villagers Required

When the 'kill' is heavy, about 30 to 40 men are required to lift it by means of props to a height of 15 feet from ground level. It is not difficult to collect this number for 'Soori Phanda'. Villagers are generally willing to help in the destruction of a cattle-lifter or a man-eater.

Any portable mechanical equipment which can easily lift a heavy 'kill', upto 15 feet off the ground, would be a great improvement. Owing to the labour and trouble involved in the crude devices usually employed, 'Soori Phanda' is now practised rather sparingly.

How to Set 'Soori Phanda'

In the preceding section all parts of 'Soori Phanda' are described in detail. The operation of the 'Soori Phanda' is described here.

Having located the 'kill' of a cattle-lifter, the first step is to search for the possibility of other 'kills' in the same locality. When a tiger can get his food on the ground he is naturally not going to take the trouble to climb the tree and oblige the 'Soori Phanda' trappers. In the Baiga demonstration carried out for the benefit of the writer the cattle-lifter was given no inducement to climb the tree, as the Baigas had neglected to remove the two other kills in the same locality.

The second step is to find two ideal trees, forking at a height of about 15 feet from the ground and as close to the 'kill' as possible. If such are not available it is best to bring from a distance two props with forks. If the ground between the two selected trees supports tall grass or bushes in the right position, do not clear the ground. If not, then branches of green leaves for camouflage between the spears may be used. Stick a green leaf or two on the tip of every spear-head; camouflage the rope tying the kill with the Bargha pole. Take a pole of Bargha or Baranga (*Kydia calycina*) 15 to 18 inches in girth and slightly longer than the distance between the two selected trees. This should be barked and the cambium layer removed. Fix the pole horizontally in the fork of the two trees or props by means of 'bakkal' (rope made out of mahul fibre). It would be unfortunate if the special species is not found for the horizontal pole within a reasonable distance in the forest, but perhaps a pole of some other species, well coated with grease or fat may serve equally.

With the help of a slinging pole and a set of forked props, raise the 'kill' until it reaches the level of Bargha pole. One or two men are required to climb up to its level with the help of a crude ladder in order to suspend the 'kill' from its middle. The slinging pole and props supporting it should be withdrawn after this and the men asked to come down.

Just below the dangling 'kill' arrange to plant about 16 spears in the ground in a space about 4 feet x 4 feet. This spot is irregular in shape and the spears are neither equidistant from each other nor are they arranged in regular lines or rows. The Baigas look up to the dangling kill while fixing position for digging holes for erecting each spear. As mentioned previously, spear-shafts are of two different sizes, the shorter and longer ones being planted alternately in the ground in holes 15 in. to 24 in. deep just below the carcass. The spears are firmly fixed with their tips pointing skyward.

If the head and neck of the kill hangs too close to the ground, it must be drawn up with ropes or removed because a tiger is tempted to jump and grab at the nearest part of the 'kill' from the ground and in doing so he gets the shortest drop on the spears—merely wounding the animal. The higher the drop, the deeper the penetration of spears is what must be kept in view. The minimum drop should be about 8 feet.

Camouflaging is the last operation. After this the men return to their village. When a tiger visits the place and finds its 'kill' hanging on a pole, he surveys the neighbourhood with great caution. When fully satisfied of the coast being clear, he climbs one of the trees supporting the pole. He tries to approach the 'kill' by walking on the pole, and is baffled and irritated when he finds it slippery. Instinctively he clings to the tree with one of his forepaws, while with the other forepaw he either tries to grab the 'kill' or strike with force so as to bring it down to the ground. Whether the action is for grabbing or for striking down, it is difficult to say but in either case, the hanging 'kill' swings away from the tiger. In his attempts the tiger soon over-reaches himself, loses his balance and drops on the labyrinth of spears below like a sack of potatoes. Being heavy and powerful animals, tigers often carry away spears some distance from the 'Soori Phanda' but are usually found dead within a few hundred yards. Panthers succumb within 30 to 60 yards and are sometimes even pinned down where they drop.

Near Daldal Forest Village, a panther which apparently succeeded in reaching the middle of Bargha pole dropped plumb on his bottom. A spear entered his rectum and ploughed right through his body. He was impaled in this position and his rampant dead body was found on the following day. The Baigas were in roars of laughter while describing the fate of this panther!

There are no authenticated instances of man-eaters being destroyed by 'Soori Phanda'. The reason for this is not far to seek. When a person dies a violent death, the dead body is not interred unless a responsible police officer has examined the same and a police panchnama held. On rare occasions there are cases of actual murder and the murderer tries to pass it off as death due to attack of a wild animal. Therefore police officers have to take extra precautions and obtain clear evidence that the victim was actually so killed. These formalities react on the mind of aborigines, and when a person is killed by a man-eater, anxiety to preserve the body is uppermost in their mind. When the man-eater of Talaidabra forest village killed Jungi Bhootia, the villagers searched for his dead body and when it was recovered, they strapped it on a branch of a tree 11 feet above the ground. This was done not for baiting 'Soori Phanda' but to preserve the body until it was examined by a police party. But the man-eater climbed the tree, chewed the 'bakkal' ropes, brought the body down and left nothing of Jungi's remains, except a few bones for identification by a police official.

Man-eaters Easier to Trap than Cattle-lifters

Man-eaters are mostly wounded or disabled animals. They secure their unnatural food with great difficulty, and having secured their prey, they are desperate. Maiku Gond, a forest villager of Bindawal, Bilaspur Division, was killed by the Talaidabra man-eater at 9 a.m. on 10th April 1949. Villagers of Bindawal, with the help of other villagers recovered his body within a few hours, giving no time to the man-eater for sufficient feed. Maiku's body was strapped to two standing trees 15 feet above ground level. Not being expert in setting 'Soori Phanda', they fixed spears and spikes of green bamboos below the body. The man-eating tigress climbed the tree the same night, and due to sagging and swinging bamboo poles, she lost her balance and fell on a green bamboo spike, and escaped with a superficial wound in the neck. On 12th April Maiku's body was brought down at the suggestion of the Range Officer, Kota, as bait for the man-eater. The tigress visited the place at night, dragged the body about 300 yards and ate a portion of it. On 13th April a beat was organised but the man-eater did not turn up. However, the Range Officer sat up over the kill the same night and shot the animal at about 8 p.m. in the act of holding the prey in her jaws. This instance shows that the man-eater visited and revisited her human kill four times, and she did not leave the locality for 3½ days (84 hours).

There is a belief that a man-eater becomes habitually addicted to human flesh. The writer, therefore, strongly advocates the use of 'Soori Phanda' for man-eaters in the same manner as employed for cattle-lifters, and would suggest that police officers should not insist on examination of the corpse until the negative result of 'Soori Phanda' becomes known. The corpse of a human victim is easier to handle than a bovine carcass and ensures better success if used as bait for a man-eater.

Whether the tips of the spear-heads are coated with poison or not, is a moot point, and Baigas speak on this subject with mental reservation. The writer feels the application of poison is perfectly justified and legitimate when 'Soori Phanda' is set for a man-eater.

An Apology for 'An Ignoble Death for a Noble Animal'?

Tradition and the finer points of sportsmanship perhaps demand that such a noble animal as the tiger—the king of the forest—should not end his life by such an ignoble death.

As long as a tiger is a game-killer, he is a gentleman; he never interferes with man or his property. And so long as a tiger remains a gentleman he has our admiration and respect. But cattle-lifters and man-eaters adopt new ways of life, (perhaps compelled to do so by man himself) which make them a source of danger and menace to man and his live-stock. Such creatures are in no way better than 'Goondas' among human society, and surely nobody feels any compunction about how a Goonda ends his life.

The writer, therefore, feels no qualms of conscience for the method suggested for the destruction of cattle-lifters and man-eaters. To the tiger-shooting enthusiast, the writer owes no apology for sending the cattle-lifter and man-eater to the scaffold of 'Soori Phanda' to put an end to their nefarious career.

ACKNOWLEDGMENT

The writer of this article is indebted to Shri C. E. Hewetson, I. F. S., O. B. E., Officiating Conservator of Forests, Eastern Circle, Central Provinces for going through the draft and making useful and necessary modifications. In fact, his words of encouragement induced the writer to prepare this article—a task beyond the province of a practical forester.

Lt. Col. R. W. Burton, I.A. (Retd.) to whom this article was sent by the editors, comments as follows:

There is much to interest members in Mr Daver's article.

In various parts of the world primitive peoples have evolved a number of ingenious ways of destroying animals for one reason or another—for food, for clothing, for gain, and to be rid of pests. This 'greasy pole' method now described is somewhat elaborate and would, as suggested by Mr Hewetson in his foreword, almost certainly fail in case of most tigers, whether those 'educated' through experience or naturally wary. Only in quite exceptional circumstances would it be likely to succeed.

When beyond Myitkyina in the Upper Irrawaddy country in 1931 the writer came to know that the people of the 'Triangle', as the area between the Mali and 'Nmai' rivers is termed, destroy tigers by a 'spiked bamboo' method. When it is noticed that a tiger has the habit of using one of the cane and bamboo bridges he is given a 'hangman's drop' of some 40 feet more or less on to spiked bamboos fixed in the ravine bed beneath the place where a few slats have been suitably adjusted for the occasion. The fall is almost always fatal and a somewhat ignominious end to a splendid beast not particularly harmful to the human beings of those parts.

A device of strewing leaves covered with sticky substance was used in the days of the Emperor Akbar for dealing with tigers which it was desired to kill or capture. 'Sport in Indian Art' by Col. T. H. Hendley, C.I.E. in the *Journal of Indian Art and Industry*, (1915), and *J.B.N.H.S.* Vol. 25. pp. 491 and 753 may be seen in this regard. On p. 753 it is recorded that the device had survived up to the year 1890-91 in the Sambalpur forests of the Chattisgarh division of the Central Provinces. This might be a more effective and less troublesome method than the 'Soori Phanda' of the Baigas of the Maikal Hills; that is to say if it is considered necessary to increase the destruction of tigers. The birdlime is spread on broad leaves around the kill or along a path and the tiger struggles and rolls about in order to free it from his paws, claws and body. The leaves stick to the face in such number as to blind the animal which can be approached and killed or netted while in this helpless and worried state.

It may be of interest to remark that a birdlime method for killing sloth bears was in use a hundred years ago, and may be even at the present day, by the people at the foot of the hills in the Tinnevelly District of South India as related by General E. F. Burton in his book, 'An Indian Olio' (1890). Powerful Poligar dogs tackle the bear, and the hunters, armed with long spears cross-barred close to the two-edged blades run to the spot where two or three of the party having ten foot bamboos smeared with birdlime at the ends dash up on either side, poke the bear in the ribs and adroitly twist the ends in the long hair thus holding the animal fast on either flank for the spearmen to complete the kill by repeated spear thrusts.

Man-eaters and Cattle-lifters

'With the departure of shikar-minded people from India the cattle-lifters have increased.' It is true that an increase of cattle killing by tigers has taken place since August 1947 in far apart areas, but the cause of the increase may be not so much the departure of shikar-minded people as the very great increased lawless killing of deer by the people themselves. That, in all probability, is the main reason for increase of cattle-lifting by tigers and to some extent of man-eaters also. When the balance of nature between the larger carnivora and their natural prey is unduly disturbed, tigers are bound to increasingly turn upon the cattle, and learn also to kill the herdsmen who continually drive them off their kills.

Heavy Economic Loss to Cattle Owners

It may possibly be more pecuniarily economic to the Government to restore rewards for the destruction of tigers, but the long view loss to the Government and the people of India through the immense destruction of wild life which is proceeding almost unchecked in most parts of the country is incalculable. It is possible that through the giving of rewards and other methods of destroying tigers, were a campaign against them to be instituted, their population would be reduced to vanishing point. It has indeed taken place in many areas to the knowledge of the writer during his sixty years experience of this country. When the unchecked slaughter of deer has resulted in the disappearance of these animals, and the tigers also have gone, it will be seen whether the people of the forests and their vicinity are better off or worse off, for the pig and other crop raiders will remain. All interested in these matters are asked to read and study the several Wild Life preservation articles by the present writer and others in the Society's Journals:—e.g. Vol. 47, pp. 778–784, Vol. 47, pp. 602–622, Vol. 48, pp 283–287, Vol. 48, pp 290–299, Vol. 48, pp 588–591.

Increase of wild dogs

As, in the opinion of the writer, the restoration of tiger rewards will be a retrograde step, so—it has been found—the discontinuance in some areas of rewards for destruction of wild dogs has been proved a great mistake. As has been remarked in the literature referred to above, rewards for destruction of wild dogs should be substantial and continually in force where these pests exist. Wild life preservation would be considerably assisted by their extinction.

Increase of gun licences and decrease of herbivora
This has been going on for a number of years, and for the past two years the deer population is being in many areas reduced to vanishing point. The time has arrived when a complete embargo for a period of years should be enforced as to killing of deer throughout the country; also all trapping of game animals and birds should be ended.

The contents of Mr Daver's article once again emphasize the need for a Wild Life organisation in all Provinces and States.'

Originally published in the Journal of the Bombay Natural History Society, 52, *1954.*

PART V

Man-Eaters and Rogues

'Most tigers are great cowards. They are nothing more than gigantic cats, and we all know that cats are not very courageous animals. A man knows that a tiger could kill him as easily as a cat could kill a mouse, but fortunately most tigers do not know this and have an instinctive dread of man. I could hardly go out a day without finding fresh footmarks, yet I know I was comparatively safe.'

GEN. DOUGLAS HAMILTON *c.* 1892

'You do not in sequestered places talk of tiger at all, because if you do you may attract its attention. Woods as well as walls have ears. Any name—jackal, dog, creature—is to be preferred, and is used, not slightingly, but out of respect. For the tiger is not supposed to know that the person who spoke of a jackal and desired its restriction in reality meant the royal beast.'

R. E. VERNEDE, *c.* 1911

JIM CORBETT

The Panar Man-eater

The Temple Tiger and More Man-Eaters of Kumaon *was Corbett's fourth book and is not among his better known works. The author only lived on for another year in east Africa where he moved after 1947. By then he had won fame for his writings on Kumaon in north India, his home for decades, and the place where he hunted, angled, shot and tracked the deer and the tiger. Corbett's is a name that is closely associated with the tiger, an animal he hunted and filmed, admired and respected. But it is his leopard stories that are among his best: the Panar man-eater even tried to climb the tree atop which Corbett sat, gun in hand!*

While I was hunting the Champawat man-eater in 1907 I heard of a man-eating leopard that was terrorizing the inhabitants of villages on the eastern border of the Almora district. This leopard, about which questions were asked in the House of Commons, was known under several names and was credited with having killed four hundred human beings. I knew the animal under the name of the Panar man-eater, and I shall therefore use this name for the purpose of my story.

No mention is made in Government records of man-eaters prior to the year 1905 and it would appear that until the advent of the Champawat tiger and the Panar leopard, man-eaters in Kumaon were unknown. When therefore these two animals—who between them killed eight hundred and thirty-six human beings—made their appearance Government was faced with a difficult situation for it had no machinery to put in action against them and had to rely on personal appeals to sportsmen. Unfortunately there were very few sportsmen in Kumaon at that time who had any inclination for this new form of sport which, rightly or wrongly, was considered as hazardous as Wilson's solo attempt—made a few years later—to conquer Everest. I myself was as ignorant of man-eaters as Wilson was of Everest and that I succeeded in my attempt, where he apparently failed in his, was due entirely to luck.

When I returned to my home in Naini Tal after killing the Champawat tiger I was asked by Government to undertake the shooting of the Panar leopard. I was working hard for a living at the time and several weeks elapsed before I was able to spare the time to undertake this task, and then just as I was ready to start for the outlying area of the Almora district in which the leopard was operating I received an urgent request from Berthoud, the Deputy Commissioner of Naini Tal, to go to the help of the people of Muktesar where a man-eating tiger had established a reign of terror. After hunting down the tiger, an account of which I have given, I went in pursuit of the Panar leopard.

As I had not previously visited the vast area over which this leopard was operating, I went via Almora to learn all I could about the leopard from Stiffe, the Deputy Commissioner of Almora. He kindly invited me to lunch, provided me with maps, and then gave me a bit of a jolt when wishing me good-bye by asking me if I had considered all the risks and prepared for them by making my will.

My maps showed that there were two approaches to the affected area, one via Panwanaula on the Pithoragarh road, and the other via Lamgara on the Dabidhura road. I selected the latter route and after lunch set out in good heart—despite the reference to a will—accompanied by one servant and four men carrying my luggage. My men and I had already done a stiff march of fourteen miles from Khairna, but being young and fit we were prepared to do another long march before calling it a day.

As the full moon was rising we arrived at a small isolated building which, from the scribbling on the walls and the torn bits of paper lying about, we assumed was used as a school. I had no tent with me and as the door of the building was locked I decided to spend the night in the courtyard with my men, a perfectly safe proceeding for we were still many miles from the man-eater's hunting grounds. This courtyard, which was about twenty feet square, abutted on the public road and was surrounded on three sides by a two-foot-high wall. On the fourth side it was bounded by the school building.

There was plenty of fuel in the jungle behind the school and my men soon had a fire burning in a corner of the courtyard for my servant to cook my dinner on. I was sitting with my back to the locked door, smoking, and my servant had just laid a leg of mutton on the low wall nearest the road and turned to attend to the fire, when I saw the head of a leopard appear over the wall close to the leg of mutton. Fascinated, I sat motionless and watched—for the leopard was facing me—and when the man had moved away a few feet the leopard grabbed the meat and bounded across the road into the jungle beyond. The meat had been put down on a big sheet on paper, which had stuck to it, and when my servant heard the rustle of paper and saw what he thought was a dog running away with it he dashed forward shouting, but on realizing that he was dealing with a leopard and not with a mere dog he changed direction and dashed towards me with even greater speed. All white people in the East are credited with being a little mad—for other reasons than walking about in the midday sun—and I am afraid my good servant thought I was a little more mad than most of my kind when he found I was laughing, for he said in a very aggrieved voice, 'It was your dinner that the leopard carried away and I have nothing else for you to eat.' However, he duly produced a meal that did him credit, and to which I did as much justice as I am sure the hungry leopard did to his leg of prime mutton.

Making an early start next morning, we halted at Lamgara for a meal, and by evening reached the Dol dak bungalow on the border of the man-eater's domain. Leaving my men at the bungalow I set out the following morning to try to get news of the man-eater. Going from village to village, and examining the connecting footpaths for leopard pug-marks, I arrived in the late evening at an isolated homestead consisting of a single stone-built slate-roofed house, situated in a few acres of cultivated land and surrounded by scrub jungle. On the footpath leading to this homestead I found the pug-marks of a big male leopard.

As I approached the house a man appeared on the narrow balcony and, climbing down a few wooden steps, came across the courtyard to meet me. He was a young man, possibly twenty-two years of age, and in great distress. It appeared that the previous night while he and his wife were sleeping on the floor of the single room that comprised the house, with the door open for it was April and very hot, the man-eater climbed on to the balcony and getting a grip of his wife's throat started to pull her head-foremost out of the room. With a strangled scream the woman flung an arm round her husband who, realizing in a flash what was happening, seized her arm with one hand and, placing the other against the lintel of the door, for leverage, jerked her away from the leopard and closed the door. For the rest of the night the man and his wife cowered in a corner of the room, while the leopard tried to tear down the door. In the hot unventilated room the woman's wounds started to turn septic and by morning her suffering and fear had rendered her unconscious.

Throughout the day the man remained with his wife, too frightened to leave her for fear the leopard should return and carry her away, and too frightened to face the mile of scrub jungle that lay between him and his nearest neighbour. As day was closing down and the unfortunate man was facing another night of terror he saw me coming towards the house, and when I had heard his story I was no longer surprised that he had run towards me and thrown himself sobbing at my feet.

A difficult situation faced me. I had not up to that time approached Government to provide people living in areas in which a man-eater was operating with first-aid sets, so there was no medical or any other kind of aid nearer than Almora, and Almora was twenty-five miles away. To get help for the woman I would have to go for it myself and that would mean condemning the man to lunacy, for he had already stood as much as any man could stand and another night in that room, with the prospect of the leopard returning and trying to gain entrance, would of a certainty have landed him in a madhouse.

The man's wife, a girl of about eighteen, was lying on her back when the leopard clamped its teeth into her throat, and when the man got a grip of her arm and started to pull her back the leopard—to get a better purchase—drove the claws of one paw into her breast. In the final struggle the claws ripped through the flesh, making four deep cuts. In the heat of the small room, which had only one door and no windows and in which a swarm of flies were buzzing, all the wounds in the girl's throat and on her breast had turned septic, and whether medical aid could be procured or not the chances of her surviving were very slight; so, instead of going for help, I decided to stay the night with the man. I very sincerely hope that no one who reads this story will ever be condemned to seeing and hearing the sufferings of a human being, or of an animal, that has had the misfortune of being caught by the throat by either a leopard or a tiger and not having the means—other than a bullet—of alleviating or of ending the suffering.

The balcony which ran the length of the house, and which was boarded up at both ends, was about fifteen feet long and four feet wide, accessible by steps hewn in a pine sapling. Opposite these steps was the one door of the house, and under the balcony was an open recess four feet wide and four feet high, used for storing firewood.

The man begged me to stay in the room with him and his wife but it was not possible for me to do this, for, though I am not squeamish, the smell in the room was overpowering and more than I could stand. So between us we moved the firewood from one end of the recess under the balcony, clearing a small space where I could sit with my back to the wall. Night was now closing in, so after a wash and a drink at a near-by spring I settled down in my corner and told the man to go up to his wife and keep the door of the room open. As he climbed the steps the man said, 'The leopard will surely kill you, Sahib, and then what will I do?' 'Close the door', I answered, 'and wait for morning.'

The moon was two nights off the full and there would be a short period of darkness. It was this period of darkness that was worrying me. If the leopard had remained scratching at the door until daylight, as the man said, it would not have gone far and even now it might be lurking in the bushes watching me. I had been in position for half an hour, straining my eyes into the darkening night and praying for the moon to top the hills to the east, when a jackal gave its alarm call. This call, which is given with the full force of the animal's lungs, can be heard for a very long distance and can be described as 'pheaon', 'pheaon', 'repeated over and over again as long as the danger that has alarmed the jackal is in sight. Leopards when hunting or when approaching a kill move very slowly, and it would be many minutes before this one—assuming it was the man-eater—covered the half mile between us, and even if in the meantime the moon had not risen it would be giving sufficient light to shoot by, so I was able to relax and breathe more freely.

Minutes dragged by. The jackal stopped calling. The moon rose over the hills, flooding the ground in front of me with brilliant light.

No movement to be seen anywhere, and the only sound to be heard in all the world the agonized fight for breath of the unfortunate girl above me. Minutes gave way to hours. The moon climbed the heavens and then started to go down in the west, casting the shadow of the house on the ground I was watching. Another period of danger, for if the leopard had seen me he would, with a leopard's patience, be waiting for these lengthening shadows to mask his movements. Nothing happened, and one of the longest nights I have ever watched through came to an end when the light from the sun lit up the sky where, twelve hours earlier, the moon had risen.

The man, after his vigil of the previous night, had slept soundly and as I left my corner and eased my aching bones—only those who have sat motionless on hard ground for hours know how bones can ache—he came down the steps. Except for a few wild raspberries I had eaten nothing for twenty-four hours, and as no useful purpose would have been served by my remaining any longer, I bade the man good-bye and set off to rejoin my men at the Dol dak bungalow, eight miles away, and summon aid for the girl. I had only gone a few miles when I met my men. Alarmed at my long absence they had packed up my belongings, paid my dues at the dak bungalow, and then set out to look for me. While I was talking to them the Road Overseer, whom I have mentioned in my story of the Temple Tiger, came along. He was well mounted on a sturdy Bhootia pony, and as he was on his way to Almora he gladly undertook to carry a letter from me to Stiffe. Immediately on receipt of my letter Stiffe dispatched medical aid for the girl, but her sufferings were over when it arrived.

It was this Road Overseer who informed me about the human kill that took me to Dabidhura, where I met with one of the most interesting and the most exciting *shikar* experiences I have ever had. After that experience I asked the old priest of the Dabidhura temple if the man-eater had as effective protection from his temple as the tiger I had failed to shoot, and he answered, 'No, no, Sahib. This *shaitan* [devil] has killed many people who worshipped at my temple and when you come back to shoot him, as you say you will, I shall offer up prayers for your success morning and evening.'

No matter how full of happiness our life may have been, there are periods in it that we look back to with special pleasure. Such a period for me was the year 1910, for in that year I shot the Muktesar man-eating tiger and the Panar man-eating leopard, and in between these two—for me—great events, my men and I set up an all-time record at Mokameh Ghat by handling, without any mechanical means, five thousand five hundred tons of goods in a single working day.

My first attempt to shoot the Panar leopard was made in April 1910, and it was not until September of the same year that I was able to spare the time to make a second attempt. I have no idea how many human beings were killed by the leopard between April and September, for no bulletins were issued by Government and beyond a reference to questions asked in the House of Commons no mention of the leopard—as far as I am aware—was made in the Indian Press. The Panar leopard was credited with having killed four hundred human beings, against one hundred and twenty-five killed by the Rudraprayag leopard, and the fact that the former received such scant publicity while the latter was headline news throughout India was due entirely to the fact that the Panar leopard operated in a remote area far from the beaten track, whereas the Rudraprayag leopard operated in an area visited each year by sixty thousand pilgrims ranging from the humblest in the land to the highest, all of whom had to run the gauntlet of the man-eater. It was these pilgrims, and the daily bulletins issued by Government, that made the Rudraprayag leopard so famous, though it caused far less human suffering than the Panar leopard.

Accompanied by a servant and four men carrying my camp kit and provisions, I set out from Naini Tal on 10 September on my second attempt to shoot the Panar leopard. The sky was overcast when we left home at 4 a.m. and we had only gone a few miles when a deluge of rain came on. Throughout the day it rained and we arrived at Almora, after a twenty-eight-mile march, wet to the bone. I was to have spent the night with Stiffe, but not having a stitch of dry clothing to put on I excused myself and spent the night at the dak bungalow. There were no other travellers there and the man in charge very kindly put two rooms at my disposal, with a big wood fire in each, and by morning my kit was dry enough for me to continue my journey.

It had been my intention to follow the same route from Almora that I had followed in April, and start my hunt for the leopard from the house in which the girl had died of her wounds. While I was having breakfast a mason by the name of Panwa, who did odd jobs for us in Naini Tal, presented himself. Panwa's home was in the Panar valley, and on learning from my men that I was on my way to try to shoot the man-eater he asked for permission to join our party, for he wanted to visit his home and was frightened to undertake the journey alone. Panwa knew the country and on his advice I altered my plans and instead of taking the road to Dabidhura via the school where the leopard had eaten my dinner, I took the road leading to Pithoragarh. Spending the night at the Panwa Naula dak bungalow, we made an early start next morning and after proceeding a few miles left the Pithoragarh road for a track leading off to the right. We were now in the man-eater's territory where there were no roads, and where the only communication was along footpaths running from village to village.

Progress was slow, for the villages were widely scattered over many hundreds of square miles of country, and as the exact whereabouts of the man-eater were not known it was necessary to visit each village to make inquiries. Going through Salan and Rangot *pattis* (*patti* is a group of villages), I arrived late on the evening of the fourth day at Chakati, where I was informed by the headman that a human being had been killed a few days previously at a village called Sanouli on the far side of the Panar river. Owing to the recent heavy rain the Panar was in flood and the headman advised me to spend the night in his village, promising to give me a guide next morning to show me the only safe ford over the river, for the Panar was not bridged.

The headman and I had carried on our conversation at one end of a long row of double-storied buildings and when, on his advice, I elected to stay the night in the village, he said he would have two rooms vacated in the upper story for myself and my men. I had noticed while talking to him that the end room on the ground floor was untenanted, so I told him I would stay in it and that he need only have one room vacated in the upper storey for my men. The room I had elected to spend the night in had no door, but this did not matter for I had been told that the last kill had taken place on the far side of the river and I knew the man-eater would not attempt to cross the river while it was in flood.

The room had no furniture of any kind, and after my men had swept all the straw and bits of rags out of it, complaining as they did so that the last tenant must have been a very dirty person, they spread my groundsheet on the mud floor and made up my bed. I ate my dinner—which my servant cooked on an open fire in the courtyard—sitting on my bed, and as I had done a lot of walking during the twelve hours I had been on my feet it did not take me long to get to sleep. The sun was just rising next morning, flooding the room with light, when on hearing a slight sound in the room I opened my eyes and saw a man sitting on the floor near my bed. He was about fifty years of age, and *in the last stage of leprosy*. On seeing that I was awake this unfortunate living death said he hoped I had spent a comfortable night in his room. He went on to say that he had been on a two-days' visit to friends in an adjoining village, and finding me asleep in his room on his return had sat near my bed and waited for me to awake.

Leprosy, the most terrible and the most contagious of all diseases in the East, is very prevalent throughout Kumaon, and especially bad in the Almora district. Being fatalists the people look upon the disease as a visitation from God, and neither segregate the afflicted nor take any precautions against infection. So, quite evidently, the headman did not think it necessary to warn me that the room I had selected to stay in had for years been the home of a leper. It did not take me long to dress that morning, and as soon as our guide was ready we left the village.

Moving about as I have done in Kumaon I have always gone in dread of leprosy, and I have never felt as unclean as I did after my night in that poor unfortunate's room. At the first stream we came to I called a halt, for my servant to get breakfast ready for me and for my men to have their food. Then, telling my men to wash my groundsheet and lay my bedding out in the sun, I took a bar of carbolic soap and went down the stream to where there was a little pool surrounded by great slabs of rock. Taking off every stitch of clothing I had worn in that room, I washed it all in the pool and, after laying it out on the slabs of rock, I used the remainder of the soap to scrub myself as I had never scrubbed myself before. Two hours later, in garments that had shrunk a little from the rough treatment they had received, I returned to my men feeling clean once again, and with a hunter's appetite for breakfast.

Our guide was a man about four foot six inches tall with a big head crowned with a mop of long hair; a great barrel of a body, short legs, and few words. When I asked him if we had any stiff climbing to do, he stretched out his open hand, and answered, 'Flat as that.' Having said this he led us down a very steep hill into a deep valley. Here I expected him to turn and follow the valley down to its junction with the river. But no. Without saying a word or even turning his head he crossed the open ground and went straight up the hill on the far side. This hill, in addition to being very steep and overgrown with thorn bushes, had loose gravel on it which made the going difficult, and as the sun was now overhead and very hot, we reached the top in a bath of sweat. Our guide, whose legs appeared to have been made for climbing hills, had not turned a hair.

There was an extensive view from the top of the hill, and when our guide informed us that we still had the two high hills in the foreground to climb before reaching the Panar river Panwa, the mason, who was carrying a bundle containing presents for his family and a greatcoat made of heavy dark material, handed the coat to the guide and said that as he was making us climb all the hills in Kumaon he could carry the coat for the rest of the way. Unwinding a length of goathair cord from round his body the guide folded up the coat and strapped it securely to his back. Down and up we went and down and up again, and then away down in a deep valley we saw the river. So far we had been going over trackless ground, without a village in sight, but we now came on a narrow path running straight down to the river. The nearer we got to the river the less I liked the look of it. The path leading to the water and up the far side showed that there was a ford here, but the river was in flood and the crossing appeared to me to be a very hazardous one. The guide assured us, however, that it was perfectly safe to cross, so removing my shoes and stockings I linked arms with Panwa and stepped into the water. The river was about forty yards wide and from its broken surface I judged it was running over a very rough bed. In this I was right, and after stubbing my toes a few times and narrowly avoiding being washed off our feet we struggled out on the far bank.

Our guide had followed us into the river and, on looking back, I saw that the little man was in difficulties. The water which for us had been thigh deep was for him waist deep and on reaching the main stream, instead of bracing his back against it and walking crab fashion, he very foolishly faced up stream with the result that he was swept over backwards and submerged under the fast-running current. I was barefoot and helpless on the sharp stones, but

Panwa—to whom sharp stones were no obstacle—threw down the bundle he was carrying and without a moment's hesitation sprinted along the bank to where, fifty yards farther down, a big slab of rock jutted into the river at the head of a terrifying rapid. Running out on to this wet and slippery rock Panwa lay down, and as the drowning man was swept past, grabbed him by his long hair and after a desperate struggle drew him on to the rock. When the two men rejoined me—the guide looking like a drowned rat—I complimented Panwa on his noble and brave act in having saved the little man's life, at great risk to his own. After looking at me in some surprise Panwa said, 'Oh, it was not his life that I wanted to save, but my new coat that was strapped to his back.' Anyway, whatever the motive, a tragedy had been averted, and after my men had linked arms and crossed safely I decided to call it a day and spend the night on the river bank. Panwa, whose village was five miles farther up the river, now left me, taking with him the guide, who was frightened to attempt a second crossing of the river.

Next morning we set out to find Sanouli, where the last human kill had taken place. Late in the evening of that day we found ourselves in a wide open valley, and as there were no human habitations in sight, we decided to spend the night on the open ground. We were now in the heart of the man-eater's country and after a very unrestful night, spent on cold wet ground, arrived about midday at Sanouli. The inhabitants of this small village were overjoyed to see us and they very gladly put a room at the disposal of my men, and gave me the use of an open platform with a thatched roof.

The village was built on the side of a hill overlooking a valley in which there were terraced fields, from which a paddy crop had recently been harvested. The hill on the far side of the valley sloped up gradually, and a hundred yards from the cultivated land there was a dense patch of brushwood, some twenty acres in extent. On the brow of the hill, above this patch of brushwood, there was a village, and on the shoulder of the hill to the right another village. To the left of the terraced fields the valley was closed in by a steep grassy hill. So, in effect, the patch of brushwood was surrounded on three sides by cultivated land, and on the fourth by open grass land.

While breakfast was being got ready, the men of the village sat round me and talked. During the second half of March and the first half of April, four human beings had been killed in this area by the man-eater. The first kill had taken place in the village on the shoulder of the hill, the second and third in the village on the brow of the hill, and the fourth in Sanouli. All four victims had been killed at night and carried some five hundred yards into the patch of brushwood, where the leopard had eaten them at his leisure, for—having no firearms—the inhabitants of the three villages were too frightened to make any attempt to recover the bodies. The last kill had taken place six days before, and my informants were convinced that the leopard was still in the patch of brushwood.

I had purchased two young male goats in the village we passed through earlier that day, and towards evening I took the smaller one and tied it at the edge of the path of brushwood to test the villagers' assertion that the leopard was still in the cover. I did not sit over the goat, because there were no suitable trees near by and also because clouds were banking up and it looked as though there might be rain during the night. The platform that had been placed at my disposal was open all round, so I tied the second goat near it in the hope that if the leopard visited the village during the night it would prefer a tender goat to a tough human being. Long into the night I listened to the two goats calling to each other. This convinced me that the leopard was not within hearing distance. However, there was no reason why he should not return to the locality, so I went to sleep hoping for the best.

There was a light shower during the night and when the sun rose in a cloudless sky every leaf and blade of grass was sparkling with raindrops and every bird that had a song to sing was singing a joyful welcome to the day. The goat near my platform was contentedly browsing off a bush and bleating occasionally, while the one across the valley was silent. Telling my servant to keep my breakfast warm, I crossed the valley and went to the spot where I had tied up the smaller goat. Here I found that, some time before the rain came on, a leopard had killed the goat, broken the rope, and carried away the kill. The rain had washed out the drag-mark, but this did not matter for there was only one place to which the leopard could have taken his kill, and that was into the dense patch of brushwood.

Stalking a leopard, or a tiger, on its kill is one of the most interesting forms of sport I know of, but it can only be indulged in with any hope of success when the conditions are favourable. Here the conditions were not favourable, for the brushwood was too dense to permit of a noiseless approach. Returning to the village, I had breakfast and then called the villagers together, as I wanted to consult them about the surrounding country. It was necessary to visit the kill to see if the leopard had left sufficient for me to sit over and, while doing so, I would not be able to avoid disturbing the leopard. What I wanted to learn from the villagers was whether there was any other heavy cover, within a reasonable distance, to which the leopard could retire on being disturbed by me. I was told that there was no such cover nearer than two miles, and that to get to it the leopard would have to cross a wide stretch of cultivated land.

At midday I returned to the patch of brushwood and, a hundred yards from where he had killed it, I found all that the leopard had left of the goat—its hooves, horns, and part of its stomach. As there was no fear of the leopard leaving the cover at that time of day for the jungle two miles away, I tried for several hours to stalk it, helped by bulbuls, drongos, thrushes, and scimetar-babblers, all of whom kept me informed of the leopard's every movement. In case any should question why I did not collect the men of the three villages and get them to drive the leopard out on to the open ground, where I could have shot it, it is necessary to say that this could not have been attempted without grave danger to the beaters. As soon as the leopard found he was being driven towards open ground, he would have broken back and attacked anyone who got in his way.

On my return to the village after my unsuccessful attempt to get a shot at the leopard, I went down with a bad attack of malaria and for the next twenty-four hours I lay on the platform in a stupor. By the evening of the following day the fever had left me and I was able to continue the hunt. On their own initiative the previous night my men had tied out the second goat where the first had been killed, but the leopard had not touched it. This was all to the good, for the leopard would now be hungry, and I set out on that third evening full of hope.

On the near side of the patch of brushwood, and about a hundred yards from where the goat had been killed two nights previously, there was an old oak tree. This tree was growing out of a six-foot-high bank between two terraced fields and was leaning away from the hill at an angle that made it possible for me to walk up the trunk in my rubber-soled shoes. On the underside of the trunk and about fifteen feet from the ground there was a branch jutting out over the lower field. This branch, which was about a foot thick, offered a very uncomfortable and a very unsafe seat for it was hollow and rotten. However, as it was the only branch on the tree, and as there were no other trees within a radius of several hundred yards, I decided to risk the branch.

As I had every reason to believe—from the similarity of the pug-marks I had found in the brushwood to those I had seen in April on the path leading to the homestead where the girl was killed—that the leopard I was dealing with was the Panar man-eater, I made my men cut a number of long blackthorn shoots. After I had taken my seat with my back to the tree and my legs stretched out along the branch, I made the men tie the shoots into bundles and lay them on the trunk of the tree and lash them to it securely with strong rope. To the efficient carrying out of these small details I am convinced I owe my life.

Several of the blackthorn shoots, which were from ten to twenty feet long, projected on either side of the tree; and as I had nothing to hold on to, to maintain my balance, I gathered the shoots on either side of me and held them firmly pressed between my arms and my body. By five o'clock my preparations were complete. I was firmly seated on the branch with my coat collar pulled up well in front to protect my throat, and my soft hat pulled down well behind to protect the back of my neck. The goat was tied to a stake driven into the field thirty yards in front of me, and my men were sitting out in the field smoking and talking loudly.

Up to this point all had been quiet in the patch of brushwood, but now a scimetar-babbler gave its piercing alarm call followed a minute or two later by the chattering of several white-throated laughing thrushes. These two species of birds are the most reliable informants in the hills, and on hearing them I signalled to my men to return to the village. This they appeared to be very glad to do, and as they walked away, still talking loudly, the goat started bleating. Nothing happened for the next half-hour and then, as the sun was fading off the hill above the village, two drongos that had been sitting on the tree above me, flew off and started to bait some animal on the open ground between me and the patch of brushwood. The goat while calling had been facing in the direction of the village, and it now turned round, facing me, and stopped calling. By watching the goat I could follow the movements of the animal that the drongos were baiting and that the goat was interested in, and this animal could only be a leopard.

The moon was in her third quarter and there would be several hours of darkness. In anticipation of the leopard's coming when light conditions were not favourable, I had armed myself with a twelve-bore double-barreled shot gun loaded with slugs, for there was a better chance of my hitting the leopard with eight slugs than with a single rifle bullet. Aids to night shooting, in the way of electric lights and torches, were not used in India at the time I am writing about, and all that one had to rely on for accuracy of aim was a strip of white cloth tied round the muzzle of the weapon.

Again nothing happened for many minutes, and then I felt a gentle pull on the blackthorn shoots I was holding and blessed my forethought in having had the shoots tied to the leaning tree, for I could not turn round to defend myself and at best the collar of my coat and my hat were poor protection. No question now that I was dealing with a man-eater, and a very determined man-eater at that. Finding that he could not climb over the thorns, the leopard, after his initial pull, had now got the butt ends of the shoots between his teeth and was jerking them violently, pulling me hard against the trunk of the tree. And now the last of the daylight faded out of the sky and the leopard, who did all his human killing in the dark, was in his element and I was out of mine, for in the dark a human being is the most helpless of all animals and—speaking for myself—his courage is at its lowest ebb. Having killed four hundred human beings at night, the leopard was quite unafraid of me, as was evident from the fact that while tugging at the shoots, he was growling loud enough to be heard by the men anxiously listening in the village. While this growling terrified the men, as they told me later, it had the opposite effect on me, for it let me know where the leopard was and what he was doing. It was when he was silent that I was most terrified, for I did not know what his next move would be. Several times he had nearly unseated me by pulling on the shoots vigorously and then suddenly letting them go, and now that it was dark and I had nothing stable to hold on to I felt sure that if he sprang up he would only need to touch me to send me crashing to the ground.

After one of these nerve-racking periods of silence the leopard jumped down off the high bank and dashed towards the goat. In the hope that the leopard would come while there was still sufficient light to shoot by, I had tied the goat thirty yards from the tree to give me time to kill the leopard before it got to the goat. But now, in the dark, I could not save the goat—which, being white, I could only just see as an indistinct blur—so I waited until it had stopped struggling and then aimed where I thought the leopard would be and pressed the trigger. My shot was greeted with an angry grunt and I saw a white flash as the leopard went over backwards, and disappeared down another high bank into the field beyond.

For ten or fifteen minutes I listened anxiously for further sounds from the leopard, and then my men called out and asked if they should come to me. It was now quite safe for them to do so, provided they kept to the high ground. So I told them to light pine torches, and thereafter carry out my instructions. These torches, made of twelve to eighteen inches long splinters of resin-impregnated pine-wood cut from a living tree, give a brilliant light and provide the remote villages in Kumaon with the only illumination they have ever known.

After a lot of shouting and running about, some twenty men each carrying a torch left the village and, following my instructions, circled round above the terraced fields and approached my tree from behind. The knots in the ropes securing the blackthorn shoots to the tree had been pulled so tight by the leopard that they had to be cut. After the thorns had been removed men climbed the tree and helped me down, for the uncomfortable seat had given me cramp in my legs.

The combined light from the torches lit up the field on which the dead goat was lying, but the terraced field beyond was in shadow. When cigarettes had been handed round I told the men I had wounded the leopard but did not know how badly, and that we would return to the village now and I would look for the wounded animal in the morning. At this, great disappointment was expressed. 'If you have wounded the leopard it must surely be dead by now.' 'There are many of us, and you have a gun, so there is no danger.' 'At least let us go as far as the edge of the field and see if the leopard has left a blood trail.' After all arguments for and against going to look for the leopard immediately had been exhausted I consented against my better judgement to go as far as the edge of the field, from where we could look down on the terraced field below.

Having acceded to their request, I made the men promise that they would walk in line behind me, hold their torches high, and not run away and leave me in the dark if the leopard charged. This promise they very willingly gave, and after the torches had been replenished and were burning brightly we set off, I walking in front and the men following five yards behind.

Thirty yards to the goat, and another twenty yards to the edge of the field. Very slowly, and in silence, we moved forward. When we reached the goat—no time now to look for a blood trail—the farther end of the lower field came into view. The nearer we approached the edge, the more of this field became visible, and then, when only a narrow strip remained in shadow from the torches, the leopard, with a succession of angry grunts, sprang up the bank and into full view.

There is something very terrifying in the angry grunt of a charging leopard, and I have seen a line of elephants that were staunch to tiger turn and stampede from a charging leopard; so I was not surprised when my companions, all of whom were unarmed, turned as one man and bolted. Fortunately for me, in their anxiety to get away they collided with each other and some of the burning splinters of pine—held loosely in their hands—fell to the ground and continued to flicker, giving me sufficient light to put a charge of slugs into the leopard's chest.

On hearing my shot the men stopped running, and then I heard one of them say, '*Oh, no*. He won't be angry with us, for he knows that this devil has turned our courage to water.' Yes, I knew, from my recent experience on the tree, that fear of a man-eater robs a man of courage. As for running away, had I been one of the torch-bearers I would have run with the best. So there was nothing for me to be angry about. Presently, while I was making believe to examine the leopard, to ease their embarrassment, the men returned in twos and threes. When they were assembled. I asked, without looking up, 'Did you bring a bamboo pole and rope to carry the leopard back to the village?' 'Yes,' they answered eagerly, 'we left them at the foot of the tree.' 'Go and fetch them,' I said, 'for I want to get back to the village for a cup of hot tea.' The cold night-wind blowing down from the north had brought on another attack of malaria, and now that all the excitement was over I was finding it difficult to remain on my feet.

That night, for the first time in years, the people of Sanouli slept, and have continued to sleep, free from fear.

Excerpted from The Temple Tiger and More Man-Eaters of Kumaon *(Mumbai: Oxford University Press, 1954).*

JIM CORBETT

The Man-eating Leopard of Rudraprayag

By the time Corbett shot the leopard of Rudraprayag in May 1956, it had killed as many as 125 people and its name evoked fear in a huge swathe of territory in Kumaon. Yet, in his book, the only full-length narrative that he ever wrote, Corbett clarified, 'Here was no fiend... Here was only an old leopard, who differed from others of his kind in that his muzzle was grey and his lips lacked whispers.' The excerpt describes the countdown to the right when the leopard finally met his end.

My Night of Terror

For several days after my experience on the pine-tree I lost touch with the man-eater. He did not return to the broken ground and I found no trace of him, or of the female who had saved his life, in the miles of forest I searched on the high ground above the cultivated land. In these forests I was more at home, and if the leopards had been anywhere in them I should have been able to find them, for there were birds and animals in the forest that would have helped me.

The female, being restless, was quite evidently straying far from her home when she heard me call from the top of the pine-tree, and on being joined by the male had gone back to her own area, accompanied by the mate I had helped her to find. The male would presently return alone, and as the precautions now being taken by the people on the left bank were making it difficult for him to procure a human kill, he would probably try to cross over to the right bank of the Alaknanda, so for the next few nights I mounted guard on the Rudraprayag bridge.

There were three approaches to the bridge on the left bank, the one from the south passing close to the bridge chowkidar's house, and on the fourth night I heard the leopard killing the chowkidar's dog; a friendly nondescript little beast that used to run out and greet me every time I passed that way. The dog seldom barked, but that night it had been barking for five minutes when suddenly the bark ended in a yelp, followed by the shouting of the chowkidar from inside his house, after which there was silence. The thornbushes had been removed from the archway and the bridge was open, yet though I lay with finger on trigger for the rest of the night the leopard did not try to cross.

After killing the dog and leaving it lying on the road, the leopard, as I found from his tracks next morning, came to the tower. Five more steps in the direction in which he was going would have brought him out on the bridge, but those five steps he did not take. Instead he turned to the right, and after going a short distance up the footpath towards the bazaar, he returned and went up the pilgrim road to the north. A mile up the road I lost his tracks.

Two days later I received a report that a cow had been killed the previous evening, seven miles up the pilgrim road. It was suspected that the cow had been killed by the man-eater, for the previous night—the night the dog had been killed—the leopard had tried to break open the door of a house close to where, the next evening, the cow had been killed.

On the road I found a number of men waiting for me who, knowing that the walk up from Rudraprayag would be a hot one, had very thoughtfully provided a dish of tea. While we sat in the shade of a mango-tree and smoked, and I drank the dish of tea, they told me that the cow had not returned with the herd the previous evening, and that it had been found between the road and the river when a search had been made for it that morning. They also told me of the many hairbreadth escapes each of them had had from the man-eater during the past eight years. I was very interested to learn from them that the leopard had only adopted his present habit of trying—and in many cases succeeding—to break open the doors of houses three years previously, and that before he had been content to take people who were outside their houses, or from houses the doors of which had been left open. 'Now,' they said, 'the Shaitan has become so bold that sometimes when he has not been able to break down the door of a house, he has dug a hole through the mud wall, and got at his victims in that way.'

To those who do not know our hill-people, or understand their fear of the supernatural, it will seem incredible that a people renowned for their courage, and who have won the highest awards on the field of battle, should permit a leopard to break open a door, or to dig a hole in a wall of a house, in which in many cases there must have been men with axes, kukris, or even in some cases firearms at hand. I know of only one case in all those eight long years in which resistance was offered to the man-eater, and in that case the resister was a woman. She was sleeping alone in a house, the door of which she had left unfastened; this door, as in the case of the door of the house occupied by the woman who escaped with a lacerated arm, opened inwards. On entering the room the leopard seized the woman's left leg, and as it dragged her across the room the woman's hand came in contact with a *gandesa*—a tool used for chopping chaff for cattle—and with this the woman dealt the leopard a blow. The leopard did not release his hold, but backed out of the room, and as it did so either the woman pushed the door to, or else this happened accidentally. Whichever it may have been, with the woman on one side of the door and the leopard on the other, the leopard exerted its great strength and tore the limb from the woman's body. Mukandi Lal, at that time Member for Garhwal in the United Provinces Legislative Council, who was on an electioneering tour, arrived in the village the following day and spent a night in the room, but the leopard did not return. In a report to the Council Mukandi Lal stated that seventy-five human beings had been killed by the leopard in the course of that one year, and he asked the Government to launch a vigorous campaign against the man-eater.

Accompanied by one of the villagers to show me the way, and by Madho Singh, I went down to the kill. The cow had been killed in a deep ravine a quarter of a mile from the road and a hundred yards from the river. On one side of the ravine there were big rocks with dense brushwood between, and on the other side of the ravine there were a few small trees, none of which was big enough to sit in. Under the trees, and about thirty yards from the kill, there was a rock with a little hollow at the base of it, so in the hollow I decided to sit.

Both Madho Singh and the villager objected very strongly to my sitting on the ground, but as this was the first animal kill I had got since my arrival at Rudraprayag in a place where it was reasonable to expect the leopard to come at an early hour—about sundown—I overruled their objections, and sent them back to the village.

My seat was dry and comfortable, and with my back to the rock and a small bush to conceal my legs I was confident the leopard would not see me, and that I should be able to kill it before it was aware of my presence. I had provided myself with a torch and a knife, and with my good rifle across my knees I felt that in this secluded spot my chances of killing the leopard were better than any I had yet had.

Without movement and with my eyes on the rocks in front of me I sat through the evening, each second bringing the time nearer when the undisturbed and unsuspecting leopard would to a certainty return to his kill. The time I had been waiting for had come, and was passing. Objects near at hand were beginning to get blurred and indistinct. The leopard was a little later in coming than I had expected him to be, but that was not worrying me, for I had a torch, and the kill was only thirty yards from me, and I would be careful over my shot and make quite sure that I did not have a wounded animal to deal with.

In the deep ravine there was absolute silence. The hot sun of the past few days had made the dead leaves on the bank on which I was sitting as dry as tinder. This was very reassuring, for it was now dark and whereas previously I had depended on my eyes for protection I now had to depend on my ears, and with thumb on the button of the torch and finger on trigger I was prepared to shoot in any direction in which I heard the slightest sound.

The non-appearance of the leopard was beginning to cause me uneasiness. Was it possible that from some concealed place among the rocks he had been watching me all these hours, and was he now licking his lips in anticipation of burying his teeth in my throat?—for he had long been deprived of human flesh. In no other way could I account for his not having come, and if I were to have the good fortune to leave the ravine on my feet, my ears would have to serve me now as they had never served me before.

For what seemed like hours I strained my ears and then, noticing it was getting darker than it should have been, I turned my eyes up to the sky and saw that a heavy bank of clouds was drifting across the sky, obscuring the stars one by one. Shortly thereafter big drops of rain started to fall, and where there had been absolute and complete silence there was now sound and movement all round—the opportunity the leopard had been waiting for had come. Hastily taking off my coat I wound it round my neck, fastening it securely in place with the sleeves. The rifle was now useless but might help to cause a diversion, so transferring it to my left hand I unsheathed my knife and got a good grip of it with my right hand. The knife was what is called an Afridi stabbing knife, and I devoutly hoped it would serve me as well as it had served its late owner, for when buying it from the Government store at Hangu on the North-west Frontier, the Deputy Commissioner had drawn my attention to a label attached to it and to three notches on the handle, and said it had figured in three murders. Admittedly a gruesome relic, but I was glad to have it in my hand, and I clutched it tight while the rain lashed down.

Leopards, that is ordinary forest leopards, do not like rain and invariably seek shelter, but the man-eater was not an ordinary leopard, and there was no knowing what his likes or dislikes were, or what he might or might not do.

When Madho Singh was leaving he asked how long I intended sitting up, and I had answered 'Until I have shot the leopard', so I could expect no help from him, and of help I was at that time in urgent need. Should I go or should I remain were the questions that were exercising me, and the one was as unattractive as the other. If the leopard up to then had not seen me it would be foolish to give my position away, and possibly fall across him on the difficult ground I should have to negotiate on my way up to the pilgrim road. On the other hand to remain where I was for another six hours—momentarily expecting to have to fight for my life with an unfamiliar weapon—would put a strain on my nerves which they were not capable of standing; so getting to my feet and shouldering the rifle, I set off.

I had not far to go, only about five hundred yards, half of which was over wet clay and the other half over rocks worn smooth with bare feet and the hooves of cattle. Afraid to use the torch for fear of attracting the man-eater, and with one hand occupied with the rifle and the other with the knife, my body made as many contacts with the ground as my rubber-shod feet. When I eventually reached the road I sent a full-throated cooee into the night, and a moment later I saw a door in the village far up the hillside open and Madho Singh and his companion emerge, carrying a lantern.

When the two men joined me Madho Singh said he had had no uneasiness about me until the rain started, and that he had then lit the lantern, and sat with his ear against the door listening. Both men were willing to accompany me back to Rudraprayag, so we set out on our seven-mile walk, Bachi Singh leading, Madho Singh carrying the lantern following, and I bringing up the rear. When I returned next day I found the kill had not been touched, and on the road I found the tracks of the man-eater. What time had elapsed between our going down the road and the man-eater following us, it was not possible to say.

When I look back on that night, I look back on it as my night of terror. I have been frightened times without number, but never have I been frightened as I was that night when the unexpected rain came down and robbed me of all my defences, and left me for protection a murderer's knife.

Leopard Fights Leopard

After following us to Rudraprayag the leopard went down the pilgrim road through Golabrai, past the ravine up which he had gone a few days previously, and then up a rough track which the people living on the hills to the east of Rudraprayag use as a short cut on their way to and from Hardwar.

The pilgrimage to Kedarnath and Badrinath is seasonal, and the commencement of the pilgrimage and its duration depend in the one case on the melting, and in the other on the falling, of snow in the upper reaches of the high mountains in which these two shrines are situated. The High Priest of Badrinath temple had a few days previously sent the telegram that is eagerly awaited by good Hindus throughout the length and breadth of India, announcing that the road was open, and for the past few days pilgrims in small numbers had been passing through Rudraprayag.

During the past few years the man-eater had killed several pilgrims on the road, and it appeared to be his more or less regular habit while the pilgrim season lasted to go down the road to the extent of his beat, and then circle round through the villages on the hills to the east of Rudraprayag, and rejoin the road anything up to fifteen miles above Rudraprayag. The time taken for this round trip varied, but on an average I had seen the leopard's tracks on the stretch of road between Rudraprayag and Golabrai once in every five days, so on my way back to the Inspection Bungalow I selected a place from where I could overlook the road, and for the next two nights sat in great comfort on a hayrick, without however seeing anything of the leopard.

I received no news of the man-eater from outlying villages for two days, and on the third morning I went down the pilgrim road for six miles to try to find out if he had recently visited any of the villages in that direction. From this twelve-mile walk I returned at midday, and while I was having a late breakfast two men arrived and reported that a boy had been killed the previous evening at Bhainswara, a village eighteen miles south-east of Rudraprayag.

The intelligence system introduced by Ibbotson was working splendidly. Under this system cash rewards, on a graduated scale, were paid for information about all kills in the area in which the man-eater was operating. These rewards, starting with two rupees for a goat and working up to twenty rupees for a human being, were keenly contested for, and so ensured our receiving information about all kills in the shortest time possible.

When I put ten rupees into the hands of each of the men who had brought me news about the boy, one of them offered to accompany me back to Bhainswara to show me the way, while the other said he would stay the night at Rudrapryag as he had recently had fever and could not do another eighteen miles that day. I finished breakfast while the men were telling me their tale, and a little before 1 p.m. I set off, taking only my rifle, a few cartridges, and a torch with me. As we crossed the road near the Inspection Bungalow and started up the steep hill on the far side of it, my companion informed me we had a very long way to go, adding that it would not be safe for us to be out after dark, so I told him to walk ahead and set the pace. I never—if I can help it—walk uphill immediately after a meal, but here I had no option, and for the first three miles, in which we climbed four thousand feet, I had great difficulty in keeping up with my guide. A short stretch of comparatively flat ground at the end of the three miles gave me back my wind, and after that I walked ahead and set the pace.

On their way to Rudraprayag the two men had told the people in the villages they had passed through about the kill, and of their intention to try and persuade me to accompany them back to Bhainswara. I do not think that anyone doubted that I would answer to the call, for at every village the entire population were waiting for me, and while some gave me their blessings, others begged me not to leave the district until I had killed their enemy.

My companion had assured me that we had eighteen miles to go, and as we crested hill after hill with deep valleys between I realized I had undertaken to walk against time eighteen of the longest and hardest miles I had ever walked. The sun was near setting when, from the crest of one of these unending hills, I saw a number of men standing on a ridge a few hundred yards ahead of us. On catching sight of us some of the men disappeared over the ridge, while others came forward to meet us. The headman of Bhainswara was among the latter, and after he had greeted me, he cheered me by telling me that his village was just over the crest of the hill, and that he had sent his son back to get tea ready.

The 14th of April 1926 is a date that will long be remembered in Garhwal, for it was on that day that the man-eating leopard of Rudraprayag killed his last human victim. On the evening of that day a widow and her two children, a girl aged nine and a boy aged twelve, accompanied by a neighbour's son aged eight, went to a spring a few yards from Bhainswara village to draw water for the preparation of their evening meal.

The widow and her children occupied a house in the middle of a long row of homesteads. These homesteads were double-storied, the low-ceilinged ground floor being used for the storage of grain and fuel, and the first floor for residences. A veranda four feet wide ran the entire length of the building, and short flights of stone steps flanked with walls gave access to the veranda, each flight of steps being used by two families. A flagged courtyard, sixty feet wide and three hundred feet long, bordered by a low wall, extended along the whole length of the building.

The neighbour's son was leading as the party of four approached the steps used by the widow and her children, and as the boy started to mount the steps he saw an animal, which he mistook for a dog, lying in an open room on the ground floor adjoining the steps; he said nothing about the animal at the time, and the others apparently did not see it. The boy was followed by the girl, the widow came next, and her son brought up the rear. When she was half-way up the short flight of stone steps, the mother heard the heavy brass vessel her son was carrying crash on the steps and go rolling down them; reprimanding him for his carelessness, she set her own vessel down on the veranda and turned to see what damage her son had done. At the bottom of the steps she saw the overturned vessel. She went down and picked it up, and then looked round for her son. As he was nowhere in sight she assumed he had got frightened and had run away, so she started calling to him.

Neighbours in adjoining houses had heard the noise made by the falling vessel and now, hearing the mother calling to her son, they came to their doors and asked what all the trouble was about. It was suggested that the boy might be hiding in one of the ground-floor rooms, so as it was now getting dark in these rooms, a man lit a lantern and came down the steps towards the woman, and as he did so he saw drops of blood on the flagstones where the woman was standing. At the sound of the man's horrified ejaculation other people descended into the courtyard, among whom was an old man who had accompanied his master on many shooting expeditions. Taking the lantern from the owner's hand, this old man followed the blood trail across the courtyard and over the low wall. Beyond the wall was a drop of eight feet into a yam field; here in the soft earth were the splayed-out pug marks of a leopard. Up to that moment no one suspected that the boy had been carried off by a man-eater, for though everyone had heard about the leopard it had never previously been within ten miles of their village. As soon as they realized what had happened the women began screaming and while some men ran to their houses for drums, others ran for guns—of which there were three in the village—and in a few minutes pandemonium broke out. Throughout the night drums were beaten and guns were fired. At daylight the boy's body was recovered, and two men were dispatched to Rudraprayag to inform me.

As I approached the village in company with the headman, I heard the wailing of a woman mourning her dead. It was the mother of the victim, and she was the first to greet me. Even to my unpractised eye it was apparent that the bereaved mother had just weathered one hysterical storm and was heading for another, and as I lack the art of dealing with people in this condition I was anxious to spare the woman a recital of the events of the previous evening; but she appeared to be eager to give me her version of the story, so I let her have her way. As the story unfolded itself it was apparent that her object in telling it was to ventilate her grievance against the men of the village for not having run after the leopard and rescued her son 'as his father would have done had he been alive'. In her accusation against the men I told her she was unjust, and in her belief that her son could have been rescued alive, I told her she was wrong. For when the leopard clamped his teeth round the boy's throat, the canine teeth dislocated the head from the neck and the boy was already dead before the leopard carried him across the courtyard, and nothing the assembled men—or anyone else—could have done would have been of any use.

Standing in the courtyard drinking the tea that had thoughtfully been provided for me, and noting the hundred or more people who were gathered round, it was difficult to conceive how an animal the size of a leopard had crossed the courtyard in daylight without being seen by any of the people who must have been moving about at that time, or how its presence had gone undetected by the dogs in the village.

I climbed down the eight-foot wall that the leopard carrying the boy had jumped down, and followed the drag across the yam field, down another wall twelve feet high, and across another field. At the edge of this second field there was a thick hedge of rambler roses four feet high. Here the leopard had released his hold on the boy's throat, and after searching for an opening in the hedge and not finding one, he had picked the boy up by the small of the back and, leaping the hedge, gone down a wall ten feet high on the far side. There was a cattle track at the foot of this third wall and the leopard had only gone a short distance along it when the alarm was raised in the village. The leopard had then dropped the boy on the cattle track and gone down the hill. He was prevented from returning to his kill by the beating of drums and the firing of guns which had gone on all night in the village.

The obvious thing for me to have done would have been to carry the body of the boy back to where the leopard had left it, and to have sat over it there. But here I was faced with two difficulties—the absence of a suitable place in which to sit, and my aversion to sitting in an unsuitable place.

The nearest tree, a leafless walnut, was three hundred yards away, and was therefore out of the question, and quite frankly I lacked the courage to sit on the ground. I had arrived at the village at sundown; it had taken a little time to drink the tea, hear the mother's story, and trail the leopard, and there was not sufficient daylight left for me to construct a shelter that would have given me even the semblance of protection; therefore if I sat on the ground I should have to sit just anywhere, not knowing from what direction the leopard would come, and knowing full well that if the leopard attacked me I should get no opportunity of using the one weapon with which I was familiar, my rifle, for when in actual contact with an unwounded leopard or tiger it is not possible to use firearms.

When after my tour of inspection I returned to the courtyard, I asked the headman for a crowbar, a stout wooden peg, a hammer, and a dog chain. With the crowbar I prised up one of the flagstones in the middle of the courtyard, drove the peg firmly into the ground, and fastened one end of the chain to it. Then with the help of the headman I carried the body of the boy to the peg, and chained it there.

The working of the intangible force which sets a period to life, which one man calls Fate and another calls Kismet, is incomprehensible. During the past few days this force had set a period to the life of a breadwinner, leaving his family destitute; had ended in a very painful way the days of an old lady who after a lifetime of toil was looking forward to a few short years of comparative comfort; and now, had cut short the life of this boy who, by the look of him, had been nurtured with care by his widowed mother. Small wonder then that the bereaved mother should, in between her hysterical crying, be repeating over and over and over again, 'What crime, Parmeshwar, has my son, who was loved by all, committed, that on the threshold of life he has deserved death in this terrible way?'

Before prising up the flagstone, I had asked for the mother and her daughter to be taken to a room at the very end of the row of buildings. My preparations completed, I washed at the spring and asked for a bundle of straw, which I laid on the veranda in front of the door of the house vacated by the mother.

Darkness had now fallen. Having asked the assembled people to be as silent during the night as it was possible for them to be and sent them to their respective homes, I took up my position on the veranda, where by lying prone on my side and heaping a little straw in front, I could get a clear view of the kill without much chance of being seen myself.

In spite of all the noise that had been made the previous night, I had a feeling that the leopard would return, and that when he failed to find his kill where he had left it, he would come to the village to try to secure another victim. The ease with which he had got his first victim at Bhainswara would encourage him to try again, and I started my vigil with high hopes.

Heavy clouds had been gathering all the evening, and at 8 p.m., when all the village sounds—except the wailing of the woman—were hushed, a flash of lightning followed by a distant roll of thunder heralded an approaching storm. For an hour the storm raged, the lightning being so continuous and brilliant that had a rat ventured into the courtyard I should have seen and probably been able to shoot it. The rain eventually stopped but, the sky remaining overcast, visibility was reduced to a few inches. The time had now come for the leopard to start from wherever he had been sheltering from the storm, and the time of his arrival would depend on the distance of that place from the village.

The woman had now stopped wailing, and in all the world there appeared to be no sound. This was as I had hoped, for all I had to warn me that the leopard had come were my ears, and to help them I had used the dog chain instead of a rope.

The straw that had been provided for me was as dry as tinder and my ears, straining into the black darkness, first heard the sound when it was level with my feet—something was creeping, very stealthily creeping, over the straw on which I was lying. I was wearing an article of clothing called shorts, which left my legs bare in the region of my knees. Presently, against this bare skin, I felt the hairy coat of an animal brushing. It could only be the man-eater, creeping up until he could lean over and get a grip of my throat. A little pressure now on my left shoulder—to get a foothold—and then, just as I was about to press the trigger of the rifle to cause a diversion, a small animal jumped down between my arms and my chest. It was a little kitten, soaking wet, that had been caught out in the storm and, finding every door shut, had come to me for warmth and protection.

The kitten had hardly made itself comfortable inside my coat, and I was just beginning to recover from the fright it had given me, when from beyond the terraced fields there was some low growling which gradually grew louder, and then merged into the most savage fight I have ever heard. Quite evidently the man-eater had returned to the spot where the previous night he had left his kill, and while he was searching for it, in not too good a temper, another male leopard who looked upon this particular area as his hunting-ground, had accidentally come across him and set on him. Fights of the nature of the one that was taking place in hearing of me are very unusual, for carnivora invariably keep to their own areas, and if by chance two of a sex happen to meet, they size up each other's capabilities at a glance, and the weaker gives way to the stronger.

The man-eater, though old, was a big and a very powerful male, and in the five hundred square miles he ranged over there was possibly no other male capable of disputing his rule, but here at Bhainswara he was a stranger and a trespasser, and to get out of the trouble he had brought on himself he would have to fight for his life. And this he was undoubtedly doing.

My chance of getting a shot had now gone, for even if the man-eater succeeded in defeating his attacker, his injuries would probably prevent him from taking any interest in kills for some time to come. There was even a possibility of the fight ending fatally for him, and here would indeed be an unexpected end to his career: killed in an accidental encounter by one of his own kind, when the combined efforts of the Government and the public had failed, over a period of eight years, to accomplish this end.

The first round, lasting about five minutes, was fought with unabating savagery, and was inconclusive, for at the end of it I could still hear both animals. After an interval of ten or fifteen minutes the fight was resumed, but at a distance of two to three hundred yards from where it had originally started; quite evidently the local champion was getting the better of the fight and was gradually driving the intruder out of the ring. The third round was shorter than the two that had preceded it, but was no less savage, and when after another long period of silence the fight was again resumed, the scene had receded to the shoulder of the hill, where after a few minutes it died out of hearing.

There were still six hours of darkness left; even so I knew my mission to Bhainswara had failed, and that my hope that the fight would be fought to a finish and would end in the death of the man-eater had been short-lived. In the running fight into which the contest had now degenerated the man-eater would sustain injuries, but they were not likely to reduce his craving for human flesh or impair his ability to secure it.

The kitten slept peacefully throughout the night, and as the first streak of dawn showed in the east I descended into the courtyard and carried the boy to the shed from where we had removed him, and covered him with the blanket which previously had been used for the purpose. The headman was still asleep when I knocked on his door. I declined the tea, which I knew would take some time to make, and assured him that the man-eater would never again visit his village; and when he had promised to make immediate arrangements to have the boy carried to the burning-ghat, I set off on my long walk back to Rudraprayag.

No matter how often we fail in any endeavour, we never get used to the feeling of depression that assails us after each successive failure. Day after day over a period of months I had left the Inspection Bungalow full of hope that on this particular occasion I would meet with success, and day after day I had returned disappointed and depressed. Had my failures only concerned myself they would not have mattered, but in the task I had undertaken those failures concerned others more than they concerned me. Bad luck—for to nothing else could I attribute my failures—was being meted out to me in ever-increasing measure, and the accumulated effect was beginning to depress me and give me the feeling that I was not destined to do what I had set out to do. What but bad luck had made the man-eater drop his kill where there were no trees? And what but bad luck had made a leopard who possibly had thirty square miles in which to wander, arrive at a particular spot in those thirty miles just as the man-eater, not finding his kill where he had left it, was quite conceivably on his way to the village where I was waiting for him?

The eighteen miles had been long yesterday but they were longer today, and the hills were steeper. In the villages I passed through the people were eagerly awaiting me, and though I only had bad news they did not show their disappointment. Their boundless faith in their philosophy, a faith strong enough to remove mountains, and very soothing to depressed feelings, that no human beings and no animals can die before their appointed time, and that the man-eater's time had not yet come, called for no explanation, and admitted of no argument.

Ashamed or the depression and feeling of frustration that I had permitted to accompany me throughout the morning, I left the last village—where I had been made to halt and drink a cup of tea—greatly cheered, and as I swung down the last four miles to Rudraprayag I became aware that I was treading on the pug marks of the man-eater. Strange how one's mental condition can dull, or sharpen, one's powers of observation. The man-eater had quite possibly joined the track many miles farther back, and now, after my conversation with the simple village-fold—and a drink of tea—I was seeing his pug marks for the first time that morning. The track here ran over red clay which the rain had softened, and the pug marks of the man-eater showed that he was walking at his accustomed pace. Half a mile farther on he started to quicken his pace, and this pace he continued to maintain until he reached the head of the ravine above Golabrai; down this ravine the leopard had gone.

When a leopard or tiger is walking at its normal pace only the imprints of the hind feet are seen, but when the normal pace is for any reason exceeded, the hind feet are placed on the ground in advance of the forefeet, and thus the imprints of all four feet are seen. From the distance between the imprints of the fore and the hind feet it is possible to determine the speed at which an animal of the cat tribe was travelling. The coming of daylight would in this instance have been sufficient reason for the man-eater to have quickened his pace.

I had previously had experience of the man-eater's walking capabilities, but only when ranging his beat in search of food. Here he had a better reason for the long walk he had undertaken, for he was anxious to put as great a distance as possible between himself and the leopard who had given him a lesson in the law of trespass.

Excerpted from The Man-eating Leopard of Rudraprayag *(Mumbai: Oxford University Press, 1948).*

KENNETH ANDERSON

The Man-eater of Segur

By the time of his death in 1974, Anderson had established himself as a well-known raconteur of man-eater hunts in southern India. Of course, the incidence of attacks on people in peninsular India has been very low in the twentieth century. No wonder some of his claims were treated with some scepticism. But as this memoir of the wildlife-rich Segur plateau shows, nothing can detract from his ability to tell a story well and to evoke a sense of fear, suspense and danger.

The hamlet of Segur is situated at the foot of the north-eastern slopes of the well-known Nilgiri Mountains in South India, or 'Blue Mountains', which is what the word 'Nilgiri' literally means. On the summit of this lovely range stands the beautiful health resort of Ootacamund, at an average elevation of 7,500 feet above mean sea level. This 'Queen of Hill Stations', as it is affectionately termed, is the focal point of visitors from all the length and breadth of India. Ootacamund has a charm of its own, with a climate that allows the growth of all types of English flowers and vegetables to perfection. The all-prevailing scent of the towering eucalyptus ('blue-gum') trees, which is wafted across the station from the surrounding plantations, mingled with that of the fir and pine, makes memories of 'Ooty' unforgettable, and with its cool climate gives a welcome change to the visitor from the heat and enervating temperature of the sweltering plains far below.

A steep ghat-road, 12 miles in length, leaves Ootacamund, and after passing through graceful, rolling downs, where fox-hunting was once a pastime of the English residents—the local jackal taking the place of the fox—drops sharply downwards to Segur, at the foot of the range, where dense tropical forests prevail. The road is so steep in places to be almost unusable, except to motor cars of fairly high power. At Segur the road bifurcates, one branch running north-westwards through dense forest, past the hamlet of Mahvanhalla and the village of Masinigudi, to meet the main trunk road, linking the cities of Bangalore and Mysore with Ootacamund, at the forest chowki of Tippakadu. The other bifurcation leads eastwards through equally dense forest, along the base of the Blue Mountains, to the forest bungalow of Anaikutty, nine miles away. Two perennial streams water this area, the Segur River and the Anaikutty River, both descending in silvery cascades from the Blue Mountains; here their waters run through giant tropical forests to join those of the Moyar River, some fifteen miles away. The Moyar River, or 'Mysore Ditch', as it is known, forms the boundary between the native state of Mysore on the north, the district of Coimbatore on the north-east, and the Nilgiri Range with its foot-hills on the south, both Coimbatore and the Nilgiris forming part of the Province of Madras. The evergreen forests of the Malabar-Wynaad extend to the west and south-west.

All the areas mentioned are densely wooded, hold game preserves on the Mysore, Malabar, and Nilgiri sides, and are the habitat of large numbers of wild elephant, bison, tiger, panther, sambar, spotted-deer and other animals.

The forest-bungalow of Anaikutty is built on a knoll, past which run the swirling waters of the Anaikutty river. In summer it is an unhealthy place, full of malarial mosquito, and the origin of many fatal cases of 'black-water' fever. In the winter it is a paradise. The mornings are fresh and sunny, ideally suited to long hikes through the forest. The afternoons are moderately cool. The evenings chilly, while with nightfall an icy-cold wind descends from the mountain tops to take the place of the warmer air rising from the surrounding forests. The nights are then so intensely cold that one is invariably confined to the bungalow itself, in all the rooms of which fire-places have been provided. Here a truly Christmassy feeling prevails, before a blazing fire of forest logs, while the party discusses the latest stories of the 'abominable snow-man', or ghost stories are related in hair-raising detail. Even within the compound of this bungalow elephant and tiger and panther roam, their screams, roars and grunts reminding the sleepy inmate, snugly tucked below double-blankets, with the glow of a fire by his bedside, that he is still in the midst of a tropical jungle.

These forests are a favourite resort of mine, and in them I have met with several little adventures that are still memorable after these many years.

To relate just one. Wild-dogs, which hunt in packs in India, varying in numbers from three to thirty, sometimes invade these forests from over the boundaries of Mysore and Coimbatore, where they are very numerous, more so in the latter district. These packs are very destructive to all forms of deer, particularly sambar and spotted-deer, which they hunt down inexorably, and tear limb from limb while still alive. On several occasions, both in these and other jungles, I have come across sambar actually being chased by these dogs. The method adopted is that, while a few dogs chase the animal, others break away in a flanking movement, to run ahead of their quarry and ambush it as it dashes past them. When hunting, they emit a series of yelps in a very high pitch, resembling the whistling cry of a bird rather than that of a dog. The quarry is brought to earth after being attacked by these flankers, which bite out its eyes, disembowel it, hamstring or emasculate it, in their efforts to bring it down. I once saw a sambar pursued by wild-dogs, dash into a pool of water to try and protect itself. It had been disembowelled and trailed its intestines behind it for the distance of twenty feet. Sambar are extraordinarily hardy, and sometimes are literally eaten alive by these dogs, before being killed.

One evening, at about 5 o'clock, I was a mile from the bungalow, interesting myself in an unusual species of ground orchid that sprouted from the earth in a spray of tiny star-shaped flowers. Suddenly I heard a medley of sounds whose origin I could not at first define. There were cries, yelps, and long-drawn bays, interspersed with grunts and 'whoofs' that puzzled me. Then I knew that the noise was that of wild dogs, which seemed to be attacking a pig or a bear. Grabbing my rifle, I ran in the direction of the din. I may have covered a furlong, when around the corner dashed a tigress, encircled by half-a-dozen wild dogs. Concealing myself behind the trunk of a tree, I watched the unusual scene.

The dogs had spread themselves around the tigress, who was growling ferociously. Every now and again one would dash in from behind to bite her. She would then turn to attempt to rend asunder this puny aggressor, when a couple of others would rush in from another direction. In this way she was kept going continually, and I could see she was fast becoming spent.

All this time the dogs were making a tremendous noise, the reason for which I soon came to know, when, in a lull in the fray, I heard the whistling cry of the main pack, galloping to the assistance of their advance party. The tigress must have also heard the sound, for in sudden, renewed fury, she charged two of the dogs, one of which she caught a tremendous blow on its back with her paw, cracking its spine with the sharp report of a broken twig. The other just managed to leap out of danger. The tigress then followed up her momentary advantage by bounding away, to be immediately followed by the five remaining dogs. They were just out of sight when the main pack streamed by, in which I counted twenty-three dogs, as they galloped past me without the slightest interest in my presence. Soon the sounds of pursuit died away, and all that remained was the one dead dog.

During the affair I had been too interested, and too lost in admiration at the courage of the dogs, to fire at either the tigress or her attackers.

Next morning I sent out scouts to try to discover the result of the incident. They returned about noon, bringing a few fragments of tiger-skin, to report that the dogs had finally cornered their exhausted quarry about five miles away and had literally torn the tigress to pieces. As far as they could gather, five dogs had been killed in the final battle, after which the victors had eaten the tigress, and even the greater portions of their own dead companions.

Three distinct tribes of natives inhabit these areas. There are the Badagas, descendants of long-ago invaders from the state of Mysore, themselves fleeing from Mahommedan conquerors from the north. These Badagas have now become rich; they own lands, vast herds of cattle and semi-wild buffaloes, with tremendously long, curved horns. Next come the Kesavas, the greatest in numbers but laziest in disposition, who work under the Badagas as herdsmen and tillers of the soil. Lastly, come the Karumbas, comparatively few in number, the original inhabitants of the land and now entirely jungle-men; they live on wild honey, roots, hunting and the trapping of small animals and birds. These Karumbas make excellent trackers, and like the Sholagas of the Coimbatore District and Pujarees of Salem, are true children of nature, who are born and live in the forests till the day of their death.

Having given the reader a little idea of the country in which the adventure took place, I shall lose no further time in telling the story of the 'Man-eater of Segur'.

This tiger was reported to have come originally from the jungles of the Silent Valley Forest Block in the District of Malabar-Wynaad, below the extreme opposite, or south-western face, of the Blue Mountains. This area is infested with elephant, of which it holds the record for 'rogues', and with bison. As a rule it does not hold many tiger. A few human kills took place in the Silent Valley and then ceased entirely, to recommence at Gudalur, some twenty miles from Tippakadu, at Masinigudi, and finally in all the areas between Segur and Anaikutty. How and why the tiger came so far from its place of origin, encircling the greater portion of the Nilgiri Mountains in doing so, nobody knows.

It was midsummer and the tiger had been particularly active, killing at Segur and, within a week, at Anaikutty, when I arrived. The last victim had been a Kesava herdsman, tending his herd of semi-wild buffaloes some two miles from Anaikutty along the lower reaches of the river, as it wended its way towards the Moyar. In this instance, the man-eater had stalked and attacked the man, completely ignoring the surrounding grazing buffaloes. It had killed him, and was perhaps carrying him away, when the buffaloes had become aware of its presence.

These animals, as I have already stated, are only semi-domesticated, and extremely dangerous to a stranger, especially if he happens to be dressed in the unusual mode—to them—of a European, when they frequently charge on sight. Seeing the tiger, they had evidently attacked him *en masse*, and succeeded in driving him off, leaving the dead man where he had been dropped.

That night, neither buffaloes nor herdsman returned to the kraal, so early the next morning a search party set out to discover the reason for their absence. It did not take them long to find the dead herdsman, surrounded by his herd of placidly grazing buffaloes, which had effectively prevented the tiger from returning to its prey. An examination of the corpse, the pug-marks of the tiger, and the pursuing hooves of the charging buffaloes, revealed the sequence of happenings as I have related them.

The kill that had taken place a week earlier at Segur had been that of a woman, as she went down to the Segur River with her water-pot to fetch the daily supply of water for her family. In this case the tiger had succeeded in carrying off its victim, the only evidence of the occurrence being the mute testimony of the broken water-pot, the pugs of the tiger in the soft mud that bordered the river, a few drops of blood, the torn saree, and a few strands of human hair that had become entangled in the bushes as the tiger made off with its prey.

I visited both localities and by careful measurement and examination of the pugs, which were very clear on the river bank at Segur, determined that the tiger was a smallish-sized male of considerably less than adult age.

Reports had it that the tiger very frequently traversed the ten miles of forest road between Segur and Anaikutty, his pugs being seen along this track, especially in the vicinity of both places. As a preliminary, I therefore decided to sit up along this road at differently selected places, and without wearying the reader with details, put this plan into practice, spending thus a whole week, alternately in the vicinity of Segur and Anaikutty, without seeing any signs of the tiger.

The seventh night I spent on a tree a mile-and-half from Segur. Returning the next morning to the Anaikutty forest bungalow, where I had established my headquarters, I was informed that a Karumba, who had left the previous morning to gather wild honey from the combs of the giant rock-bee that abound by the hundred in the region of a place called 'Honey Rock', about four miles from the lower foot-hills of the Nilgiri, had not yet returned.

Assembling a group of twenty persons as a search party, I accompanied them to this wild and densely wooded spot. Splitting the party into four groups of five men each, as protection against possible attack by the man-eater, we searched till noon, when one of the men discovered the body of the dead Karumba lying in a nala, and brought me the news. I went to the spot and found the man had been killed by being bitten through the throat. Beside him lay his empty kerosine-tin, in which he had been gathering honey, all of which had split on the ground to form a feast for a colony of black-ants, which covered the tin in a black mass.

Thinking at first that the tiger had killed him, and wondering why it had not devoured him in that quiet, secluded spot, we cast around for tracks, but soon discovered that the killer had been a female sloth-bear, accompanied by its cub. The human-like imprints of the mother's feet, and the smaller impressions of the cub's, were clearly to be seen in the soft sand that formed the bed of the nala.

Evidently the she-bear had been asleep with its cub, or perhaps about to cross the nala, when the Karumba, in his search for honey, had suddenly come upon it. The sloth-bear has a very uncertain temperament at the best of times, being poor of sight and hearing, so that humans have often been able to approach them very closely before being discovered, when in the fright and excitement of the moment, they will attack without any provocation. Undoubtedly this is what happened, when the she-bear, surprised, frightened and irritated, and in defence of her young which she fancied was in danger, had rushed at the Karumba, bitten him through the throat, severing his jugular, and then made off as fast as she could.

I did not wish to spend time in hunting a bear which, after all, had only killed in defence of its cub, and was for returning to the bungalow at Anaikutty, when the four Karumbas that were included in the party, urged me to track down and shoot the animal, which they felt would be a menace to them when they, in turn, came to the same place for honey. More to please them than because I had any heart in the venture, I therefore sent the rest of the party back and followed these four men on the trail of the she-bear.

The tracks led along the nullah and then joined a stream, down which Mrs Bruin and her baby had ambled for some distance before breaking back into the jungle. Thence she had climbed upwards towards the many rocks and caves that gave them shelter, where hundreds of rock-bee hives hung from every conceivable rock-projection in long, black masses, sometimes attaining a length of over five feet, a width of a yard, and a thickness of over a foot.

The ground we were now traversing was hard and stony, and to my unskilled eyes presented no trace of the bears' passing. But the Karumbas were seldom at fault, and their powers of tracking really worth witnessing during the long and tiring walk that followed. For more than two hours they led me up hill and down dale, and across deep valleys and stony ridges. An overturned rock or stone, a displaced leaf, or the slightest marks of scraping or digging, showed them where the bears had travelled.

At last we approached the mouth of a cave, high up over a projecting rock from which hung twenty-three separate rock-bee hives. This cave, they declared, was the home of the she-bear and her cub.

Standing ten yards from the entrance, and to one side, I instructed the men to hurl stones into the interior, which they did with unabated vigour for quite twenty minutes. But not a sound did we hear. Then the Karumbas made a torch of grasses, which we ignited and threw into the cave, but still nothing happened. Finally they made five similar torches. Lighting one of these, I followed the Karumba.

The cave was comparatively small inside and was obviously empty, the occupant and her youngster having left at hearing our approach, if not before. But I was glad of this, because, as I have said, I had no quarrel with this animal, and would have regretted having to shoot it and leave the baby an orphan.

We had lit our third torch, and were just about to leave, when one of the Karumbas came upon an interesting curiosity of the Indian forest, of which I had heard but never seen, and the existence of which I consequently never believed.

He found in a corner what the natives call 'bear's bread', or 'Karadi roti', as it is named in Tamil. This was a roughly circular mass, about ten inches across, an inch thick, and of a dirty blackish-yellow, sticky consistency. Female bears are reported to seek out the fruits of the 'Jak' tree, a large fruit with a rough, thorny exterior; wild 'Wood-apple' fruit, the size of large tennis balls and with hard shells; and pieces of honey-comb, including bees, comb and honey. Each of these ingredients is first eaten by the she-bear in turn, the whole being then vomited in the cave in a mass, which the she-bear allows to harden into a cake, as reserve food for her young.

Due possibly to its long hair, the bear is marvellously impervious to the stings of the rock-bee, which prove fatal to a human being if suffered in large numbers. It can climb trees and rocks with astounding ability in search of wild fruit and honey-combs. A cave, inhabited by a she-bear with her young, is reported to contain sometimes a dozen such cakes. In this case we had found only one, but inasmuch as I had never come across or seen one before, I was immensely interested. As evidence of the wholesomeness of this 'bread', the four Karumbas offered it to me for immediate consumption, but I declined with thanks. They then divided it equally between themselves, and ate it with evident gusto.

We were now sitting at the entrance of the cave. Whether the flames of the torch, the smoke, or the sound of human voices disturbed them, I do not know, but suddenly a few black objects buzzed around us, and as we sprang to our feet to make off, the rock-bees from the hive immediately above, grossly disturbed and angry, descended upon us in an avalanche.

It was every man for himself, and as I grabbed my rifle and sprinted down the hillside, the agile Karumbas outdistanced me very quickly. What I lacked in speed I made up, however, by being clothed, the bees being able to register some twenty stings on my neck and hands and other exposed parts, as compared with about forty that I counted on each of the bare skins of the Karumbas, when we slowed up half-a-mile away. It was indeed a comical ending to a tragic but interesting morning, and we returned to Anaikutty a far wiser, but very sore and smarting party.

I set out at 8 p.m. that night, motoring slowly to Segur in my Studebaker, a Karumba acting as assistant by my side and flashing the beams of my 'sealed-beam' spotlight along the jungle on both sides. Meeting nothing, we continued for six miles along the north-western road, past Mahvanhalla and Masinigudi, and finally the four miles to the forest chowki at Tippakadu. Here we allowed an hour to pass before returning, this time encountering, three miles from Tippakadu, a herd of bison along the roadside.

These animals crashed away as the car approached closer. Between Masinigudi and Segur we met several spotted-deer, and just after taking the turn to Anaikutty a large bull elephant in the centre of the road, his tusks gleaming sharply white in the powerful beams of the spotlight. A gentle toot of the horn sufficed to send him scurrying on his way, and finally we reached the Anaikutty bungalow without seeing a trace of the man-eater.

The following morning dawned bright and fine, and I set out with my Karumba guide across the forest towards the Moyar River, nine miles to the north. Again we encountered no trace of the tiger, but came across the pugs of an exceptionally large forest-panther as the land began to dip sharply to the basin of the Moyar. Judging by its tracks, it was indeed a big animal, approaching the size of a small tigress, and would have made a fine trophy, had I the time to pursue it. We returned to the bungalow in the late afternoon, tired and somewhat disappointed, after our long and fruitless walk.

Again that night we motored to Tippakadu and back, encountering only a solitary sambar, when returning, at the river crossing before the Anaikutty forest bungalow.

With dawn I undertook another hike with my Karumba, this time in a south-easterly direction and towards the Nilgiris. We found a dead cow-bison in the bush four miles away, an examination of the carcase revealing that the animal had died of rinderpest. The Forest Department had reported some two months earlier that this epidemic had spread from the cattle of the Badagas, whose herds it afflicted in a contagious and epidemic form, to the wild animals, especially bison, and here lay proof of the statement.

That afternoon a report came in from Mahvanhalla that the tiger had taken a woman near the bridge by which the main road to Tippakadu crosses the Mahvanhalla Stream before it joins the Segur River. Motoring to the spot I was shown the place where, the previous evening, the tiger, which had been lurking on the banks of the stream, had attacked the woman, who had been among the herdsmen watering the large number of mixed cattle and buffaloes kraaled at Mahvanhalla.

Apparently nobody had actually witnessed the incident, the woman having been a little apart from the rest of the party assembled near the bridge and hidden by the bend the little stream takes just before it passes under road. She had screamed shrilly and silence had followed. The remaining graziers, five in number, had hastened to Mahvanhalla, gathered reinforcements in the form of six others, including the husband of the unfortunate woman, and turned to the spot to look for her. They had found the basket she had been carrying, and close by in the soft earth, the pug-marks of a tiger; then the whole party had returned to Mahvanhalla. At dawn the next day, the four men who now reported themselves had set forth by footpath to cover the ten miles to Anaikutty and report the incident to me, it having become known that I was in the area to shoot the tiger.

We went to the place where the woman had been attacked, and with the expert help of my Karumba tracker were soon able to pick up the trail of the tiger and its victim. Before following I dismissed the party from Mahvanhalla, together with the bereaved husband, who made me promise faithfully that I would bring back at least a few bones of his beloved spouse to satisfy the requirements of a cremation ceremony.

Almost without faltering, my Karumba guide followed what was to me the completely invisible trail left by the tiger. The man-eater was evidently making towards a high hill, an out-spur of the Nilgiri

Range, that ran parallel to the road on the west at a distance of about two miles. Years ago I had partly climbed this very hill in search of a good bison head, and knew its middle slopes were covered with a sea of long spear-grass which gradually thinned out as the higher, and more rocky, levels were reached. In that area, I felt we had little chance of finding the tiger or its prey.

Nevertheless my stout little guide continued faultlessly, and within a mile of the foot of the hill came across the woman's saree, caught in the undergrowth. Shortly afterwards the tiger appeared to have changed its mind, in that its trail veered off to the right, parallel to the hill, which now was quite near, and back again towards the bed of the Mahvanhalla Stream. Still following, we eventually reached the stream, which here ran through a deep valley. Scattered bamboo-clumps grew in increasing numbers down this declivity, and the shrill squeal of vultures and the heavy flapping of their wings soon heralded the close of our search. The remains of the woman lay below a clump of bamboo, eaten by the tiger, and the vultures had finished what had been left over from his feast. We found the head lying apart, the eyes picked out of their sockets by the great birds, which had also devoured most of the flesh from the face. This ghastly remnant, together with her hands and feet, her glass bangles and silver anklets, we gathered together and wrapped in grass, and in fulfilment of my promise, the Karumba very reluctantly carried the bundle to her husband. Sitting up at the spot seemed a waste of time, as the tiger would not return to such scattered remnants.

That night I continued my hunting by car but without success, and next day procured a young buffalo, which I tied close to the bend in the stream at Mahvanhalla, where the woman had been attacked. The chances of the tiger visiting the same spot being remote, I decided not to sit up over the live buffalo, but to await a kill, should it occur.

In this surmise I was wrong, for before noon next day runners came to Anaikutty to report that the buffalo had been killed and partly eaten by a tiger. By 3 p.m. my machan was fixed, and I sat overlooking the dead bait, at a height of some fifteen feet.

A large red-martin, a big species of mongoose that inhabits the lower slopes of the Nilgiris, was the first visitor to put in an appearance at the kill. He came at about 5.30 p.m., at first nibbled cautiously, and then began to gobble chunks of the raw meat. By

6 o'clock he had filled himself to bursting-point, and made off to pass the night in dreamless contentment, or with a heavy attack of indigestion.

As dusk fell the shadowy forms of three jackals slunk forward. With the greatest temerity they approached the dead animal, sniffed it while glancing around apprehensively, and made off in frightened rushes, only to return each time, as hunger and the demands of a voracious appetite urged them on. At about the fourth attempt they finally settled down to eat, but had hardly taken a few mouthfuls when there was a rush and a scamper and they were gone.

This was surely the coming of the tiger, I thought, and sure enough a slinking, grey shape flitted into the open and cautiously shambled up to the kill. No tiger would so shamefacedly approach his own kill, however, and the hyaena—for hyaena it was—began all over again the cautious and frightened approaches of his smaller cousins, the jackals.

Several times he sniffed at the meat, the while he glanced furtively to right and left. Several times he shambled away, to scurry around in a wide circle and see if the coast was clear, before cautiously slinking up again to take a hasty mouthful. Then off he went in another wide circle to make sure that the rightful owner of the kill was not in the vicinity to catch him red-handed. This continued for quite half-an-hour, before the hyaena settled down to a serious meal.

He had been eating for ten minutes when a tiger called nearby, '*A-oongh, Aungh-ha, Ugha-ugh, O-o-o-n-o-o-n*', was four times repeated in the silent night air, and the hyaena whisked away as if by magic and did not return.

I sat in readiness, momentarily expecting the appearance of the king of the jungle. But the hours dragged on and he did not come. Whatever may have been the reason, that tiger gave the kill a wide berth that night, and the false dawn found me shivering with the intense cold that had now set in despite the summer season. I remained till day-light, to descend from my machan, cramped and still, thoroughly disgusted and disappointed with the world at large and with the tiger in particular. What caused him to approach so close to his kill that night without actually putting in an appearance will ever remain a mystery.

I rested the next day and followed my usual nightly procedure of motoring along the road to Tippakadu and back, but without seeing any sign of the tiger. This time a small panther jumped off a roadside culvert exactly at the turning point on the road to Anaikutty, and crouched in the ditch, its bright eyes gazing into the spot-light as the car passed by. I let the little brute alone, to pass its days in happiness in the beautiful forests where it rightly belonged.

The next week proved uneventful, and I began to think of giving up the chase and returning to Bangalore, but eventually I decided to wait three more days before departing. One of these three days passed uneventfully, when, at midday the second day, a Badaga boy, the son of a rich cattle-owner at Segur, was carried away by the tiger while taking the midday meal to his father, who was with the cattle, and other Kesava herdsmen.

I was soon at the spot with my Karumbas, to follow up the trail. The tiger had carried the boy across the Segur River and into the jungle to the north. Again we followed, without delay, and this time found the body hidden in a nala and only half-eaten. Unfortunately, the father arrived at the scene, and wanted to remove the body, and an hour was lost in argument to persuade him to let it remain and give me a chance of sitting up.

In the vicinity of the body there was no suitable tree or rock in which to conceal myself, and eventually it was decided to move the corpse some fifty feet towards a bamboo clump, on the top of which an unstable machan was erected. To reach this, I had to climb up the notches of a cut bamboo stem, only to find the machan one of the worst I had ever sat on in my life, it swayed alarmingly with every current of wind, and my slightest movement caused the bamboos to creak ominously below me. Besides, I did not have a good view of the body, which was over thirty yards away. The bamboo stems growing around me completely obstructed any view at a close range.

The beginning of my vigil was most uninteresting, and no living thing put in an appearance beyond a peacock, which alighted higher up the nullah bank from the place where the corpse was lying. From there it walked slowly downwards, till it suddenly caught sight of the prone, human form in the nullah where we had left it, when with a great flapping of wings it sailed away above my head, its flowing tail glinting a greenish-red in the rays of the setting sun, for all the world like a comet flapping through the forest.

At 9 p.m. I became aware of the presence of the tiger by the low moan he emitted from near the spot where he had originally left the corpse. I cursed myself for having shifted it, but realized that this had had to be done, as there was absolutely no shelter for me at that spot.

Finding his kill had been moved, the tiger then growled several times. After all, we had shifted it a bare fifty feet, and from where he stood the tiger would undoubtedly see it in its new position and come towards it; or so I hoped. But moving the kill had been fatal, raising within the tiger a deep suspicion as to why the man it had left dead, and had partly eaten, had now moved away.

It is extraordinary how very cautious every man-eater becomes by practice, whether a tiger or a panther, and cowardly too. Invariably, it will only attack a solitary person, and that, too, after prolonged and painstaking stalking, having assured itself that no other human being is in the immediate vicinity. I believe there is hardly any case on record where a man-eater has attacked a group of people, while many instances exist where timely interference, or aid by a determined friend or relative, has caused a man-eater to leave his victim and flee in absolute terror. These animals seem also to possess an astute sixth sense and be able to differentiate between an unarmed human being and an armed man deliberately pursuing them, for in most cases, only when cornered will they venture to attack the latter, while they go out of their way to stalk and attack the unarmed man.

This particular tiger was definitely possessed of a very acute sixth sense, for it guessed something was amiss. Instead of openly approaching its kill, as I had hoped it would, it then began to circle the whole area, plaintively moaning at intervals, as if in just complaint against the meanness of fate at having moved the kill. Around and around it travelled for quite an hour, till it finally decided that the kill was forbidden fruit, and the last I heard was its plaintive moaning receding to southwards, as it made for the sheltering hills.

This last episode, with its attendant failure, caused me to redouble my determination to bag this most astute animal, and to postpone my departure, if need be, for another fortnight, to enable me to do so.

Gossip with the Karumbas now suggested that the tiger might be met at nights along the many cart-tracks that branched into the forest from Anaikutty, Segur and Mahvanhalla, rather than along the main roads on which I had been motoring for several nights.

The Man-eater of Segur

As these cart tracks were unmotorable, I hired a bullock cart for the next week, and determined to spend each night in it meandering along every possible track in the vicinity of the three places. The driver of this cart, a Kesava, was an unusually doughty fellow, and my two Karumba scouts were to accompany me to suggest the most likely tracks.

The next three nights we spent in this fashion, encountering only sambar and spotted-deer, and on the third night, an elephant, which gave us some anxious moments. It was where the track crossed the Anaikutty River, two miles from the village, that we first met him. He had been standing under a mighty 'Muthee' tree, as motionless as a rock, and quite unnoticed by us. In the cart we carried the car-battery, which I had detached together with the 'sealed-beam' spotlight, to be used in emergency only. For ordinary illumination, and for picking up eyes in the jungle, we were using two torches, a seven-cell and a five-cell respectively.

As I was saying, we were crossing the river, and the cart was about midway in the stream, the bulls struggling valiantly to pull the huge wheels through the soft sand, when the elephant, alarmed and annoyed by the torches, let forth a piercing scream, like the last trump of doom, and came splashing at us through the water. Switching on the 'sealed-beam', I caught him in its brilliant rays about thirty yards away. The bright beam brought him to a halt, when he commenced stamping his feet in the water and swinging his great bulk and trunk from side to side, undecided whether to charge or to make off. We then shouted in unison, and focussed all lights on him, and with a parting scream of rage he swirled around and shuffled off into the black forest, a very angry and indignant elephant indeed.

We continued these bullock-cart prowls for the next three nights, but without success.

On the morning of the seventh day, the tiger killed the son of the Forest Guard stationed at Anaikutty, a lad aged eighteen years. At 9 a.m., and in bright sunlight, he had left his hut in the village and gone a short distance up the path leading to the river and to the bungalow where I lay sleeping after my nights in the cart. He had gone to call his dog, which was in the habit of wandering between the village and the bungalow, less than a mile away, because of some scraps which I daily gave it after each meal. That boy was never seen alive again.

His father, the Forest Guard, thinking he was with me at the bungalow, took his absence for granted. When noon came, and it was time for the midday meal, the youth had not yet appeared and the Guard decided to come to the bungalow and fetch him. By the roadside, within a furlong of the river, he came across his son's cap. This alarmed him, and he called aloud to the boy. Receiving no answer, he ran the remaining distance to the bungalow and awoke me, to tell me what he had found.

Feeling something had befallen the lad, I picked up the rifle and accompanied the anxious father to the spot where the cap was still lying. Looking around, we found his slipper under a bush ten yards away and realized that the worst had happened. Remaining at the spot, I told the now-weeping guard to hurry back to the village and summon the Karumba trackers.

Within a quarter-hour these men had joined me and we started on the trail.

The boy appeared to have struggled and had probably cried aloud, although none had been there to hear him, for within a few yards of where we found the slipper, the tiger had apparently laid him down and bitten him savagely. This was made clear by the sudden spurts of blood that smeared the dried grass for the area of quite a square yard. The tiger had then proceeded with its prey, leaving a trail of blood on the ground, which gradually petered out as the blood began to coagulate.

The river here turned north-west and we found that the tiger was making in its direction. Pressing forward, we soon reached the thick jungle that clothed the river-banks, where the blood-trail once again became evident in the occasional red smear that marked the leaves of the undergrowth as the tiger, holding its prey in its mouth, had pushed through.

In the soft mud of the river-bank we saw the fresh pugmarks of the killer, which passed across the shallow water to the opposite bank. Here it led up the slope to the shelter of a clump of jungle-plum bushes, before which we found the lad's remains. He had been almost half-devoured and was a ghastly sight. The reason for the sudden spurts of blood on the trail we had followed now became apparent. The boy had obviously been still alive when carried off by the tiger, as we had already surmised, and must have screamed and struggled in an effort to get free. Annoyed the tiger had thrown him down and dealt the boy a smashing blow across the skull with its forepaw. The forehead had been crushed inwards, like a squashed egg-shell, while the sharp talons had half-torn the scalp away, leaving the bare bone of the skull exposed to view. One of the eyes had also been gouged out by a claw, and hung from its socket.

The grief of that poor father was truly heart-rending to watch, as he prostrated himself at the feet of his only son, kissed the poor mangled remains, and called aloud to heaven and earth for vengeance, while heaping dust, that he scratched from the dry ground, on to his own head.

Shaking him roughly to bring him to his senses, and telling him earnestly that the more noise he made and the more time he wasted correspondingly lessened our chances of shooting the tiger, we succeeded in reducing his cries to a whimper.

A medium-sized 'Jumlum' tree overshadowed the plum-bush beneath which the boy lay, and on this tree I determined to erect my machan. I sent the father and the two Karumbas back to the bungalow, instructing the latter to bring my portable 'charpoy' machan, and pending their return climbed up into the 'Jumlum' tree, both for my own protection, and to get a shot if the tiger should suddenly put in an appearance.

The two Karumbas were back in under an hour with the machan, which they securely and efficiently tied in the lower branches of the 'Jumlum' tree at the usual height of a little over fifteen feet. They had also brought my water-bottle, gun-torch and blanket, as instructed, and soon after noon I climbed into the machan in high hopes of securing a shot.

Evening came, and nightfall, without a sign of the tiger. Then followed a sudden rain-storm, such as sometimes occurs in the midst of a dry summer in India. Lightning flashes illuminated the forest as bright as day, vividly revealing the corpse beneath the plum-tree below; thunder crashed, reverberating against the adjacent hills; and the rain literally descended in torrents, preceded by a sharp shower of hailstones.

Within a minute I was soaked to the skin, and as the downpour continued, before my very eyes the river rose with the mass of water that rushed down from the surrounding hills, till I judged it to be unfordable.

This dreadful state of affairs continued till well past midnight, when the rain died down to a thin, cold drizzle. Then a horribly chill wind set in, blowing down from the Nilgiris to the sodden forest. The drizzle finally ceased, the wind continued and, as evaporation began in my soaking clothes and blankets, the cold became intense and unbearable.

Gladly would I have faced a dozen man-eaters and returned to the bungalow, but this was impossible owing to the swollen river, which still remained unfordable. The waters swirled by, carrying with them uprooted tree-trunks, stripped branches, and debris of all kinds, including dead logs that had been lying for years along the river bank. To attempt a crossing of that raging torrent, in pitch darkness, would have been to invite death by drowning, if not by a blow across the head from a racing tree-trunk.

To add to my difficulties, the acute cold, the exposure, the lack of proper sleep and my now generally exhausted condition, brought on a sharp attack of malaria. The onset of the attack was in the usual form of ague, which caused my teeth to chatter like castanets, followed within an hour by high fever to the verge of delirium. I lost all interest in shooting and the tiger, and how I passed the remainder of that never-to-be-forgotten night without falling off the machan, I shall never know.

It was past 8 o'clock next morning before my Karumbas returned, to find me almost unconscious and still in a high fever. Somehow they got me across the river and to the bungalow, where I spent the next forty-eight hours in bed in the grip of successive attacks of malarial chills and fever, which only abated on the third day with the assiduous use of paludrine.

How I did not contract pneumonia as an after-effect of this terrible experience is a question to which only Providence knows the answer.

On the third day, when I began to take a little more interest in life, my men told me that the tiger had returned, either during or after the raging storm, and removed its victim while I was huddled, so ill, in the tree above it. This gave me an added reason for gratitude to Providence, for had I fallen off that tree in delirium, I might easily have become a fresh victim.

I took it easy for the next two days to give myself a chance of recovering completely from my bout of fever, and also to await news of a fresh kill, which was bound to occur, sooner or later. The third day I spent in procuring three buffalo baits, tying one each at Anaikutty, Segur and Mahvanhalla. They were all alive the following morning, so on the next day I resumed my peregrinations, roaming the forest around Anaikutty in the morning in the hope of meeting the tiger accidentally, and driving by car to Tippakadu and back by night.

Two days later, the run of bad luck I had been experiencing over the past three weeks suddenly changed for the better. That morning, for a change, I decided to follow the course of the Segur River for some miles downstream; driving with my Karumbas to Segur village, I left the car at a large banana plantation and began to put this plan into effect.

Hardly a mile downstream is a swampy area, much inhabited by bison in years gone by, and still locally known as 'Bison Swamp'. A half-mile beyond this, a large patch of dense bamboo jungle covers both banks of the stream. These bamboos have always been a favourite haunt of elephant and sambar, tiger occasionally passing through on their way down from the hills.

It was 8.30 a.m., and we had just entered the bamboos, when a sambar doe belled loudly from the opposite bank, to be taken up almost immediately by the hoarser cry of a stag. The Karumbas and I sank to the ground among the rushes that grew profusely along the river edge at this spot. The two sambar repeated their calls in quick succession, and it was obvious that something had alarmed them.

At first I thought they had seen or winded us, but I dismissed the idea with the realisation that what breeze existed was blowing upstream, from the sambar to us. Also, they could not have spotted us, being, as they were, some distance within the bamboos, from where we were quite out of sight. Our progress had been very silent and cautious, so the only conclusion to be drawn from the continued strident calls of the sambar was that they had seen or winded a tiger or panther, as no other human beings would be about in such a lonely place, due to the panic created by the man-eater.

We lay in the rushes for almost ten minutes when, with a loud clatter over the loose stones in the river-bed, a sambar stag, closely followed by a doe, dashed across, to disappear among the bamboos on the same bank as that on which we were hiding.

Not a move came from any of us as the anxious moments passed and then, silently, gracefully and boldly, a tiger stepped out of cover from the opposite bamboos and glided down the steeply declining bank to the river's edge. Without hesitation he walked into the river, ignored the cobbled stones, and when the water had reached his chest, he stopped and commenced lapping.

Taking careful aim, I fired behind his left shoulder. He sprang backwards, emitting a coughing-grunt, and then rolled over on his side, facing away from me and towards the bank from which he had just come. Running forwards out of concealment, I advanced some forty yards, from where I could just see a part of the side of his face, the rest having sunk below the water. Here I waited quite fifteen minutes, to put in another shot if need be, but it proved unnecessary, for the tiger was dead.

I have mentioned that I had taken careful measurements of the man-eater's pugs, which I compared with those of the specimen now before me, to find they corresponded exactly. Thus I knew that at last, after many tiring efforts and exasperating failures, I had shot the man-eater of Segur.

The reason for his man-eating propensity also became apparent, in that the animal had only one eye. The remaining eyeball had shrunk to nothing. When skinning this tiger I took particular trouble to investigate the cause of the loss of that eye, and upon digging out what remained of the organ with a knife, I found a gun-shot slug embedded in the socket.

Here was clear reason why the animal had become a man-eater. Someone, in all probability a poacher armed with a muzzle-loader, had fired at the animal's face, in the far-away Silent Valley of the Malabar-Wynaad. A slug had entered the eye and blinded him. Desperate, in pain, and hampered by the loss of his eye, the tiger had found it difficult to hunt his normal prey, and so had taken his revenge upon the species that had been responsible for the loss of his eye.

Excerpted from Nine Man-eaters and One Rogue *(London: Allen and Unwin, 1955).*

KENNETH ANDERSON

The Crossed Tusker of Gerheti

Kenneth Anderson was primarily a hunter of tigers and leopards, but in this case, a marauding tusker was his chosen quarry. Anderson lived, worked, and hunted in a region which, unlike Corbett's patch of the country, had large mega-herbivores: the gaur and the elephant. It was no mean feat to track and shoot a rogue elephant. Elephants normally do not attack people but there have been individual males who regularly come into conflict with human beings due to their crop-raiding activities.

This is the story of another elephant that earned the name of a 'rogue', and was proscribed by the Government of Madras by notification through the Collector and the District Forest Officer of Salem.

The events I am going to relate took place quite a time ago. As usual with rogue elephants, no one knew just what caused this elephant to start molesting human beings. The forest guard then stationed at Gerhetti stated that, one night about a month before the rogue began his depredations, he had heard two bull elephants fighting in the forest. According to his story, the contest had raged off and on for over three hours, and had taken place in the vicinity of a water-hole situated just about half a mile in front of the forest bungalow.

Next day he had found the jungle trampled down and great splashes of blood were everywhere in evidence of the punishment that had been inflicted. Judging by the account he gave me, and from the pandemonium that had raged, it must have been a mammoth struggle. Possibly the rogue, as we came to know him, had been the elephant that had got the worst of that fight and from this moment had begun to vent his spleen on all and sundry.

Another explanation might have been that the rogue was just an ordinary bull elephant in a state of 'musth', a periodical affliction that affects all bull elephants and lasts for about three months, during which time they become extraordinarily dangerous.

A third possibility was that this elephant had been wounded by one of the many poachers that are to be found in the forests of Salem District. These gentry sit up over water-holes and salt licks to shoot deer that visit such spots during the hot and dry summer months. Generally, when a poacher sees anything more formidable than a harmless deer, he keeps very quiet or slinks away if he feels the going is good. Yet even amongst poachers we find a few that are 'trigger-happy'. They discharge their muskets at any animal that puts in an appearance, and it may have been that one of these adventurers had wounded the bull and started him on his career as a rogue.

It may even have been that a simple peasant, guarding his crops by night, had shot at him with his match-lock. Elephants are fond of destroying crops that grow close to the forest.

Whatever it was that had originally upset him, the rogue of Gerhetti started his career quite suddenly, and for the short time he held sway in the fastness of the jungle where he lived, he became a terror, bringing all traffic, both bullock cart and pedestrian, to an end within an area of about 400 square miles.

Gerhetti is the name given to a tiny hamlet comprising some five or six huts about two miles off the track leading from Anchetty to Pennagram in the Salem North Forest Division. The country here is very hilly, and thick bamboo jungle grows to a distance of about three miles from both banks of a rocky stream known as Talvadi Brook, which joins the Cauvery about fifteen miles south-west of the spot I am telling you about. This bamboo jungle nearly always harbours herds of elephant, and quite often three or four independent elephants, which, although not rogues, are very carefully avoided by the jungle folk.

Another stream, called Gollamothi, flows almost parallel with the Talvadi rivulet, about 12 miles north of it, and joins the Gundalam River, itself another tributary of the Cauvery. These three rivers, with the hills that surround them on all sides, and the thick bamboo jungle that abounds, makes an ideal habitation for any elephant, and it was here that the rogue started his career as a killer.

The Crossed Tusker of Gerheti

It began like this. With the midsummer heat, the Gollamothi stream had dried up, except for one or two isolated pools of water which had managed to survive, being formed between huge rocks that cropped up on the river bed, and fed by sub-soil percolation. One of these pools was known to hold fish of some size, perhaps six to eight inches in length, and one afternoon two men from the village of Anchetty, five miles away, decided they would go to this pool and net some fish in the restricted area that had resulted from rapid evaporation.

So they arrived at the pool and cast their nets. Soon they had made a considerable catch. They then put the fish into their baskets and lay down under a shady tree by the side of the water-hole to enjoy a brief siesta.

It was about five when one of them awakened. The sun had just sunk behind the top of a hill that jutted out to the west of the pool, but it was still quite bright. As he sat up beside his sleeping companion, something caused the man to look behind him, where he saw the slate-grey bulk of an elephant descending the southern bank of the Gollamothi on its way to the pool.

The man reached out and vigorously shook his companion, to whom he whispered in Tamil *'Anai Varadhu'*, which means 'an elephant approaches'. Then he got up and ran to the northern bank and into the forest. His companion, suddenly aroused from sleep, did not quite grasp the significance of the warning, and as he sat looking around and wondering what had happened to his friend, the elephant was upon him.

The man who escaped told me he heard the screams of the friend he had left behind, mingled with the shrill trumpeting of the enraged elephant. Then there was silence. Naturally, he had not waited to hear more. Two days later, when the search party from Anchetty came to look for the remains, they found a pulpy mass of broken flesh and bones decaying in the hot sun. There was evidence that the elephant had first placed his foot upon the man and then had literally torn him apart with his trunk. He had carried one leg to a spot ten yards away, where he had beaten it against the gnarled trunk of a jumblum tree before finally throwing it away among the rocks.

That was the rogue's first victim. His second attempt was upon a herd-boy who was driving his herd at sunset to the cattle patti at Gundalam. This boy, being young and agile, had fled along the sands of the dry stream, hotly pursued by the vicious elephant. Finding he was losing ground, the boy had the sense to run up the steep side of the hill where the rocks were very slippery and small loose boulders abounded. This had enabled him to maintain his lead.

In his mad rush to escape, the boy cut his bare feet literally to ribbons on the protruding sharp stones, while his body was lacerated in a hundred places by the thorns and shrubs that sought to hold him back. But he had kept on running, and managed to escape the elephant by climbing on to a high rock that protruded about two hundred yards up the hill side.

He told me afterwards that when the bull reached the rock, he walked around it several times, trumpeting and attempting to reach his victim with his trunk. But the boy kept his head, and moved around with the elephant, keeping as far as possible from the tip of that dreaded trunk. He told me that the top of the rock was only about twenty-five square feet in area and that the snake-like tip of the killer's trunk had sometimes been within a foot of his ankles. Nevertheless he had managed to avoid it, and after an hour of this game the elephant suddenly lost interest in his victim and wandered away. Still the boy had been too frightened to come down and had spent the night on top of the rock, only getting down next morning after the sun had risen high in the heavens, when he felt that the elephant was nowhere in the vicinity.

After this there had been a lull for a month, when the few folk who lived in the area began to feel that the elephant had perhaps departed to other regions, or alternatively, if it had been in musth, that the musth season had elapsed and that its condition had returned to normal.

But they had been far too optimistic, for exactly five weeks after unsuccessfully chasing the herd-boy this elephant attacked two wayfarers as they were journeying through the forest to the village of Natrapalyam, which lies about eight miles south of Anchetty. These men had been suddenly chased by the rogue and had begun to run along the forest path with the animal in hot pursuit about 100 yards behind them. One of these men was about thirty years old and the other some ten years older. Age soon began to tell as the older man began to lag, his breath coming in sobbing gasps. He knew that a terrible death was behind him, and he tried his best to keep running. Unfortunately, he had quite lost his head, and made no attempt to circumvent the animal, as he might have done, by perhaps climbing a tree, or by getting behind a rock or even by throwing down a part of his clothing as he ran. This last action might have served to delay the attacker for a few minutes. For when chased by an elephant, it is advisable as a last recourse to shed some part of one's clothing; when the elephant reaches it and catches the strong human smell, he will invariably stop to tear it to ribbons. In the precious seconds thus gained, the victim has a chance of making good his escape.

But this unfortunate man simply ran on till he could run no more. The elephant overtook him as he lay sprawled on the path, his body heaving to the gasps of his tortured lungs. Soon their services on this earth were ended, for the elephant picked him up in his trunk and dashed repeatedly against a wayside boulder, beating him to pulp before finally tossing him aside into the jungle.

As a result of these incidents, petitions had been forwarded, through Forest authorities, to the Collector of Salem District to proscribe this animal as 'rogue'; which means that permission was granted for the elephant to be shot by any game-licence holder in the district. Normally, the elephant is strictly protected in India. Red-tapism, as anywhere in the world, is a slow process, and this is particularly the case in India, so that three months or more elapsed before action was taken to issue the necessary order. A further month's delay occurred before all game-licence holders in the district were notified.

Meanwhile the 'rogue' was rather busy. He attacked a bullock-cart that was laden with sandalwood cut by the Forest Department. The driver of this cart, and the forest guard who was accompanying the sandalwood, escaped by running into the forest, but the cart was smashed to pieces and one of the bulls slightly injured.

Not long afterwards, the rogue did rather an unusual thing. Cattle, let loose by graziers in the forest, generally scatter over a fairly wide area. One of these animals evidently strayed too near the rogue, which attacked and broke the beast's back.

A further short lull was followed by the news that he had killed a 'Poojaree', one of a jungle tribe of this area, as he was returning with honey from the forest for the contractor who had bought the right from the Forest Department to collect all the wild honey in this particular division.

Then the official notification reached me. As a rule, I take no pleasure in elephant-shooting, as I have a very soft corner for these big and noble animals. Secondly, I feel it is comparatively tame sport, as apart from the large target presented by the animal itself, elephants invariably give away their position by the noise they make in the undergrowth when feeding. It is then a comparatively simple matter to get up-wind of the quarry and stalk to within a short distance of it. All that is required is a little experience in knowing where to place one's feet, to avoid stepping on dried leaves or twigs that crackle and so give away the stalker's position.

Another very important aspect of 'still hunting' is the ability to 'freeze', to become absolutely still in whatever position you may be at that moment, if the quarry looks your way. This may prove a little awkward at times, when in a half-crouched position, but to straighten up or squat down would be fatal, as the slightest movement involved in doing so will give the stalker away. The thing to do is to remain absolutely and completely 'dead', even for as much as ten minutes in the same half-crouching position. I can assure you that this can sometimes be extremely tiring.

By these methods I have often stalked elephants to within a few yards and watched them grazing peacefully, without their being in the least degree aware of my presence. But the slightest whiff of my scent, or the slightest crackle underfoot, would have sent them thundering away. An elephant has surprisingly poor sight, and if you are dressed in military khaki-green and you keep absolutely still, it will often look your way without ever becoming aware of your presence.

For these reasons, as I have said, I do not like shooting elephants. Also, many of the so-called rogues are not rogues at all. As I have mentioned before, poachers and cultivators are in the habit of firing at elephants and often wound them in the process, when they become embittered against the human race. Again, many of the incidents reported against the so-called rogues never occurred, for people interested in shooting an elephant sometimes concoct tales and urge the villagers to write exaggerated reports in order to induce the Collector to proscribe the rogue. Collectors as a rule go into such matters very carefully and thoroughly before issuing orders; but sometimes, with all these precautions, elephants are killed which are not rogues at all in the real sense.

So I did not pay much heed to the notification till it was followed, about three weeks later, by a letter from Ranga, my shikari who lived at the town of Pennagram. He wrote to report that the elephant had killed a Poojaree woman four days previously at a spot called Annaibidhamadhuvu, which lay about seven miles from Gerhetti. The literal interpretation of this name is 'the pool into which the elephant fell'. It is a natural pool, formed by steep rocks on the bed of Annaibidhahalla river, a sub-tributary of the Chinnar river, which itself is a tributary of the Cauvery, that largest of South Indian rivers. Moreover, this pool is deep and never dries up, and I am told that many years ago an elephant, while reaching for water with its trunk, fell in and could not get out again because of the steep and slippery rocks that ringed it. Elephants are good swimmers, and have herculean strength and endurance, but they also have great bulk; so this poor creature, after swimming around and around in the pool for three days continuously, slowly sank and drowned before quite a large crowd of people who had come all the eleven miles from Pennagram to witness the 'tamasha'.

After receiving Ranga's letter, and as some time had passed since I had last seen him, I got leave of absence for four days and motored down to Pennagram. I picked him up and covered the 18 miles or more of terribly bad forest road that leads to the Gerhetti forest bungalow, passing Annaibidhamadhuvu on the way.

The forest guard at Gerhetti told me that the rogue was in the vicinity, as well as a herd of about ten elephants. All these animals were in the habit of drinking at the water-hole in front of the bungalow, and as there were several animals in the herd of about the same size as the rogue, it would be difficult to know which was which. This precluded any possibility of following with any degree of certainty any particular set of tracks.

Further, the description of this animal in the notification was very vague; it merely stated that the measurement around the circumference of the forefoot had been 4' 10", which made the elephant approximately 9' 8" tall, as twice the circumference of the forefoot is the approximate height. The colour was reported as black, but all elephants are black after they have washed, but they soon cover themselves with sand or earth. Sand in the forest is of different hues, varying from red to brown, grey, and almost black. So this was no distinguishing factor either. The only feature that appeared to identify the elephant was that the two tusks, which were reported to be over three feet long, met and crossed near their tips.

You will therefore understand that the last factor was the only one by which I would be able to identify this particular animal, for it was doubtful if he would give permission to approach and measure his height! But to see if the tusks were crossed meant getting a frontal or head-on view of the elephant, as at an angle tusks may appear to cross without actually doing so. I certainly did not want to shoot the wrong animal, apart from the immense amount of trouble and official explanation that would follow.

This meant that I could select any set of tracks that came up to the measurement of the rogue and follow them till I came upon the animal that had made them, then manoeuver for a frontal view of the animal to see whether he possessed the hallmark of crossed-tusks. If he did, he was my elephant. If he did not, I would have to start all over again by going back to the pool and following another set of tracks. It must be remembered that I had only four days at my disposal, and of these four days one had already passed in picking up Ranga, coming to Gerhetti and making the necessary inquiries.

At about ten that night I heard the sound of elephants feeding in the vicinity of the pool. This was undoubtedly the herd, and the rogue would not be with them. So I went to sleep again.

At dawn I started with Ranga and the Forest Guard on my plan of following up one of the sets of tracks. The margin of the water-hole was fairly ploughed up by a mass of footprints of all sizes, where the herd had watered the night before. These included the tracks of some very young elephants, which could hardly have been over three feet tall.

Circumventing the pool, I found three sets of tracks which came near to the size of the rogue. Two of these three sets, I noticed, had been made on the same side of the pools the herd had watered, while the remaining set had been made by an elephant which had approached the water from quite another direction. I therefore argued that this third animal might be the rogue, and we began to track him in single file. Ranga went in front, following the tracks; I followed, covering him with my double-barreled 450/400 Jeffries; the Forest Guard came behind me, his duty being to guard against an attack from the rear in case it happened that the elephant had gone round in a semi-circle, and was now grazing behind us.

For a short distance after leaving the water the ground was covered by long spear-grass and clearly revealed the passage of the elephant the night before, being trampled flat in all directions. Then the spear-grass gave way to the usual thorny growth of lantana and wait-a-bit thorn. Here also it was comparatively easy to see where our quarry had passed, but our own passage became more difficult by reason of the thorns that plucked and tore at our clothing.

Yet the tracks were clear and we made fairly good time for about a mile, when we reached the base of a small hill. The slopes of this hill were covered with heavy bamboo growth, and the elephant had passed through this, climbing the hill as he went. He had also stopped to feed on the tender shoots that spring from the end of the fronds of bamboo, as was clearly evident by the havoc he had created in the mass of broken bamboo stems we met along this trail. Here he had passed a considerable quantity of dung, and as we reached the top of the hill and went down the other side, it was evident that the elephant had fed until the early hours of that morning; for the dung was fresh and had not had time to cool.

From here onwards our passage became laboriously slow. The dense bamboo completely surrounded us and a careless step by any one of us resulted in a rustle or sharp crackle, depending on whether we trod upon the leaves of bamboo fronds.

I touched Ranga's elbow and motioned to him to stand still, the guard and myself doing the same. We listened for over ten minutes for the familiar sounds of a feeding elephant, or the deep rumble that issues from his cavernous stomach in the process of digestion. But the forest was comparatively silent except for the cheery calls of grey jungle cocks everywhere around us, and the distant whoops of langur monkeys on the opposite hillside.

Evidently our quarry was resting, or had perhaps passed further down into the deep valley that lay before us. As the latter seemed more likely, we proceeded on tip-toe, slowly and carefully in his wake. We had all to keep our eyes down to make sure we did not trample on anything that would betray our presence. Another half mile brought us to the valley, where the undergrowth was extremely dense; wildplum, wood-apple, and mighty tamarind trees grew profusely everywhere, making visibility beyond 15 to 20 yards impossible. Another 200 yards brought us to the rocky bed of a small tributary of the Gollamothi river.

The elephant had here skirted the bed of the stream and crossed it at a sandy spot 50 yards further down. The opposite bank, up which he had then climbed, was fairly steep, so that we were now faced with the prospect of having the elephant above us, which is hardly the best way of meeting a 'rogue'.

We went forward very slowly indeed, and as silently as was humanly possible. Our quarry had stopped feeding, and was now on the move again. We soon saw that he appeared to be making for another valley that lay beyond the spur of the hill up which we were now climbing. Having reached this conclusion, we began to move faster, but had gone only another quarter-mile when we heard the rumbling sound made by the elephant's stomach in the process of digesting his heavy meal.

I sent Ranga and the Forest Guard up a stout tamarind tree and crept forward alone in the direction of the sound. The elephant was in a small depression, densely wooded by bamboo. Evidently he was resting, or perhaps lying down, as there were no sounds of his feeding. By this time the rumbling had also ceased.

Very carefully, almost inch by inch, I went down into the depression. Then, stopping for a moment, I gathered a little soft earth in my hand and held it up before me, letting it drop in order to see from which direction any current of air might be blowing. The earth fell straight, indicating that there was hardly any breeze in the depression. This was a handicap, as there was a chance of the elephant smelling me in the still air.

So I went forward, still more carefully, if that were possible. The bamboos towered above me, and I peeped around each clump as I came abreast of it. A few more yards of this sort of progress and I saw what appeared to be a slate-grey boulder before me. It was the elephant, lying on the ground, and as my bad luck would have it, facing in the opposite direction.

I could now do one of two things: either make a detour and try to come upon the elephant from the front, where I might see his tusks and identify him as the rogue before shooting, or, much simpler, rouse him from where I stood. He would undoubtedly turn around to face the disturbance, so that I could then identify him and shoot before he could know what was happening.

Deciding on this second and easier course, I slipped partly behind a clump of bamboo, then softly whistled. The elephant took no notice. Perhaps he was deeply asleep, or thought the sound had come from some forest bird. Then I clicked my tongue loudly. This had the desired effect, for the elephant scrambled to its feet and spun around to face me.

He was a magnificent tusker, quite ten feet tall, and his ivory tusks gleaming magnificently in the early morning sunlight. But they were wide apart, not crossed in the least. I had spent my time tracking the wrong elephant.

The pachyderm looked at me in amazement for quite half a minute, his small eyes contemplating the creature who had disturbed his slumbers. I could almost read the thoughts that were passing through his brain. His first reaction, after surprise, was annoyance, and he moved forward a pace or two in a threatening attitude. I gave another sharp whistle, at the sound of which his courage ebbed away, and he turned tail and bolted into the forest, the crashing sound of his retreat dying away in the distance.

By the time I returned to the spot where I had left Ranga and the Forest Guard, they had already climbed down from the tamarind tree, guessing, by the sounds they had clearly heard, that I had found an animal which was not the rogue we were after. The three of us then trudged dejectedly back to the water-hole, not only disappointed, but annoyed at the time we had wasted.

As previously related, there were two other tracks of approximately the same size. They had been made in the mud of the pool and nothing could be gained by measuring them with my tape to determine which came nearest to the notified dimensions of the rogue; soft mud exaggerates the track of any animal. Ranga followed one, and the guard and myself the other, with the understanding that we would return to the water-hole in fifteen minutes for further consultation.

It was not long before I could see that the animal I was following has been one of the regular herd, for the broken undergrowth revealed the presence of the feeding cows and young that had accompanied him. He was obviously not the rogue, and in exactly fifteen minutes by my watch I turned and made my way back to the water-hole. Ranga, having no watch, had not yet arrived, so I sat down to a quiet pipe and sip of hot tea from the flask carried by the Forest Guard. After about ten minutes, he came to report that the elephant had made a detour a quarter of a mile from the water-hole, had moved around in a semi-circle and passed through a strip of jungle that led to a hill in exactly the opposite direction, behind the bungalow.

This news seemed promising, so we were up and away. Nor was it long before we came to the spot whence Ranga had returned to report. It soon became evident that our new quarry was a traveller, for he had hardly stopped to feed, other than pluck an occasional small stem of succulent young leaves. That elephant led us on and on, over the hill behind the forest bungalow, over the next two hills, and then in almost a straight-line to the Talvadi stream.

In all we covered well over four miles before reaching the bed of that stream, when we found that the elephant had turned south-west and was moving directly down the Talvadi river itself. I knew the Cauvery River lay within a distance of 15 miles, and I began to feel our quarry had suddenly made up his mind to reach the big river. Once he did this, and particularly if he swam across to the opposite bank, it would be hopeless to follow him, as the terrain there is not only extremely dense, but leads on and on as unbroken forest and hill country to the Niligiri and Biligirirangan Mountains, over a hundred miles away.

So we passed on with all possible speed, casting discretion to the winds, but our elephant had had a lead of several hours, and judging by the long and determined strides he had taken, he had been bent upon travelling.

The soft sand of the river-bed was now scalding hot under the midday sun. It hampered my walking and trickled into my boots by means that only fine river or sea sand knows. Every now and again the stream bed became rocky, and for long stretches the fine sand gave way to a succession of rounded, water-worn boulders. In such spots the elephant had pushed through the undergrowth of the banks to avoid the boulders, and we did the same, bent double to dodge the dangling lines of creepers, and pouring with perspiration from our exertions.

Fifteen miles of such walking brought us near the confluence of Talvadi and the Cauvery. A few hundred yards from the big river, the Talvadi stream is crossed by the rough track leading from Uttaimalai village to Biligundlu. The elephant had changed direction here and had followed the track towards Uttaimalai for another two miles, before turning southwards again towards a swamp that borders the big river. This swamp, known as Kartei Palam, which means Bison Hollow, was well known to me. Years before it had been a regular haunt for bison herds that swam across the Cauvery from the Coimbatore bank to the Salem side. At this time of the year the swamp was fairly dry, except in places, but lush grass grew everywhere, while shady clumps of trees dotted the whole area.

We now met with signs that the elephant had begun feeding, and as we made our way towards the centre of the swamp mounds of fresh dung showed that the animal was not far away.

The ground also became boggy, and once more I sent Ranga and the guard back to minimize the squelching sounds that were bound to arise from three people walking in the mud. Progress was necessarily at a snail's pace, for I had not only to look out for the elephant, but study the ground carefully at each step, to avoid suddenly plunging waist-deep into the clinging black clay. Yet, several times I sank knee-deep, and to extricate myself I had to struggle and flounder about, making no end of a noise, before I gained a firmer footing.

Several times I stopped to listen but heard nothing, and then, without warning, there came a violent 'swoosh' of the reedy grass, and the elephant stood some twenty paces away, all dripping and covered with the sticky muck in which he had been lying. It was a big bull, with gleaming white tusks, symmetrically curved. But they were not the crossed tusks of the rogue. Disappointment and disgust so overcame me that I fairly 'shoo-ed' that poor elephant away, and when I rejoined my followers, I was in no good mood, as they could clearly see.

It was now past 4 p.m. and we had some 15 miles to retrace along the Talvadi Stream, plus another four to the Gerhetti bungalow. Alternatively, we could camp at Biligundlu and return next morning; but this would mean the loss of another half-day, out of the two days that were left to me. So I gave the order for the return march, much to the disgust of my companions, who reminded me that, as we had no light of any kind, the major portion of our journey would have to be performed in darkness, there being no moon. We might even meet the elephant! My reply, I am afraid, was terse, and consigned this elephant, and all other elephants, to a region they would find far too hot.

That return journey seemed one long succession of stumbling, slipping, slithering over rocks, or tripping over stumps, or being caught by creepers without sign or sound of the elephant. It was almost midnight before we limped into the Gerhetti bungalow, thoroughly exhausted and as fretful as children. We had been up with the dawn, walking incessantly, stalking through thorns, grass, river-sand and swamp, and had covered about 40 miles. We were ravenously hungry and thirsty too.

Next morning I was cramped and foot-sore. The Forest Guard showed a swollen ankle, which he had contrived to twist somewhere along the Talvadi stream, and begged to be excused from that day's operations. Only Ranga appeared fit, and ready for another hard day. Porridge, bacon and eggs, and strong coffee put new life into me, while a huge ball of 'ragi', which is a small foodgrain, boiled and made into a sphere almost the size of a tiny cannon-ball, washed down with coffee, would satisfy Ranga till night-fall.

We had all been too tired to hear any sounds during the night, but a visit to the water-hole now indicated the herd had returned while we had slept. There were also the fresh footmarks of two big bulls, one of which was probably the first elephant I had followed the previous day, while the other was the animal I had not followed at all. The third bull, as we well knew, we had left far away at Kartei Palam.

Nevertheless, nothing could be left to chance; so we followed the same plan as that of the previous day, tracking each of these two animals till we came up with them. By 9.30 a.m. we had come up to the first bull that we had decided to follow. He was slightly smaller than the two we had tracked the previous day, and he was not the rogue! No doubt this was the third animal of the trio whose footprints we had noted the previous day. Going back to the water-hole, we set out on the remaining track, and came upon its maker at 2.30 p.m., quietly standing under a large and shady tamarind tree. Nor was he the rogue. I readily recognized him as the first animal I had tracked the previous day and had disturbed while lying among the bamboos.

Thus it was clear that the 'rogue' was not in the immediate vicinity. Three of the four days available to me had now gone, but I was still no further forward than on the day of my arrival.

At five we were back at the bungalow, brewing a large 'degchie' of tea. Then at half-past five a party of bullock-carts arrived from Anchetty, eight miles away, to shelter for the night because of the presence of the rogue. The cartmen stated that at a spot about half-way from Anchetty, where the Gollamothi stream traversed the road, they had come upon the tracks of a large elephant which had crossed and recrossed the road at several points and was evidently hanging around not far from the ford itself.

Determined not to give up till the last moment, Ranga and I ate an early dinner and, bundling the still-tired Forest Guard into the car, motored to the ford of which the cartmen had spoken.

It was 7 p.m. when we arrived there and almost dark. The car lights revealed the tracks of the elephant where the cartmen had said. At the ford itself, with the aid of torches we made out the plate-like spoor of the elephant superimposed upon the narrow ruts made by the cart-wheels of our friends. Elephants do not wander about in day-time in hot weather, and this clearly indicated that the pachyderm had been on the road that very evening, before our arrival. Perhaps he had even heard the sound of the car, or seen our lights, and had moved off just before we came on the scene.

We quickly lowered the hood of the Studebaker. I handed my 'sealed-beam' spotlight to Ranga, whom I placed in the 'dickie' seat behind, but kept the guard in the front seat. I myself sat on the folded hood, my feet on the driving seat, with my rifle and torch at the ready. Complete darkness soon enveloped us, overshadowed as we were by the towering muthee and jumblum trees that bordered the banks of the Gollamothi, together with bamboo clumps, whose stems creaked weirdly to the jungle air-currents that blew up and down the dry sandy bed of the river.

The prospects were poor. To begin with, we did not know whether the tracks we had seen belonged to the rogue. They might, indeed, have been made by any elephant. All we knew for certain was that they had not been made by any of the three big males around Gerhetti. Secondly, there was not the slightest reason for this elephant to return to the ford he had so recently passed, for he had the whole wide jungle in which to roam. Thirdly, we all knew that ten miles to an elephant is scarcely two hours' easy ambling, and that when he is really travelling he moves much faster. Fourthly, the wind was blowing in all directions, and would carry our scent to any elephant within a quarter-of-a-mile and, if he was not the rogue, drive him off. On the other hand, there was the slender hope that, if this was the rogue, our scent might attract him.

So we waited in the pitch darkness till 8.30 p.m., and then the dull sound of a hollow log being turned over came to us from somewhere upstream. The elephant was on the move at last and, judging by the sound, was some four or five hundred yards away. Silence followed for another quarter hour, when the sharp crack of a breaking branch, much closer, indicated that the elephant was feeding and moving towards the ford as he did so.

I knew he would take thirty minutes at least to finish eating the young leaves from the ends of the branch he had just broken, so that there would be plenty of time before he came near. The car, with us inside it, would be clearly visible to him as he came around the curve of the river, and there was every possibility, if he was not the rogue, that on seeing us he would just fade quietly away into the forest. But I had not taken the wind into consideration; just then it blew strongly from us towards the elephant.

Minutes of silence followed, and then we heard a slight rustling in the undergrowth from the bank of the river nearest to the car. It was a faint sound, apparently made by a small body in the bushes. Then the ominous crack of trodden bamboo came to us suddenly. Silence again, deep and enveloping. Even the breeze seemed to have died, to allow full opportunity for the next event.

This was the ear-splitting scream of a charging bull-elephant, mingled with the crashing of bamboos and undergrowth as they collapsed before the monster that rushed towards us.

Ranga never flinched! The beam of the spot-light cut through the enveloping gloom. My own torch-beam, mingling with that of the vastly more brilliant spot-light, showed an enormous tusker, his bulk pitch black, his trunk curled upwards and inwards, with two wicked white tusks that were crossed at the tip, thundering upon us!

At 50 yards range the bullet from my right barrel took him in the throat. He stopped with the impact, screaming with rage. No doubt this was more than he had bargained for! The explosion, the pain and the lights confused him, and he half-turned into the jungle. My second bullet, aimed hastily at the temple, struck him somewhere on the side of the head. He rushed into the jungle, stumbled on his forefeet, picked himself up again as I reloaded, and disappeared in the bamboos as my third shot struck him somewhere in the body. For quite fifteen minutes we could hear heavy crashes in the jungle as the elephant reeled, collapsed and then recovered, to continue his flight.

Starting the car, I reversed and returned to Gerhetti. I slept soundly that night, Ranga awakening me at dawn with a mugful of steaming, strong tea.

By 6.30 a.m. we were on the track of the stricken rogue. Great gouts of blood had issued from the throat-wound and had sprayed through his trunk over the surrounding bushes, which had been reddened by his passage. Soon we too were red with his blood as we pushed ourselves through the undergrowth in his wake.

He had lain, or fallen down, in several places, where the greensward had been dyed a deep red. He had leaned against several tree-trunks that were still sticky with his blood. Truly, if you do not finish the job of killing an elephant, you let yourself in for a gory trail. I really pitied this poor beast, murderous killer of men though he had been.

After two mile I found him, half-kneeling, half-lying against a tree-trunk. He was so weak from loss of blood that he could scarcely move, although he clearly saw me as I walked towards him. The only sign of life were his wicked little blood-shot eyes, that gleamed and moved as they watched my approach. Fifteen yards away I raised my rifle to deliver the coup de grace. As if to salute approaching death, that game and mighty beast shivered from head to foot as he drew up his mighty bulk to its full ten feet. The trunk curled upwards, the big ears flapped outward, and he staggered two paces forward in his last charge, when the heavy 450/400 bullet crashed into his temple and he collapsed, as if pole-axed, to earth.

Although a killer, the 'crossed-tusker of Gerhetti' was a brave fighter, and I honoured him as he lay before my still-smoking muzzle—mighty in life and even mightier in his death!

Excerpted from Man-eaters and Jungle Killers *(London: Allen and Unwin, 1957).*

R. P. NORONHA

Tigers in Bastar

Though it is Kumaon that won fame, or notoriety, for fatal encounters between people and tigers, mainly due to the writings of Jim Corbett and J. E. Carrington-Turner, it was Bastar in central India that reported man-eating till much later. As Noronha, veteran hunter and civil servant, and Deputy Commissioner of Bastar, shows us, man-eating is a highly specific phenomenon. When tigers attack people as prey, it indicates a breakdown of the normal patterns of avoidance and self-defence. Noronha's experiences date to the 1950s, and what is notable is his close attachment to the cultures of tribal groups who were both neighbours and victims of the tiger.

THE MAHSEER TIGER

The Indrawan river has a short and merry life. It races through the Central India hills behind Dumarpadao, accelerates over several waterfalls, slows down in the plains, and obliterates itself without a whimper in the Mahanadi. In the hills it is alive, dancing and sparkling through cataracts and gorges and dense forest. There it teems with mahseer, and the country through which it runs is full of tigers, frequently maneaters. What makes a tiger a maneater? After forty years in the jungle, I don't know. There are the obvious causes, a wound, old age, other forms of physical incapacity. But they don't explain the tigers which become maneaters without any physical incapacity whatsoever, nor do they explain why some areas are more infested with maneaters than others; like the Sunderbans, like Kalahandi in Orissa...like Dumarpadao. Training by a maneating mother? On four occasions I was able to save the lives of the half grown cubs of maneating tigresses. I followed their subsequent careers closely, and none of them became a maneater.

There was a tiger around Dumarpadao again, with thirty-two human kills to its discredit and Caesar was after it with his friend Charan Singh, the Gond. Caesar is not his name, but he was a government servant wielding power, hence Caesar. They had been after the tiger for four days now, and the technique they used was unique. A maneater has perforce to adapt itself to the habits of its prey, in this case man. Normally it is a nocturnal animal, sleeping in the day and killing at night, when the deer family and pig are on the move. But man is not nocturnal. He sleeps by night and is on the move during daylight hours. If therefore the tiger wants human flesh, he too has to hunt by day and sleep at night. That was the secret of the method used by Charan Singh and Caesar. Once they found a human kill, they tracked the tiger without respite, all through the day. When dusk fell, they ate and slept where they were, picking up the trail again at sunrise. And so on, day after day, like a two-man pack of wild dog, remorseless, relentless. There was not much danger at night, because then the tiger too slept. Not that they relied wholly on the tiger; one of them was always awake. The close tracking kept the tiger on the move and gave him no chance to kill and eat. Sooner or later anger and hunger drove him to stalk the stalkers or to turn at bay. In either case there was a showdown, which is the objective of the shikari who is on the trail of a maneater. Or should be.

I hope I have not made the technique sound too easy. It is not; and it is not to be recommended to amateurs. You need a cool head, quick and precise shooting, excellent physical endurance; and above all, experience of the jungle. Caesar and Charan Singh had all of these. Even more important, they combined beautifully as a team.

Caesar pushed up the steep bank of a nala, overgrown with dense bush, making no effort at concealment. In contrast Charan Singh slid silently into the nala and faced their rear with his old double-barreled .500 Express at the ready. For some reason the other maneaters leave the first man in a single file alone; it is always the last that is whipped off, suddenly and silently, so silently that the others are ignorant of his fate until much later. The two men completed the crossing without incident. At the moment they were not tracking. The tracks had petered out over a rocky stretch. They were searching for tracks, following the contours which a tiger follows, guessing at his probable direction and basing the guess work on a very sound knowledge of this jungle and this tiger. Charan Singh skirted a large ber (wild plum) bush carefully, found something and held it up. Caesar peered over his shoulder. A single coarse hair, a yellow-white hair. He shook his head admiringly. The tiger had kept close to the bush, using the hard ground beside it to mask his pug marks. So much caution might mean an ambush, he thought, and his eyes roved restlessly. They caught a glimpse of movement, the rifle leaped to his shoulder. Disappointed of achieving a surprise the tiger woofed and vanished in the undergrowth, giving no chance for a shot. Resignedly Charan Singh sat on the ground, his rifle across his knees, and said, 'Come on. Give your lecture, how you checkmated the tiger, how quick you were, how lucky the tiger is to be alive. I'd much rather hear it sitting comfortably than while we're on the move.' They were old friends, these two. Caesar sat back to back with Charan Singh from force of habit (the jungle saying is—two men facing the same way are one man) and began. 'You are lucky to be alive. You would not have been alive if I had not been so alert, so quick, such a superb shikari. I apologize to the world for saving your life.'

Charan Singh threw a clod of earth backwards at him. It was fielded neatly. 'See? So quick. I think he'll stop running today.'

'Perhaps, but I think not. He's a coward like all maneaters, and he's had a fright. May be tomorrow. But what does it matter, the exercise will keep your fat down.' The injustice of the remark was so patent—it was Charan Singh who was fat and Caesar who was thin—that it silenced Caesar, at least for a while. Charan Singh's voice intruded on the stillness in conversational tones, without a trace of tension—'He's about fifteen yards directly behind you. Turn as soon as I fire.' Caesar shifted his hand unobtrusively to the grip of his rifle, saying in the same conversational tones, 'O.K. Seventeen eighteen thirty-four fifty I can't think of anything to say forty-five ninety.'

Charan Singh spoke again—'He's gone. You know, that tiger deserves admiration. If we had been beginners one of us would have been dead by now.' He got up. 'We'll find the tracks in the nala, he went that way, and he had not time to be careful.' They retraced their steps and right enough the pug marks were there, water still oozing freshly into them. They followed the tiger the whole of that day without ever getting close to him. But they were never so far as to let him rest. At dusk they slept alternately on a hillock high enough to catch the cool night breeze, the handful of gur and chana which was all they had to eat resting lightly on their stomachs. As Caesar dozed off, Charan Singh said, 'Tomorrow we must find time to bathe, or else even the tiger will smell us.' Tigers, and to some degree leopards, have a very weak sense of scent. They hunt by sight and sound.

Early next morning they resumed the tracking. A definite direction began to emerge. In spite of twisting and turning it soon became apparent that the tiger was heading for Devdhara, a spectacular horseshoe shaped waterfall, not far from where they were. Charan Singh was pleased. 'Now we can have that bath. Hathikoh is close by. I'll tell them to bring our clothes and some hot food to Devdhara. Wait for me.'

Caesar found a convenient depression at the top of an anthill and settled himself comfortably. He passed the time by checking his rifle, a very special rifle. Starting life as a perfectly respectable 10.75 mm (.423) Mauser, many indignities had been heaped on it. The barrel had been chopped down to 21 inches and a fat blob of a foresight, visible in the poorest light, installed. The back sight too had been replaced with a very wide shallow 'V', and the stock on it had been tailored to Caesar's measure. If you looked carefully you might see traces of bluing on the metalwork; the butt however was innocent of any kind of polish. The safety had been removed; with the bolt locked, that rifle was ready to fire, and no fiddling with safeties. Caesar sometimes made a concession to convention by carrying it with an empty chamber, but not often. He claimed that the trigger guard was safety enough for a rifle in good condition. The checking done, he relaxed, as much as he ever permitted himself to relax in the jungle.

It was just after ten. The heat waves were beginning to dance in the sunlight, but it had not really warmed up as yet. He sat without movement of any kind, even when the eye flies settled in the little wells of sweat under his lashes. A sleek shiny cobra poured itself over the ground into the next patch of scrub where a cluster of ground thrushes reacted hysterically. He was pleased; snakes were always lucky for him on a tiger shoot. Then he heard the splashy footsteps of a peacock on dry leaves, so similar to the sound of a careless tiger. The peacock came down an incline on his left and stood motionless at the edge of a tiny clear place in the bushes, looking about him with exaggerated caution, as if his clumsy feet had not just warned the whole jungle of his arrival. A large solitary langur, the common black-faced monkey of Central India, followed him down the incline at an awkward gallop, and the peacock disappeared into the scrub. Caesar sat still, partly from habit and partly because it was the easiest way to keep cool in the pervading heat. Now there was a cobra, a peacock and a langur, all in the same small patch of scrub, not to mention the ground thrushes. He wondered what would happen. Perhaps they would keep out of each other's way, perhaps...

The return of the monkey put an end to speculation. He came out of the scrub walking slowly, stiffly, on two legs, a paw stretched out, the cobra coiled round the arm, gripped firmly just behind its head. Back in the scrub the thrushes shrieked and the peacock confusedly gave his alarm call for a tiger. The langur picked out a suitable rock, squatted on top of it and slowly, systematically, started to rub the snake's head into a pulp against the stone. He interrupted his labours after a minute and gravely lifted his hand to examine the results of his work. The coils round his arm were beginning to loosen. He resumed the rubbing and continued until they fell away entirely and the cobra was hanging limply from his paw. Then he threw it violently from him and returned to the scrub at the same awkward gallop. Everything was as it had been before. Caesar blinked. Had he seen what he had seen or was it an illusion? He descended the anthill and went to the rock. It was not fantasy. The cobra, almost headless, lay a few feet away.

Charan Singh's voice interrupted him. 'You're more than usually stupid today. If I had been that tiger I would be tasting you by now—and regretting my choice of meat.'

Guiltily Caesar accepted the rebuke. He gestured to the snake. 'Monkey rubbed its head off, I saw the whole thing.'

Charan Singh was immediately interested. 'You're lucky. All the jungle people say that is how a monkey kills a snake, but very few have actually seen it being done.' 'Did you ever see it?' 'Dozens of times. And now I'm a bigger liar than you! Come on, let's go to Devdhara.'

At Devdhara, the Indrawan spills over a rocky lip and falls sixty or seventy feet, scooping out a horseshoe shaped depression which is very deep and about two acres in area. It emerges at the other end of the horseshoe, much chastened, a wiggly skimpy little river, hardly thirty yards across, flowing over a rocky bottom that keeps it crystal clear. For a quarter of a mile the high banks are covered with dense undergrowth, broken only by game paths made by animals coming down to drink. There are mahseer in the horseshoe shaped part of the river, great massive twenty and thirty pounders, but they do not grow to that size by being stupid. They *can* be caught—rarely. It is the exit part of the river which provided Caesar's favourite fishing. Sporty four to eight pounders that felt like whales on the light line that was essential. They were all surface feeders and a little bait fish had to struggle along not more than three feet below the surface if it was to find any takers. A heavy line dragged the bait down too deep. The minnow, hooked lightly through the tail, was floated slowly downstream without a sinker, gently guided into each likely looking eddy, and soon there would be a crashing strike and the scream of an outraged reel... Charan Singh broke sympathetically into this fisherman's day dream—'First a bath, then food, then sleep, then I give you half an hour to catch a mahseer. I told them to bring your rod and bag. And then we look for tracks downstream, minus your rod but with—I hope—a mahseer. I have a feeling the tiger is somewhere along the river today.'

They climbed down to the horseshoe and found food and clothes and Caesar's fishing tackle awaiting them. They took turns to bathe, one on guard, one in the water. Splashing mightily in the shallows, Caesar, who could not swim, asked, 'What happens if the tiger gets into the water after me?'

Charan Singh said drily, 'Then you'll learn to swim quicker than any man has learned before.' They ate and slept, the mahseer nagging at Caesar's subconscious. In less than an hour he was up and putting together a limber nine foot bamboo rod. Five minutes with dough and a tiny hook gave him half a dozen minnows. By that time Charan Singh too was awake. They took their rifles and the rod and followed the river bank on the exit side until they came to a suitable game run that gave room for strip casting. Caesar drifted the minnow down with the current, checking at every likely spot. On the second cast the rod dipped violently, the reel began to scream and he was into a six pounder, which took him ten thrilling minutes to land. Charan Singh said, 'That's my dinner. Let's go.' Caesar pleaded for one more, just one more. It was nearing five. Three casts later the rod kicked again, again the reel screamed, but this time it was no six pounder. This time, thought Caesar exultantly, it was a double figure fish, fifteen, may be twenty pounds. He used every trick of the trade to slow it, utilizing to the full the elasticity of the rod and giving side pressure whenever he could make the current work for him. After a long twenty minutes he began to recover line, pumping slowly, delicately. He felt the fish coming up, then there was another mad rush and they were back to square one. He pumped again, the line began to fatten on the nearly empty reel, and for the first time he allowed himself the luxury of hope.

Charan Singh's whisper brought him back to earth—'Tiger. Coming down the other bank to drink.' He retreated into the game run, only the rod exposed, and looked. An old, a very large tiger was coming down the opposite bank, and no one will ever know why he had not seen them. There was no old tiger in the area except the maneater. Caesar passed the rod to Charan Singh and picked up his rifle—'For God's sake don't lose that fish, just hold him,' he breathed. The 10.75 mm came smoothly up, the blob of the foresight steadied at the junction of neck and shoulder, the shot crashed out and echoed madly in the surrounding hills. The maneater sank down slowly, dead before he reached the ground. But even before that the rod was back in Caesar's hands and the fight with the mahseer was on again. Charan Singh covered the tiger for a minute, then he lowered his rifle and awaited the outcome of the battle with the mahseer, which had got its second wind while Caesar was busy with the tiger. Now he went deep and used the current for added strength. The fragile rod bucked and bent with each frantic rush, but the line held, the angler's skill too held. Another twenty minutes later the magnificent deep bodied fish was floating on its side and Charan Singh made no mistake with the landing. It took longer to reach the tiger; they had to go half a mile downstream before they could find a ford. A broken canine identified him beyond all possibility of doubt.

That night, when the tiger had been skinned and the Gonds of Hathikoh were dancing in celebration, the two men lay on string beds and remembered aching muscles which they had not known they possessed. Charan Singh said, 'You must tell everyone you caught that tiger with a fishing rod. I'll support you.'

Caesar said virtuously, 'I never tell lies. But I'm going to call him the mahseer tiger. If it hadn't been for the mahseer putting up such a fight, we wouldn't have been on the bank when he came down to drink. You haven't praised my skill, my timing, my...'

Charan Singh interrupted resignedly, 'It was a wonderful shot ...' 'What shot? I was thinking about the mahseer. Six pound line and a forty pound fish, well thirty, all right damn you, twenty, and I won't go an ounce below that!'

The Tiger that Was

There is a difference between reticence and silence. Reticence is a definite holding back, and there should be no room for it between friends. But Charan Singh was curiously reticent. 'Yes, there is also a problem. You will see for yourself.' Caesar—let us call him that, for he was a government servant and all government servants are Caesar in that they wield power—lit up at mention of a tiger. 'What problem? He won't take a kill? He breaks out of the beat? He won't return to kill?' 'No, not that sort of problem. I can show him to you quite easily. The problem is something else.' He would say no more.

Dumarpadao in late March is quite lovely. Rich dark sal trees and hills, lit up by the red blossoms of Flame of the Forest, and small black tribals, and heavy quiet afternoons when the heat bakes everything and everyone into sleep. But there was to be no sleep this afternoon. When they had eaten the noon meal, Charan Singh hitched up his dhoti, and nodded at the nearby hill. 'He's there. Do you want to see him now?'

Caesar scoffed—'He's there! I suppose he sent you word that he's ready to receive visitors!'

'He's there. And I know exactly where he lies up in the afternoon.'

There was shouting at the foot of the hill, and a flock of goats scampered noisily down. The baredi (grazier) came running to them. 'He took a goat just now. He took a goat!' he shouted. Charan Singh selected a paan with leisurely deliberation and inserted it into his mouth. 'He always takes a goat in the afternoon in that hill and you always forget that he does. So why the excitement? In any case the goat is dead by now. By the time we reach it it will have been digested. Stop behaving like a young girl who has been seduced by a bear—for the first time.' There was a gale of laughter at this unfounded libel on the Central India sloth bear which is persistently accused by the tribals of abducting attractive young women. The excitement abated. Five minutes later they started for the hill, Charan Singh in front, Caesar behind him, cradling a beat up, short barreled 10.75 mm Mauser, last of all the baredi. Caesar asked, 'No rifle?' Charan Singh grunted. 'Let be. Today we are not killing tigers, today is only for seeing.' They set off for the hill at the easy jungle pace that eats up the miles so deceptively. In those days Caesar was half Charan Singh's age, but by the time they climbed the hill he was breathing hard and trying not to be envious of the way in which the other flowed effortlessly over the ground. There was no trouble in finding the kill. The baredi took them straight to it. Caesar stared. And stared. And continued to stare.

'Well?' said Charan Singh.

'This is no tiger kill. I don't say that a tiger will never kill a goat, but no tiger killed *this* goat. Look.'

'I have looked. And I have *seen* this tiger killing other goats in exactly the same way.'

'That is so', said the baredi unexpectedly, but both men ignored him. About half the goat had been eaten, scratched into shreds and eaten. There was not a single tooth mark, only long scratches that had ripped the flesh away, scratches made by long claws. Charan Singh looked amused. 'Well?' he asked again. 'Well—nothing. If you say this is a tiger's kill, it is a tiger's kill. But this tiger I must see. It is not like any tiger I have ever known.'

'No, it is not like any tiger you have ever known. Let us go now.'

They left the goat where it lay, since Charan Singh turned down the idea of sitting over the kill or of beating.

'I tell you', he said, 'I shall show you this tiger in the act of killing. A little patience! There is also tomorrow.' And with that Caesar had to be content.

The Gonds say, 'In the beginning of all things was the jungle, and the jungle created the gods'—which makes the jungle pretty ancient. Charan Singh's forebears had been in these hills as long as the jungle could remember, a memory stretching back for thousands of years. They were the first worshippers of Lingo Pen who is older than the civilised deities of the Aryans; of Telgin and Budha Deo before whom civilisation is the plaything of a child. That night when they were staring into the fire after a full meal he said suddenly, 'Do you know what is the saddest thing in life? It is age. Our Gond gods are old and tired, now they cannot look after their creatures, they cannot even die themselves.' Then, with a sudden change of mood, he asked impishly, 'Tell me, these young strong gods of yours, can *they* die?'

Caesar grunted drowsily, 'How can gods die, o fool!'

'Then they will get old and they will be gods no longer, nor will they be able to die. What stupidity! Let us sleep now.'

They were up at dawn to collect a couple of jungle fowl for the pot. That task completed and the food cooked and eaten, they got ready for the mission of the day. Again Charan Singh refused to take his rifle. 'I shall be holding the goat' he said, and would not be drawn by Caesar's caustic 'So you have only one hand now!' It was past noon when they set out for the hill, the goat dragging reluctantly behind Charan Singh. He climbed the hill careless of noise, which was surprising for a man who had so often successfully stalked tiger. But Caesar said nothing. There was always a good reason for whatever Charan Singh did. At last they reached a spur of the hill, overlooking a small open space, surrounded by karonda bushes. It was cool here; the karonda, a bushy evergreen bearing sweet-sour plums, is always cool. Charan Singh tied the goat in the opening and the two men took their position on the spur, melting quietly into the background. The experienced shikari needs no screening, he disappears into the details of the jungle. The goat bleated once or twice, then fell silent and the vigil began. An hour passed, two hours, the shadows lengthened visibly. Charan Singh looked worried. 'He *must* come now', he muttered, as if trying to drag the tiger out by a conscious effort of will. But no tiger appeared. Instead, there was a muted, far off groan. There is no other word for it, it was not a roar, it was a groan. Charan Singh's brow cleared at once. 'Fool that I am not to have thought of it! Come, there will be no tiger today. But for tomorrow, I promise the tiger.' There is a Gondi saying—'The return journey from a successful hunt is the space of a breath; the return from failure is twenty miles.' Their return journey was a dreary twenty miles.

That night as they sat before the fire Caesar plied Charan Singh with questions, to all of which he returned evasive answers or no answers. But he said, 'The tiger did not groan because he was wounded. He had a belly ache, he is not accustomed to eating. Work it out for yourself.' Which, on the face of it, was absurd. In disgust, Caesar went to bed early and, as ever, the jungle lulled him to sleep in minutes. First the cicadas, then the coppertoned calling of a nightjar, far off the belling of an alarmed sambhur—the jungle comes to life when men are resting. Their bullocks panicked in the last quarter of the night but the questing torch beams revealed no cause for alarm. It was only at dawn that they discovered a prowling tiger had disturbed them. Charan Singh glanced at the pug marks and dismissed them with a laconic—'A young male, curious or hungry.' But he was thoughtful.

They took the goat up the hill in the afternoon and everything of the previous day was repeated. An hour later Charan Singh tensed, and Caesar slid the 10.75 mm forward silently. The jungle exploded into action. A tiger erupted from nowhere and the goat was dead between its paws, skull crushed between the massive jaws. Caesar felt Charan Singh's hand on his arm and swivelled his eyes to it. One finger moved horizontally in a negative motion. He relaxed his grip on the Mauser. The tiger—it was a young male—crouched over the goat until death was certain, jaws locked in the death grip. It is this habit of tigers that has given rise to the myth of their sucking the victim's blood. They don't; they merely want to make sure. After a good ten minutes the tiger relaxed and released the goat. Now the second act will begin, thought Caesar. The tiger got up and was swallowed by the undergrowth, they heard it sit in the screening karonda a few yards from the kill. They waited for what both of them knew must follow. A tiger likes its meat to cool before it begins to feed. After an interval it returned and the goat disappeared in less than an hour, except for the head, the hooves and a few tatters of skin. The tiger belched noisily and walked towards the river.

Charan Singh sighed. 'Again I have let you down. This is the fool that scared our bullocks last night. No more promises, but will you try again tomorrow? I have a reason for asking.' His voice was as close to pleading as it would ever get. Caesar patted his shoulder, and said simply, 'Of course.' They returned to camp in companionable silence, and later, the full bellies mitigated disappointment. Caesar said abruptly, 'If you don't tell me something about this mysterious tiger of yours, I'll burst!' Charan Singh nodded, 'I knew that was coming. I remember the question you used to ask ten years ago—for twenty-four hours a day! But perhaps you have learned enough for there to be no need of questions now.'

'I have not learned about this tiger. Tell.'

'There is little to tell. He is very old. He is very big. He wants to live and I want him to die but I cannot bring myself to kill him because I am a foolish old Gond and I keep thinking that the gods—my gods, not yours—do not want him to die as yet.'

'You sound as if you are sorry for him.'

'I am sorry for him. You will be too, when he appears.'

'When', not 'if'. That at least was reassuring. The thought enabled Caesar to sleep soundly through the night.

The next afternoon saw them plodding up the hill with another goat in tow, like the tenth retake in a badly directed film. The goat was tethered. They settled down to wait. Caesar amused himself by counting the little red ants that were climbing in single file up a leafless tree which they had used to break up their silhouettes. He lost count and rather than begin again, looked for something else to occupy him. A small perky mongoose emerged from the karonda and inspected the goat gravely, decided it was too big to tackle, and wandered off on its lawful occasions. A group of ground thrushes squabbled noisily in their progress through the undergrowth. When they made themselves visible, Caesar counted and found there were five. 'Why do they call them seven sisters?' he wondered, and checked himself before he could ask Charan Singh—he probably didn't know either. Ants, and a mongoose, and the seven sisters; you were never alone in the jungle, he decided. Or was that true? What if you were deaf and blind, *then* you would be alone. He felt a surge of pity for himself, blind and deaf, and a wave of anger swept over him against whoever or whatever made him blind and deaf, and amusement followed at how stupid he could get. It is always like that with people who spend much time in the jungle. One thought leads to another, the mind is never purged of thought, but nothing sensible ever emerges, and the thoughts themselves disappear like wisps of fog. It is a pleasant form of thinking, warranted to be headache free because there is no end result. And it is an easier form of relaxing than making the mind a blank as in yoga. His mind wandered pleasantly again—ants and a mongoose and thrushes and how horrible to be blind and deaf in the jungle and... Charan Singh's hand on his knee pressed insistently. Then he heard it, a shuffling, dragging sound in the karonda bushes, a sound he could not identify.

Reddish yellow appeared and grew in the green of karonda. Slowly, materializing like a ghost, it became a tiger. A huge tiger, an old, old tiger. A travesty of a tiger. A living skeleton from whose bones the loose skin hung like chain mail from the skeleton of a gladiator. It dragged itself over the ground with infinite effort, the breath wheezing from an half open, almost toothless mouth in painful gasps. The goat collapsed in a heap out of sheer fright. The tiger collapsed over it and the emaciated paws began to claw and scratch and tear and make the marks which had so puzzled Caesar when he saw the first kill. Beyond an initial whimper, the goat made no sound, dying silently and without a struggle, numbed by terror. Charan Singh whispered, 'Poor devil!' and he was not thinking of the goat. Caesar stared, the 10.75 mm forgotten, the pity welling out of him like tears, and his pity was not for the goat either. Now the tiger began to slobber over the goat, sucking up the flesh which his claws had shredded, like an old man slobbering over a bowl of bread and milk.

'It is not right', thought Caesar, 'it is not right, a tiger should not come to this.' But then he thought, 'After all, the old gods of Charan Singh may have come to this too—no longer able to make an end.' Charan Singh touched the 10.75 mm. Caesar shook his head, a gesture of negation saturated with regret, 'I can't. He wants to live so much.'

Charan Singh breathed in his ear. 'That is what is so sad, his wanting to live. He is no longer a tiger, what has he to live for! Be kind.' Caesar half lifted the rifle and put it down again. 'I cannot. He wants to live so much,' he repeated. The tiger licked and slurped, and they watched him, enveloped with a sense of futility.

Suddenly the matter was taken out of their hands. The karonda shook. There was a tremendous roar and a tiger leaped out and on the pathetic ruin at the goat. A sickening, worrying sound, then silence as the long canines found their target, the base of the old one's skull. There was no resistance, no struggle, just that one bite and death. And silence. The young tiger dragged his victim a few feet from the goat, growling, and returned to eat. They watched until the goat disappeared and the tiger went off to drink. Now there was nothing to see except the old, dead tiger, sprawled out in final repose. Now he was a tiger again, at last.

THE JUNGLE GODS

From Kashmir to Kanyakumari, from Kutch to Calcutta, there are vast areas of jungle in India and in them dwell the aboriginals, the forest tribes who existed before the British, before the Moghuls, before the Scythians and the Aryans. Their story goes back into the mist of the past. 'In the beginning was the jungle and the jungle was bare, so Bara Deo created us from the womb of Earth.' It's as good a theory of evolution as any. They are a cheerful people who laugh easily and savour the zest of living without inhibitions. Existence is somewhat less than a bed of roses by our standards. One scratches a living from the clearings and the bounty of the forest, but the distance between a full belly and starvation may be only a week. Always there is the menace of death in many forms—the falling tree, the epidemic, the tiger.

This is the story of an aboriginal family in Central India, and of the destiny which fulfilled itself on a drab July day. The family consisted of an old man who was completely crippled with arthritis; of his wife who was almost as old; of his son, a sturdy Gond (that was their tribe) who wore the cock's feather in his hair with gay abandon, and was a dashing blade by jungle standards; and of a girl whom he had loved, eloped with, and married, all within three days. The only earning member was the boy, Mahugu who became a great friend of mine. He cheerfully supported the entire family without giving his parents the feeling of dependence which civilisation exacts as the price of charity. To him it was not charity but the repayment of the debt he owed them for his childhood. The girl was young, and he loved her, and life was good. The year rolled along with its cycle of festivals and fairs and dances (Aboriginals are passionately fond of dancing and do it really well). Small things brought happiness—the pennyworth of bangles at the fair or the handful of tiny fish broiled for the Nava Khani, the feast of the New Eating when the first crops are cut.

In July, the rains came and the jungle was filled with exciting new things. There were the mushrooms and the wild yams and the new bamboo shoots; and the people of the village were plucking and garnering from dawn to dusk. July also brought the return of the maneater, the tiger which was famous for the punctuality with which he went on his rounds. You could almost compile a calendar from the dates of his visits to the villages in that tract, each date marked by a death. He was a heavy male, in the prime of life and there was no excuse for his turning to human flesh. When his score was near the half century mark I spent a fruitless week—all I could spare—in the previous December chasing after him. It was like chasing a ghost. The aboriginals accepted his exactions fatalistically. 'Death comes', they said, 'by his own road, and who shall cut firelines to circumscribe him?' But they offered sacrifices to the appropriate demons although they were somewhat doubtful if demons had any power over the tiger which is the undisputed master of the jungle. And the jungle, as everyone knows, is the most important part of the world. The more intelligent amongst them, including Mahugu, accepted my advice and prepared a system of drum signals by which information of a kill could reach the nearest telegraph office (40 miles) in two hours, for transmission to me.

It was four in the afternoon when the telegram arrived, and it found me in the middle of a conference which was too important to break up. The telegram said, 'Tiger killed woman. Come.' I left at six and reached the place at midnight, having done the last ten miles on foot in pouring rain.

Mahugu's mother went out with two other old women to pick mushrooms at dawn as usual. The aged are always allotted patches close to the village for this sort of work so as to save them a long walk. The three old women picked and gossiped and retailed scandal until nearly twelve, when they broke off to have their meal. They were eating when the tiger walked out of some undergrowth—the survivors were unanimous about this, he did not rush, he walked—picked up Mahugu's mother by the neck and carried her off. Reflexes in the old are slow and she did not scream until she was a good fifty yards away. Then she screamed once. The tiger closed his jaws tighter and there was no further sound.

By three o'clock the news had reached the village and the men had assembled to recover the body. It took that long because they had to be called from different points in the forest and, anyway, it requires a little time to make up your mind to chase a tiger off his kill when you have not got so much as a musket amongst you. They took drums, and the horns which provide the orchestra for their dances, and they made quite a noise. The tiger had eaten one leg up to the hip when they arrived, and he went off reluctantly. He took the game track across the hills and two miles away he met Mahugu, his wife and a little girl of the village who had accompanied them.

If you live much of your life in the forest, as I have, you come to realise that the gods of the jungle possess a queer and perverted, even sadistic, sense of humour. They are the old gods who were before Pan and Silenus, and it is not good to trifle with them or to let your vigilance relax. Always they are on the watch for such a moment. Doubtless the Greeks had a word for it, but I never was a classical scholar. You spend six hours after a wounded tiger and he dies at the end of a gallant charge which makes you wish you carried heavier insurance—you relax with a sigh of relief and on your way back to camp a falling tree does what the tiger failed to do. Or, you spend ten years of your life earning enough to buy a pair of bullocks so that you may cultivate your own land instead of working for others. You buy your bullocks and again you relax— and out of all the dozens of pairs in the village, the tiger picks on yours and you are back ten years. It is not without reason that the aboriginal prays when his child is born—'O Jungle Gods, forget him; never remember him until he is dead.'

Mahugu did not know his mother was dead. He put his arm round his wife's shoulder as they walked and she rebuked him— 'What will the child think!' They had had a good day—her basket was full of yams and fish—and he made the mistake of being thankful in his mind and of allowing himself to refrain from fear. The fourth freedom is dangerous in a jungle.

The tiger was angry and hungry. He had merely whetted his appetite on the old woman when the search party arrived. He saw Mahugu coming along the winding track and headed to cut him off, his tail rising a little in his excitement, his mouth slightly open, head close to the ground. The woman saw him first and screamed— too late. Mahugu turned and tried to get at the axe slung over his shoulder but he was down before he could touch it. In that first impact the tiger flung him ten feet and went with him, his jaws clamped at the junction of neck and shoulder, fangs deep in the chest, into the lung. The man was probably dead when he hit the ground. At least, I hope so. His wife recovered a second later and flung her basket at the tiger. Then she rushed screaming at him and he—because he was not used to this sort of thing—left Mahugu and slunk off a leap away. She took her husband's head in her lap and cried to him to awake and the tiger watched her, growling. The jungle gods must have watched too, faintly smiling. It was *their* jungle, and Bara Deo had no right to create his aboriginals in it. For a long time the woman sat and wept and then the girl called to her to come away and she went. Most marvellous of all, the tiger did nothing.

I sat on the edge of an uncomfortable string bed and heard the story without comment. What was there to say? I called some of the men I knew, who were of good nerve, and said we would look for the tiger in the morning. Then I slept like a log till dawn. We set out at sunrise and first visited the place where the old woman had been killed. The corpse was still lying there but the tiger had not returned to it in the night. From there, Mahugu's wife took over as guide and led us across the hills for a couple of miles to the spot where her husband had died. I saw the basket of yams and fish lying where she had thrown it at the tiger and once again wondered at the courage of woman. For the first time since her husband had died, she lost control of herself and would have run into the jungle to search for the corpse if we had not restrained her. We sent her back to the village and picked up the trail.

It was a fairly easy line to follow because the ground was wet. After about two furlongs the tiger had crossed a small stream, carrying the body. At the opposite bank he dragged it for a bit and then carried it again. One leg of the corpse touched the ground at places and made identification of the spoor certain. It took me an hour and a half to do the next half mile, partly because I remembered the jungle gods and prospected every likely spot before entering it, although it was very dense and I did not anticipate an attack. In my experience a tiger will not charge in very thick cover—it impedes his movements too much. At last we found a place where the kill had been put down three times in fifty yards, and knew that it must be somewhere close. A tiger puts his kill down a little before he sits to eat, so as to look back on his track and make sure he has not been followed.

I stopped and examined the area. To the left was the stream we had crossed, to the right a hill which sloped down to a narrow pass about a thousand yards away. Between the two was a patch of dense bamboo. It was raining and the bamboos offered not only shelter but dry ground. This was obviously where the tiger was. In any other kind of cover I would have tried a stalk, but you can't go through bamboo without a noise, and stalking was ruled out. The other alternative was a beat or drive. I spent the next three hours circling the ravine of bamboos and very tough going it was. The rain dripped down my neck and my rope-soled boots slipped over every rock in Asia.

I finally decided to beat and explained my plans to the men who were with me. They agreed willingly and we arranged the line of advance.

I found a rock about eight feet high and sat down just behind its apex. The beat started and nothing happened for half an hour, except the shouts of the beaters and the noise of their drums and trumpets. Suddenly, a hundred yards to my left but out of sight, a man screamed, and a tiger roared—short sharp roars that mean business. I got off my rock and ran as fast as I could towards the sound. There was a little clearing on the way and I crossed it in nothing flat. A voice called to me from a tree, 'Look out! Tiger!' I couldn't see a thing and stood very uncomfortably with my rifle poised, listening. Sometimes in the jungle, ears are better than eyes. But I couldn't hear anything either; the persistent mutter of the rain blanketed all sounds. I backed to the edge of the clearing and saw a tall rock—it must have been twenty feet high—out of the corner of my eye. Without conscious action, I found myself on the rock. 'Is anyone hurt?' I asked. The man on the tree replied, 'He nearly got me but I just managed to climb out of reach. He's still here.'

'Can you see him?'

'Yes. Look out, he's going for you.'

The tiger followed the last word very closely. His leap at the rock fell just a couple of feet short—if I'm a liar, I'm not doing it deliberately, that's what it looked like. He landed back on his feet, half turned and suddenly tried the leap again. Again he failed. He was at the edge of the clearing, going away, when my bullet caught him behind the left shoulder, raking forward into the base of the neck. He never moved after that.

I climbed off my rock and wiped my face and suddenly felt very tired. I rolled and smoked a cigarette and watched the men tying the tiger to a roughly made litter. They treated him with respect, an enemy who had won all battles but the last, and there was no laughter, no joking as there is when an ordinary tiger is dead. Later, in the village, they moistened rice and sealed his mouth with it, saying conventionally, 'Speak well of us to ghosts and all dead things, and do not come back to haunt us.' But the wife of the dead man took no part in these proceedings and sat aloof brooding, looking into a future which was without Mahugu and in which an old, old man loomed large to be fed.

And I—I thought of the jungle gods and hoped they were satisfied.

MOTHER LOVE

I once read an article by an American in the course of which he said that hunting tigers was as dangerous as shooting rabbits from a Sherman tank. I forced myself to think of the article when I heard the tigress growl. She was twenty yards away, in dense cover, and the growls sounded like nothing so much as a two-stroke motor cycle four-stroking at low revs. There's a drum in the growl, if your know what I mean.

This was in Bastar, one of the remotest parts of India, outside of Assam and the North East Frontier. Look up a large-scale map of the country and locate Raipur in what used to be the Central Provinces and is now Madhya Pradesh. Move your pencil due south for a couple of hundred miles and you'll hit Jagdalpur, the capital of Bastar. Two hundred miles from the nearest railway station, in the middle of fifteen thousand square miles of forest and hill, with the sparsest population in India, a mere nine hundred thousand. That's Bastar. I was there as the Deputy Commissioner. The tigress was there in her own right—she was the original (if not the oldest) inhabitant. Also she was a confirmed maneater.

The people of Bastar are predominantly aboriginal and most of them are optimists. It is this latter quality that accounts for the high percentage of maneaters amongst the tigers there—the tribesmen cannot resist letting off an arrow at a tiger, irrespective of the range or the target offered. Wounded tiger—maneating tiger, the terms are almost synonymous. That was what had started my tigress off on her career, an arrow in her paw. But let us leave her growling for a moment and go back to the beginning of the story.

Murnar is a jungle village, thirty miles from the nearest road of *any* kind, and the people are Murias. They live on what they can grow in the valley, and on the produce of the forest, the wild yams and mushrooms and fruit. And of course they hunt. There is a ritual hunt with bows and spears, called the Bij Parad, held in May to sanctify the new seed which will be sown in June. The animals killed are considered to be sacrifices to the Earth God. Several villages unite for this hunt and large areas of forest are combed out. Everything that appears is shot at but the shooting is erratic and the arrows don't always hit the right spot. That is how our tigress came to be wounded.

To make bad matters worse, she had cubs, a male and a female, and the strain of keeping them fed must have told on her. Less than a month after being wounded she killed her first human being. Three more kills occurred in rapid succession, and then the Murias got worried enough to come to me for help. They had tracked her into an isolated patch of jungle and were sure she had not left it.

I reached the village at about three in the afternoon and it took until four to collect enough men for a drive, or beat as it is called in India. There was no other method possible—the cover was too dense for a successful stalk. I sat behind a tree, nursing my .465, and listened to the noise of the beaters approaching me. Then suddenly, a roar and a series of roars—and dead silence. After a while I heard snatches of talk. The Murias were calling to each other, following her movements. It is very difficult to place a tiger's roar with any accuracy, but I was sure she had not been more than a couple of hundred yards from me when she roared. At last a Muria approached.

'Is anyone hurt?' I asked.

'No', he said, 'she broke back towards us but didn't attack. She's still there, she won't move. If we try to push her forward she growls and demonstrates. Come along and tackle her yourself. She won't run.' Fair enough, I thought. We headed for the spot where she had gone to ground, dog-trotting to save time. It was rapidly becoming dark by then.

This was evergreen sal forest that had been cut a few years previously, and the new shoots had grown into a tangled mass. There were virtually no big trees left—visibility was a bare ten yards. We pushed our way through the stuff, the heavy Holland's double held ready with the safety off. She growled then, and we stopped and tried to locate the sound. This is where you came in. The growls that sounded like a two-stroke motor cycle four-stroking at low revs. We couldn't locate it, it seemed to be coming from every direction at once, but we knew that it must be in front of us. When a tiger growls like that, it is giving you a very definite warning, a final ultimatum—so far and no further. A loud roar may be nothing more than a demonstration but a low growl that rumbles means business. So we waited for the charge.

We waited for what seemed like a long time, then I tried a trick that had worked for me before. I took the Muria's turban, avoiding all sudden movement, tied a stone in one end of it, and threw it forward and a little to my left, using an under arm bowling action.

A tiger will usually charge at an object thrown like that. But not our girl. She flashed out, identified the object before she was completely out of cover, and was back before I could bring the rifle to my shoulder. Under cover, she roared twice and again lapsed into the old rumble. To have followed her up into that stuff would have been a painful form of suicide. So I backed off ten paces and then ran like hell to get the beaters. We had a short half hour before the light faded completely.

I explained the plan to them hurriedly. We would form a line, with me a few yards ahead of it, and walk her up. The beaters would make as much noise as possible, but would stop dead as soon as I raised my hand.

'Understood?'

They did. We moved off. I expected that the tiger would charge out at the line of beaters as she had done before, and that would give me my chance. But she didn't. The yelling line approached the spot we had reached before and I raised my hand for silence. We stopped. The tigress had a sense of the dramatic. She waited for pin-drop silence, then she roared, jut once. Then she fell back on the old growling, the old low rev four-stroking. And never at any moment did she show so much as a whisker. We stuck it out minute after minute, hoping that her nerve would give and that she would move before it was too dark to see the sights, but nothing happened.

The Muria touched my shoulder. 'We must be out of this place before it gets dark. She's a maneater, the gods are with *them* after dark.'

He was right, at that. We conceded the victory to her and went back to the village. And so ended the first round.

The frequency of her kills diminished after this incident. She may have got scared, or, as is more likely, the increasing mobility of her cubs made greater demands on her time. Between the end of May and the middle of August there were only three human kills and the Murias came to accept her as a necessary evil. She did not leave the area but, on the other hand, neither did she bother the aboriginals unduly. I tried for her once or twice over animal kills, without success, and then, like the villagers, I also lost interest—she had become a more or less law abiding citizen of the jungle.

Towards the end of August, however, her behaviour changed. Normally a tiger is an extremely elusive animal and you can pass one at ten yards without even knowing that you have done so. But this one suddenly acquired a distaste for human company and threw an iron curtain around her jungle by the simple process of charging out at anyone who entered it. This was far more serious than killing. It meant starvation because it kept the aboriginals out of the main source of their food supply, the jungle. And of course, as was bound to happen, some of the people at whom she charged were not quick enough to avoid her, so there were more kills. The last one took place in the middle of September and rang up the curtain on the final act.

An old woman was the victim. She had gone out to collect young bamboo shoots, a much prized article of diet during the rains. The tigress tackled her in a rather messy fashion but left enough of the corpse to make sitting up worthwhile. A normal tiger usually returns to the remains of its kill for a second meal. Hiding out to await its return is called sitting up because one is usually perched on a tree. A maneater seldom returns to its kill, but there was just a chance this one would—she had cubs.

Once again the news reached me so late that it was evening before I arrived at the spot, and I only succeeded in reaching it at all because I used a small motor cycle which is mobile enough to go literally anywhere. I tied a small machan—a miniature bedstead without legs—in the solitary available tree, about six feet from the ground. The height was dictated by the tree—it was eight feet high—and not by my heroism.

It must have been nearly six when I settled down to my vigil, and she had probably seen or heard me when I was erecting the machan. I sat and suffered the mosquitoes in silence, promising myself a revengeful orgy with a flit gun when I got home.

The tigress approached her kill at about midnight, and had nearly reached it before she detected my presence. I have already said that I was in the only available tree; that was enough to make my machan stick out like a sore thumb. There was a blasphemous roar and then another and another, until I thought the jungle was full of tigers. And then a crashing of the undergrowth followed by silence. My lady friend had been and gone. I sat out the night—partly because I was too scared to face the walk through that jungle, alone and at night, with a maneater around—without results.

Just after dawn, I began to stretch my cramped legs as a preliminary to getting down from the tree, and then the tigress made her only mistake; she growled. The sound came from a point about a hundred yards to the north, and there was only one place from which it could possibly have come—a small hill with fields on three sides and the scrub jungle, in which I was sitting, on the fourth. The fields she would not cross in daylight, and the scrub jungle was quite small, it would be easy to cordon it off with a few men so that she could be held where she was until we were ready to go in after her. I stood up in the machan and shouted at the top of my voice, mainly to ensure that she remained in the hill. There was no answering roar; she had learnt belated wisdom. I let my rifle down with a piece of string and swung myself off the machan till I was hanging by my hands. That brought my feet to the ground, so the descent was not difficult. I collected my rifle and made for the village.

The cordon was in position twenty minutes after I left the machan. I had a hurried cup of tea and sent men out to collect beaters. I proposed to walk her up in line with them as I had done on the previous occasion, but this time I considered I had better chances of success with the whole day before us. I was quite sure she would not leave the hill, because the cubs were with her; if she had been alone she would never have allowed the dawn to catch her in such a vulnerable position. What had probably happened was that the cubs had followed her to the kill just before dawn and she had not dared to take them back through the thin cover of the scrub or across the fields once the dawn had broken. There was the risk of meeting an early ploughman.

I turned over the plan in my mind. We would approach *uphill* to force her to charge. There were fields on the other side, she wouldn't face them in daylight, she would *have* to break through us. Of course, the charge would be downhill, more difficult to deal with, but there was no alternative. And that damned hill was another cut coupe, the secondary growth would be terrific... Oh well, there was no other way. I took out the hand-detachable locks of my .465 and oiled them. They were perfectly all right, but there's never any harm in making sure.

It was nearly eight by the time the beaters collected. They came from three different villages and there must have been a hundred of them when we set off. None of us had any illusions about the job we were starting on—we were all jungle people and we knew better than most that we would need a lot of luck today. We formed a long, slightly curved line at the beginning of the scrub and I gave my final instructions.

'The drums and horns will stop as soon as I raise my hand. After that, *there will no no noise at all.* Got that? Absolutely no noise. If I whistle, everyone will climb a tree and stay there till I whistle again. There will be no talking.'

They understood. They were my friends and we had turned some tight corners together, we had confidence in each other. The horns (trumpets to be more correct) blasted, the drums thundered out, and every man shouted as loudly as he could. Our line moved forward slowly, with me a little in front of the centre, the double .465 hot and heavy in my hands. It's a beautifully balanced rifle but it *does* weigh 10½ pounds. After we had covered a furlong, when we had climbed a quarter of the hill, I raised my hand and the noise died. For a long ten seconds there was no sound and then I heard distinctly, 'Thunk! Thunk! Thunk!' It was a rather nasal thud, solid, not like a drum, regularly spaced. I twisted my hand in a wordless question to the Muria behind me. He breathed into my ear, 'Tiger. Tail hitting ground. Angry.'

That was the first time I heard a tiger lashing the ground with its tail. I heard it again on two subsequent occasions, but that's all, in more than thirty years of jungle life. It must be an extremely rare occurrence. I slipped a couple of paces ahead to get a sal sapling in front of me and then whistled. All the men who could find trees, climbed them but this was a cut coupe and there were not many climbable trees; quite a few beaters remained on the ground with me, including one with a trumpet in his hand. The silence was deep enough to magnify the noise of the tiger's tail hitting the ground. There was something ominous about those measured 'Thunks'. We waited. After a few seconds the old rumbling growl started, the four-stroking growl, and it came as a relief from the nervous tension that had been built up by the silence and the tail lashing. Again the seconds trickled away, timed by the rumbling of the tigress. I am quite sure that she would have lost her nerve first and moved if we had held the silence. And in that movement she would have disclosed her position, which would have given me a chance to shoot. But the Muria with the trumpet lost *his* nerve first and blew a tremendous blast that made most of us jump out of our skins. That did it. The tigress came out in a succession of standing leaps, straight for the Muria, who was hardly four feet from me. I caught the merest glimpse of her. My despairing effort to get my sights aligned was quite futile. I don't think the bead came within yards of the tigress during the time she took to cover the twenty odd paces between us. The Muria saved himself with a magnificent sideways jump that made the tigress miss by inches. She kept going, a yellow streak twisting through the dense undergrowth down the hill, followed by a chorus of excited shouts from the beaters. I never had a chance at her.

I called for silence, and the chattering ceased. The Muria gave me a sheepish grin. 'I couldn't stand it any longer' he said, frankly. 'That's all right...where did she go?' 'Oh, she must have crossed the scrub into the big tree forest by now—she's a mile from here at least.'

That was the moment she chose, with a superb sense of timing, to return. She had left her cubs behind, and she was not going to make her getaway at their cost. We were between her and the cubs and she could only reach them by breaking through us again. Which she did. With a series of terrific roars that gave warning of a fixed purpose she shot up the hill and through our line, and I was just as unsuccessful this second time in getting my sights on her as I had been the first. There was no bonus in staying where we were so I waved the Murias down from their trees and we retreated to the base of the hill.

'What now?' I asked.

'Now it's easy', said Budsan, my tracker. 'We know she won't leave her cubs. All we have to do is go round the hill and beat it downwards, towards the big tree forest which is her natural home. She'll move, but only after a series of demonstrations—that's to give the cubs time to slip away.'

The idea was sound. She would certainly demonstrate, fake charges at the beaters and I would get a shot during one of them, with any luck. But I was not so keen on killing her now, this gallant old lady who had shown her willingness to sacrifice everything for her cubs. I said as much.

'We also have cubs—there are a dozen children in my village, without fathers, because of her. She's a confirmed maneater and her cubs have tasted human flesh too—they'll be maneaters as well when they grow up. If you let her get away this time, Sahib, nothing will stop her, she'll have enough confidence to enter into the village itself. We *must* finish the job today.'

That was Somaru speaking. There was a remorseless logic in what he said. I remembered the Keskal tigress that had accounted for over three hundred people before she was shot. I remembered the young mother that *this* tigress had killed only last month, and her two infant children.

'I'm afraid you're right' I said, 'Let's go.' We detoured the hill and climbed it from the other side in silence. When we had formed our line on the top, we started the noise as before and beat down hill, pausing every fifty yards to look and listen. We were halfway down the hill when she flashed out for her first demonstration, roaring and slapping at the trees as she passed them. This time I was a little too far down the line to get a shot. We straightened out the line and removed the kinks that had formed in it where men had retreated, and pushed on. She went to ground in almost exactly the same spot from which she had charged downhill earlier in the morning, and once again we were treated to the old four-stroking rumble. It was nearly eleven by now, and very hot.

There is a considerable degree of nervous strain involved in walking up a tiger with the object of making it charge and I was pretty edgy. That must be the justification for what I proceeded to do. I stopped the beaters and put them up trees. Then I hitched the .465 forward and walked towards her cover.

I said, 'Come and get it, damn you!'

She knew this was the show-down and didn't try to evade it. There was a deep, grunting roar and out she shot, straight for me, her tail held high. I think, perhaps a tiger's charge is the ultimate in loveliness, if you live to remember its beauty. The death charge— your death or the tiger's—begins with deliberation, there is no cheap flamboyance about it. The tiger is all head and snarl, the colours of its skin seem to change as the muscles ripple under it, it *flows* like a big wave. Like a wave, too, it is inexorable.

I hit her first at the junction of neck and shoulder and she folded up under the impact of the heavy bullet. I gave her the other barrel through her head for luck, and then it was over.

The Murias climbed down from their trees and cheered wildly. I wiped my glasses and my face, and lit a cigarette. That first cigarette after you have stopped a charge tastes awfully good. They gathered round me and all of them started to talk at once. Somaru clarified the matter.

'There are still the cubs left. They won't move without the mother. We have to get them too.' He was right. Cubs which have been brought up by a maneating mother do not *invariably* become maneaters themselves, but there is an excellent chance of the taste persisting. I made it clear that I had no intention of killing the cubs if I could catch them.

'How big are they?' I asked.

'Oh, about three months,' replied Budsan.

I was relieved. They could be caught. I sent my heavy rifle back to the village and armed myself with a stout stick.

'Now listen carefully', I said. 'I want these cubs alive and unhurt. This is what we'll do. Throw a line round the scrub and close in. Every second and third man will form a pair, holding their turbans between them. As soon as you see the cubs, muffle them in the turbans. Ten rupees to anyone who gets me a cub.'

A quarter of an hour later, I was the most surprised man in Bastar, and the most uncomfortable. The male cub rushed out at me, and he wasn't three months old by a long chalk—he was nearer seven, about the size of a small panther. I lifted my stick and shouted at him, but he came on like a good'un. At the last possible moment I snatched a muzzle loader from the man next to me and was lucky enough to almost blow the cub's head off with a load of buck-shot. When I looked at him more closely I understood why the tigress had thrown an iron curtain about the jungle. He was very lame and there were festering sores in his right rear leg, the result of an encounter with a porcupine. In that condition he could not go far, and he could certainly not run. The tigress had tried to protect him in the only possible way, by keeping everyone else out of their jungles.

After that I told the Murias they could deal with the other cub themselves. They went on, and I sat under a tree until they returned with the corpse of the female cub, which they had finished off with spears. I got my last and final surprise when we reached the village. As soon as the women—and they had been the heaviest sufferers from the tigress—learnt about her injured cub, and how she had refused to escape without it, they cut short their rejoicing and, quite spontaneously, produced the cymbals which they use as an accompaniment for their songs. They circled the bodies of the tigress and her two cubs slowly, singing the lament for a mother:

'She is dead, and her children are alone,
She is dead and her milk is dry.
She is dead—she is dead—she is dead.'
But her children were dead too.

Excerpted from Animals and Other Animals *(Delhi: Sanchar Publishing House, 1992).*

A. J. T. JOHNSINGH
AND G. S. RAWAT

On Corbett's Trail: Tracking the Man-eater

Two naturalists follow Jim Corbett's tracks, mapping the route he took when going after the man-eating tiger of Thak. The hunt for that tiger took place in 1938: since then, there have been many changes in the land and its wildlife. Dr Johnsingh of the Wildlife Institute of India is not only a senior wildlife biologist, but a tireless foot-marcher through tiger land, and it is his eye for detail that is evident in this account co-authored with another biologist, G. S. Rawat.

While reading through 'The Temple Tiger' by Jim Corbett no one would miss his absorbing description of a patch of dense jungle about 3 km south of Devidhura in the Kumaon Himalaya. In the early part of this century, when Corbett pursued the Temple tiger of Devidhura, this jungle had sambar, barking deer, langur and numerous pheasants. Corbett vividly describes the fight between a black bear and the tiger he heard in a ravine, late one evening, while he waited for the tiger on a 2 m high tree branch. Waged in the hollow of a restricted area the sounds of the fight were terrifying and he was thankful that the fight was between two contestants who were capable of defending themselves and not a three cornered one in which he was involved. He missed the tiger due to 'rank bad shooting' and shot dead the screaming bear as it rushed towards him.

In late April 1993, 84 years after Corbett's visit, we were in the same forest. The oak and scrub jungle between this forest and Devidhura which Corbett has mentioned, has been lost to cultivation. In place of the single grass hut village, there were numerous masonry houses. The cloudy weather with occasional distant thunder made our walk up the ridge, still with dense forest cover, enjoyable. We walked for two hours as silently as possible, looking for pheasants and large mammals. The cooing of the Rufous turtle dove, and calls of the Great Himalayan barbet and songs of cowherd boys rang through the forest. Redbilled blue magpies flew though the trees and across the valleys. A group of White-throated laughing thrushes rummaged through the understorey in a dense patch of oak and rhododendron. There were a few wild pig diggings and a group of barking deer pellets. There was no evidence of either black bear or sambar. It was apparent that the jungle had ceased to be a tiger habitat for a long time.

Between 1907 and 1938, Jim Corbett had shot eight man-eating tigers and leopards in Kumaon hills which in total had killed nearly 900 people. Much has changed in these hills from the beginning of this century as it has happened in many other parts of the world. The bridle paths where men and their horses walked for days to reach a destination have been converted into motorable roads where buses ply now. Pati, a village mentioned as Pali by Corbett in 'The Champawat man-eater' with a populace of about 50 people in 1907 now has about 2500 people. The major victims of this change are the forest cover and wildlife.

Since boyhood, we had read Corbett's books and nourished a desire for years to see some of the places where Corbett had hunted. We decided to see areas between Kaladhungi and Tanakpur where Corbett had shot the Mukteshwar, Champawat, Chuka and Thak maneaters. From Kaladhungi to Tanakpur via Mukteshwar, Champawat and Chuka, the distance is 300 km. Time was a constraint and therefore a jeep was used wherever it was possible and a 10 day trip was planned in late April 1993.

Rhododendron was in bloom when we went to Mukteshwar. Its scarlet flowers stood conspicuous among the white, light green and rusty green new foliage of the three species of oak. Corbett shot the Mukteshwar tigress near an orchard owned by his friend Badri Shah. We were lucky to meet 65 year old Ramesh Lal Shah, a nephew of Badri, who as a boy had seen Corbett and has been looking after the orchard after Badri's death in 1925. R. L. Shah showed us the ravine where the tigress was shot. Now there are farmlands and houses on either side of the ravine and the forest cover is scanty. Shah briefed us about the decline in wildlife in and around Mukteshwar. The last tiger he saw in his farm was about six years ago. Sambar has become almost extinct although it still occurs in the 13 sq km well protected jungle around the 100-year old Indian Veterinary Research Institute and in the Ramgarh hills, 16 km south to Mukteshwar as the crow flies, which as Corbett wrote, are well wooded even now. On the way to Mukteshwar we had seen these forests where still tigers stray from the foothills and nature reigns supreme in certain pockets. However, even in Kasyalekh, five kilometres north of Ramgarh forests, the last tiger shot in the vicinity of the village was 40 years ago. We could see the pitiful remains of the tiger, head with broken teeth and mouth painted red, in the house of a friend.

A tigress known as the 'Champawat man-eater' arrived in Kumaon as a full-fledged man-eater from Nepal from where she had been driven out by a body of armed Nepalese after she had killed 200 people and during the four years she had been operating in Kumaon, had added 244 to her tally of victims. Corbett was asked by the government to track down this man-eater. In 1907 he began hunting this tigress when she killed a woman in Pati while she was cutting leaf-fodder with a few other women. When Corbett reached Pati from Nainital after three days of strenuous journey on foot and by pony it was five days after the woman had been killed. During these five days the villagers said that no one had gone beyond their own doorsteps and the insanitary conditions of the courtyard testified to the truth of their statement. Corbett saw the spot where the woman was killed in the densely wooded ravine near the village. As night approached, he seemed to court death by sitting on the road to wait for the tiger which was reported to be around the village. Corbett was only 33 years old then and inexperienced with hunting man-eating tigers. The long night was miserably cold and he was terribly frightened by his imagination of a dozen tigers advancing on him. Fortunately for him the tiger was nowhere in the vicinity of the village. When his men found him early morning he was fast asleep resting his head on his drawn up knees. Corbett could not kill the Champawat man eater near Pati but before he left the village he shot three goral, a mountain goat, in one of the nearby grass covered steep ridges. His men ate one and the villagers shared the other two.

After exploring the forest south of Devidhura we set out towards Pati. The striking feature of the habitat close to Pati was the rolling hills and steep ridges covered with grass and pine; even now an ideal goral habitat. Like Devidhura village and temple, which has undergone changes due to modernisation, Pati also sprang a surprise on us. There were numerous tea shops on either side of the road where Corbett had waited for the man-eater. Instead of the few grass covered huts, there stood a prosperous village of more than 100 concrete houses. Several villagers however, still remember the stories related to the death of the woman and the visit of Corbett. We could even track down a boy who was said to be the great grandson of the woman killed. The present village headman, 72 years old and a descendant of the headman of Corbett days, took us to the place where the man-eater had killed the woman. There was no ravine which Corbett describes. Instead there was a broad valley devoid of vegetation. Hardly 50 m from the place where the woman was killed there is a school where there were at least 50 children between six and 10 years of age.

Every villager we had interviewed said that tiger and sambar no longer occur on these hills but goral and barking deer are still common. We had doubts even about this reported wildlife abundance as the goral habitat showed signs of excessive use by cattle and people. Numerous trails criss-crossed the hills. Three kilometres east of the village there is a ridge covered with dense forest. It appeared to us that it might still hold some wildlife. One villager remarked that it is one place where sambar and black bear could be seen.

The pleasant cloudy weather we had the day before had changed. We climbed the ridge and wandered for an hour looking for animals. There was no sign of either barking deer or goral. We saw the ill-conceived aborted attempt of the Forest Department to raise Deodar trees amidst the dense pine, oak and rhododendron forest. Many rhododendron trees had been felled for firewood. While descending down to the road we met a man who was nervously herding away his goats from the ridge. He said that leopards were numerous and killed many a sheep and goat and were responsible for the decline of goral population. At night we stayed in the Dhunaghat forest rest house situated amidst pine trees in a scenic setting.

Wherever we enquired about wildlife around Champawat, we were informed about the forest patch between Khetikhan, five kilometres from Dhunaghat and Champawat. Along the road the distance between Khetikhan and Champawat was 33 km. We decided to walk the 15 km short cut through the forest and over two hills. We left Khetikhan around 0700 hrs and walked leisurely looking for birds and signs of large mammals. When we were about 5 km from Champawat an incident which had occurred almost in the same area and described by Corbett came to our mind.

Twenty men were walking along this forest road and about midday they were startled by the agonized cries of a human being coming from the valley below. Huddled together on the edge of the road they cowered in fright as the cries drew nearer and nearer and a tiger presently came into view carrying a naked woman. The woman's hair was trailing on the ground on one side of the tiger and her feet on the other and the tiger was holding her by the small of her back and she was beating her chest and calling alternately to God and man for help. Fifty metres from and in clear view of the men, the tiger passed with its burden and when the cries had died away in the distance the men continued on their way. When Corbett asked them why they did nothing, the men replied that they did nothing as they were afraid and wondered what could men do when they were so frightened.

Just on the outskirts of Champawat, on the side of Pithoragarh road is a narrow road with a picture of the tiger and information that the 'Champawat man-eater' was shot three kilometres from the road. We walked along the narrow road and eventually realized that the distance was not three kilometres but could be three miles. Fortunately a few boys whom we met on the path could take us to the exact location in a gorge where a man-eater was shot 84 years ago. The area was the most beautiful location we had seen during this entire trip. The Gaida river flowed through a deep gorge forming pools of blue waterbodies. The hills around the gorge rose steeply for 300–400 m. On the steep slopes there were stately pine trees and many of the rhododendrons were in bloom. Except for an abandoned road built about 20 years ago on the left bank of the gorge to a small hydel power house which is not functional now, the entire area looked undisturbed. The serenity in the gathering darkness was so infectious we thought that we might hear a tiger calling any time.

From Champawat, our next destination was Chuka in Ladhya valley. The drive to Chalthi, 40 km from Champawat, took us for the first 10 km through excellent oak-rhododendron forests and thereafter the continuity of the forest was broken by patches of cultivation. At about 1000 m, the vegetation drastically changed. Sal trees with its golden yellow ripe leaves dominated the vegetation and Indian Coral tree (*Erythrina indica*) with its coral red flowers replaced rhododendron.

We started walking from Chalthi at 0900 hrs towards Chuka along the Ladhya valley. Even in the morning the temperature soared to over 35°C. The hills on either side rose steeply for 300–500 m where, according to our local guide, there is abundant goral and an occasional serow. In several places we had to cross the meandering Ladhya river and therefore the going was slow. The human population in the valley was very low and distributed in 5 villages ranging in size from 2 to 10 houses. There were quite a few sambar tracks on the river bed. Extraction of dead and fallen logs was in progress and there were three mills powered by water for making wooden vessels. The mill owners who were making many more vessels than their quota permitted complained that the tree species Sanan (*Ougenis oojeinensis*) and Genthi (*Boehmeria rugulosa*) suitable for making vessels were becoming rarer and rarer in the valley.

In 1937 a tiger known as Chuka man-eater and in 1938 a tigress known as Thak man-eater terrorized the valley and the latter, in fact, almost brought to a halt the work of about 5000 labourers extracting timber from Ladhya valley. Thak and Chuka even now can be reached only on foot either from Chalthi or Thuligad. Fifty years ago access to these areas was even more difficult and how the 64 year old Corbett reached these places within a short span of time and how he outwitted these tigers can be appreciated only by those who have seen the terrain.

On reaching Chuka in the evening we went to Ladhya-Sharda junction to try and catch a mahseer. Corbett talks of splendid fishing at this junction. Although we tried hard for two hours we did not have a single bite. The evening, however, was not dull. We were alerted by the rolling of stones in the Himalayan slopes across the border in Nepal and scanning the slopes with binoculars saw five goral coming down to drink.

Before leaving for Chuka we had no idea of the village and had thought we would be able to buy some food there. On seeing Chuka, whose population has declined over the last 50 years, we realised that it was not possible to buy any food here. However, 62 year old Ummed Singh, the Headman of Chuka, who had seen

Corbett as a boy, with the usual largeheartedness of the hill people, gave us a dinner of dried goat curry and roti and a place to sleep on his terrace. We were fatigued by the long walk in the hot sun. The cloudless sky was studded with stars and the Sharda river roared nearby. There were no alarm calls from the jungle and the silence of the night was broken only by the barking of the village dogs. We had a most refreshing sleep.

Next morning around 0900 hrs, we left leisurely for Thak. The distance was about 4 km but the walk involved a steep climb through dense sal forest. On the way Ummed Singh showed us the rock on which Corbett had laid down and mimicked the call of the man-eater which was looking for a mate, lured it close to him and shot it dead. Near Thak village which has not changed over the last 50 years, we saw the giant mango tree from the roots of which issues a cold spring of clear water. The Thak man-eater and Corbett had drunk from this spring and we also quenched our thirst after the steep climb in the hot April sun.

If the climb to Thak, which is atop a hillock, was very steep the descent to Kaldhunga forest rest house, our next destination, was steeper. Ummed Singh after showing us the way to Kaldhunga bid good-bye and an hour later we reached Kaldhunga. When Corbett visited Kaldhunga in 1938 there were no human habitations as far as the eye could see and judging from the tiger and other animal calls that can be heard from the bungalow, there appeared to be an abundant stock of game in the valley. We found the bungalow unused except by some sheep herders and the view of Sharda in front of the guest house was blocked by overgrown vegetation. Due to vandalism by poachers from Nepal, the wire mesh and glass panes of several windows had been broken. As darkness gathered, we pulled out two wooden cots on to the verandah and settled for the night. There were no alarm calls that night and it appeared that all wildlife had deserted the area. Next morning, in order to avoid the heat of the day, we left Kaldhunga when there was sufficient light to walk by. We went past Purnagiri temple along the path which runs parallel to Sharda gorge and completed the 14 km walk at Thuligat at 0830 hrs. Our vehicle, waiting for us in Thuligat, took us back to Dehra Dun.

As boys when we read through the narratives of Corbett, we had visions of hills around Pati with plentiful goral, forests around Devidhura populated by sambar with antlers as big as the branch of an oak tree and Sharda full of big mahseer. Years have changed this wildlife abundance. Everywhere we saw evidence only of human abundance. On the bank of Sharda opposite Kaldhunga where sambar and barking deer were abundant fifty years ago, there exists now a village named as Sananni with about 300 houses. The Nepalese not only have shot out wildlife on their land but also frequently cross Sharda by inflated rubber tubes and decimate wildlife on the Indian side. Indian poachers also contribute their mite in eradicating fish and wildlife. Ummed Singh claims that 10 years ago he had counted 12 barking deer between Chuka and Kaldhunga and regrets he hardly sees any now.

Nevertheless, wildlife conservation in Corbett's Kumaon can be strengthened by the creation of the Sharda Biosphere Reserve covering an area of about 2000 sq km. If Nepal also contributes to this conservation effort by creating a similar Biosphere Reserve, various forms of outer Himalayan wildlife can be protected in an area of about 4000 sq km. The first step to be taken in this tract is to stop the destruction of wildlife. Poaching needs to be arrested immediately. Special efforts must be taken to confiscate all illegal guns. Legal gun owners should be requested and warned not to poach wildlife. Fishing must be regulated. Existing patches of natural vegetation in the Biosphere area must assiduously be protected and eco-restoration programmes need to be initiated in other degraded areas. Most parts of the suggested Biosphere area in the conceivable future may not again echo with tiger calls and sambar alarms. But barking deer and goral would reappear and leopards may become much more common. Sharda would again become the favoured home of the mighty mahseer.

Originally published in Blackbuck, *10:2, 1994.*

Abbreviations

BNHS	Bombay Natural History Society
DFO	Divisional Forest Officer
IBWL	Indian Board for Wildlife, formed in 1952, the key advisory body to the Indian federal government on wildlife-related matters
IUCN	International Union for the Conservation of Nature. Founded in 1948, collects scientific information on conservation.
WWF	World Wide Fund for Nature, originally known as the World Wildlife Fund, a major international fund-raising and lobbying conservation group.

Glossary

Amildar — Revenue official

arribada — Spanish term for arrival, refers to mass nesting of turtles on beaches, e.g. in Gahirmatha

arrack — Country liquor

baith — Literally, to sit

banya — Trader, moneylender, member of a mercantile caste

bandobast — Arrangements, refers here to organization of camping and beats to hunt or watch wildlife

Barasingh, Barasingha — Literally a stag with 12 tines; used for swamp deer of central and north India; also refers sometimes to the Kashmir stag

babool — Acacia

battue — A large shoot, a continental European term

bauleah — Pond or marsh

beat — Patch of forest beaten in a shoot; also, the actual beating in a shoot

ber, byr — *Zizyphus jujuba,* tree which grows in north and west India; the fruit is eaten by many animals

bhagar — Pond or marsh

bhang — Hashish

bheel — Pond or lake in eastern India; not to be confused with Bhils, a tribal people

Bhil or Bheel	A tribe in western and central India
~~Bij Parad~~	~~Annual ritual hunt of the Murias, a tribe in central India~~
budgerow	Boat-house in Bengal, used by hunters in the Ganges delta
bund	Dam, embankment
chana	Parched gram
charpoy	Bed, often made of rope and wooden frame
chars, churs, chaurs	Wet grassland
chela	Understudy; used for younger male in twosome of wild buffalo or elephant; also for a jackal that follows a tiger
chinkara, chikara	Indian gazelle
chita	Cheetah, or hunting leopard, now extinct in India
chital, cheetal, cheetul	Spotted deer, a species with a wide distribution in India
chota shikari	Junior hunter
chowkidar	Watchman, guard at a post (*chowki*)
chuprassy	Peon
cooly	Labourer
Coronda	Karonda, a fruit tree
dak	Post
darbar, durbar	Court; also used for a court official
Dawk Bungalow	Rest-house
degchi, degchie	Large utensil for cooking
deodar	Himalayan Cedar
dhoolie	A type of palanquin
Dollond	Eyeglasses
Doorgah Poojah	Annual festival of goddess Durga
filwan	The one who takes care of elephants, *née* mahout

Glossary

gaddi	Seat of power; alternately, elephant pad
ghats, ghauts	River wharf or crossing; also for hills or range of mountains (e.g. Western Ghats)
Gonds	Tribe of central India
goondas	Musclemen, scoundrels
goral	A goat-antelope of the foothills of Himalayas
goru (guru)	Teacher, mentor
Gujars	Pastoralists of the Himalayan foothills
gur	Unrefined sugar, jaggery
hangul	Kashmir stag, sub-species of red deer found in the Valley
Harpat	Kashmiri name for Himalayan black bear
hankh, hanka, honk	Beat, hunt
hookah	Pipe traditionally used to smoke tobacco
howdah	Platform with seats for riding on elephant back
hulkaras	Beaters in a shoot
jemadar	Supervising officer
jheel, jhil	Small pond, or lake in north India
Kavalai	Tamil word for a device used to lift water for irrigation
kedah, keddah, khedda	Stockade; driving elephants into a stockade; traditional catching technique
khabar, khubr	Literally news, used to refer to information on presence of game
khanat	Medieval water system, often underground
khillut	Gift or reward from a superior or ruler
khud	Literally, a hole, a depression in the ground
kill	Remains of an animal killed by a wild carnivore

Glossary

khansamah	Cook, house-keeper
koonki	Tame elephant trained for elephant-catching
Korkus	Tribe of central India
kota, kotah	Stone tower
kraal	Elephant training stockade in southern India
kukri	A curved knife
Kuruba, Karumba	Tribe in hill ranges of southern India
kutchery	Court
lathi	Staff or bambo stave
lebru	A small bullock cart
machan	Raised platform, usually on a tree to shoot or film from
madapolam	Fine white fabric or goods made from it, originally made at Madapollam in south India
mahout	Elephant trainer, often a hereditary post, highly skilled in the care and husbandry of elephants
mahseer	Freshwater fish much sought after by anglers
makhna, muckna	Male Asian elephant without tusks
maidan	Grassland, open patch or clearing
margs	paths, footpaths
mela shikar	Noosing of wild elephants from the backs of tame elephants, method mostly used in Assam
Mhowa	*Bassia latifolia;* its flowers and fruits are popular with wild animals and people alike
moonshee, munshi	Account-keeper
Mughs	Pirates off Bengal coast, often slave-raiders

Glossary

mukkam	Place of encampment, especially for travellers
Murias	Tribe of Bastar, central India
musth	Psychological disturbance in male elephant, marked by discharge of fluid from temporal glands, time of sexually aggressive behaviour
nala, nullah	Dry stream-bed
natchni, nautchni	Dancing girl
ness	Settlement of buffalo-herding pastoral people in Gir Forest, Gujarat, usually fenced in with thorns
pagi, puggee	Tracker, from 'pag' or 'pug' meaning track or footprint
pan	betel leaf
phand	noose used to catch elephants in Assam
phandi	Professional elephant-nooser, mainly in north-east India
poojari	Priest
ragi-flour	Sorghum, millet
rukh	In Kashmir, the game reserve of the Dogra rulers; in central India, government-owned grasslands
ryots	peasants
sahiblog	White men, masters, white-collar officers
sakhni	Female elephant with tusks, rare in Asian elephants
Sal	*Shorea robusta*, major timber tree in north India
salaam	Salute
sambhar, sambar, sambur	Largest Indian deer
Santhal	Tribe in eastern India
serow	Himalayan goat antelope
shaitan	Devil

shamiana	Canvas or cloth tent
shikar	Hunt; used for a particular expedition ('out on shikar'); also for ritual or ceremonial hunts ('the Viceroy's shikar'); or for catching ('elephant shikar')
shikari, shikkaree, shikarie	Hunter; professional village-based hunters; in Gir Forest, Gujarat, for game-watchers who no longer kill animals
shola	Thickly wooded valley in south Indian hills, patch of evergreen tropical forest
sola topee	Hat worn by the British to protect from strong sunlight
soondry	*Heritiera minor*, a tree which grows in the Ganges–Brahmaputra delta
sowari	Horse rider
syce	Groom
swaraj	Self-rule
tamasha	Literally, fun; a kind of theatrical display
tat	Short for *tattoo*, an Indian native pony
Tehsildar	Local official, in-charge of a *tehsil*
Terai, Tarai	Moist flat lands south of the Himalayan foothills, wet savannah in north India
thakur	A term of respect, Lord, master, especially applied to Rajput nobles
Todas	A mainly pastoral tribe in the Nilgiris
Toria, toras	Rocky outcrops on central Indian plains
tote	Home, cottage
urna	Wild
Yuvraj	Crown prince
Zamindar	Rent-receiver in eastern India or a large land-owner, extending control over a *zamindari*

Bibliography

Ali, Sálim, *The Fall of a Sparrow*, Delhi: Oxford University Press, 1985.
Anderson, K., *Nine Man-Eaters and One Rogue*, London: Allen & Unwin, 1955.
——— *Man-Eaters and Jungle Killers*, London: Allen & Unwin, 1957.
——— *The Black Panther of Sivanipalli and Other Adventures of the Indian Jungles*, London: Allen & Unwin, 1959.
Baldwin, J. H., *Large and Small Game of Bengal and the North West Provinces of India*, London: Henry S. King, 1876.
Baskaran, S. Theodore, 'Chennai's Patch of Green', *The Hindu*, 5 October, 1997.
Best, J. W., *Forest Life in India*, London: John Murray, 1935.
Braddon, E., *Thirty Years of Shikar*, London: William Blackwood, 1895.
Champion, F. W., *With a Camera in Tigerland*, London: Chatto and Windus, 1927.
——— *The Jungle in Sunlight and Shadow*, London: Chatto and Windus, 1934.
Corbett, Jim, *Man-Eaters of Kumaon*, London: Oxford University Press, 1944.
——— *The Man-Eating Leopard of Rudraprayag*, London: Oxford University Press, 1947.
——— *The Temple Tiger and More Man-Eaters of Kumaon*, London: Oxford University Press, 1954.
Dattatri, S. and R. Whitaker, 'Cobra', *Sanctuary Asia*, October 1983.
Daver, S. R., 'A Novel Method of Destroying Man-Eaters and Cattle-Lifters Without Firearms', *Journal of the Bombay Natural History Society*, vol. 49, 1951, pp 54–66.
Davidar, E. R. C., *Cheetal Walk, Living in the Wilderness*, Delhi: Oxford University Press, 1997.
Dharmakumarsinh, R. S., *Birds of Saurashtra, India*, published by the author, Bombay, 1951.
——— 'Following the Lion's Trail: The Lion-trackers of Mytiala', *The India Magazine*, March 1986, pp 166–72.
——— 'Gulam Hussain Bazdar, The Falconer', *The India Magazine*, June 1983, pp 40–9.
——— *Reminiscences of Indian Wildlife*, Delhi: Oxford University Press, 1998.

Divyabhanusinh, *The End of a Trail, The Cheetah in India*, Delhi: Banyan Books, 1995.

Dunbar Brander, A. A., *Wild Animals in Central India*, London: Edward Arnold, 1923, 1931.

Fletcher, F. W. F., *Sport in the Nilgiris and in Wynad*, London: MacMillan, 1911.

Forsyth, James, *The Highlands of Central India, Notes on Their Forests, and Wild Tribes, Natural History and Sports*, London: Chapman and Hall, 1879.

Fry, C. B., *Life Worth Living, Some Phases of an Englishman*, London: The Pavillion Library, 1939, 1986.

Gee, E. P., *The Wildlife of India*, London: Collins, 1964.

Hamilton, Douglas, *Records of Sport in Southern India*, London: R. H. Porter, London, 1892.

Hardinge, Lord, *On Hill and Plain*, London: John Murray, 1933.

Johnsingh, A. J. T., 'Dhole: Dog of the Indian Jungle', *Sanctuary Asia*, 1984.

Johnsingh, A. J. T., and G. S. Rawat, 'On Jim Corbett's Trail', *Blackbuck, Journal of the Madras Naturalists Society*, vol. 10, no. 2, 1994.

Khacher, Lavkumar, 'A Fine Effort—The Lions of Gir', WWF—Indian Quarterly, 2nd Quarter, 1979.

Kirkpatrick, K. M., 'Aboriginal Methods Employed in Killing and Capturing Game', *Journal of the Bombay Natural History Society*, vol. 52, 1954, pp. 285–90.

Krishnan, M., *Jungle and Backyard*, Delhi: National Book Trust, 1961.

——— *Nights and Days, My Book of India's Wildlife*, Delhi: Vikas, 1985.

——— 'Fights to the Death', *The Statesman*, 11 December 1983.

——— 'Our Wildlife: A Great Legacy Dissipated', *Illustrated Weekly of India*, 24 August 1980.

——— 'Did-he-do-it?', *The Statesman*, 11 July 1983.

Manfredi, P., ed., *In Danger*, Delhi: Ranthambore Foundation, 1997.

Mervin Smith, A., *Sport and Adventure in the Indian Jungle*, London: Hurst and Blackett, 1904.

Mundy, G. C., *Pen and Pencil Sketches in India, Journal, A Tour in India*, London: John Murray, 1858.

Noronha, R. P., *Animal and Other Animals*, Delhi: Sanchar, 1992.

Pollock, F. T., and W. S. Thom, *Wild Sports of Burma and Assam*, London: Hurst and Blackett, 1900.

Rice, William, *Tiger Shooting in India; Being an Account of Experiences of Hunting Expeditions on Foot in Rajpootana, During the Hot Seasons from 1850 to 1854*, London: Smith, Elder and Company, 1854.

Roussellet, Louis, *India and its Native Princes. Travels in Central India and the Presidencies of Bombay and Bengal*, London, 1882.

Sanderson, G. P., *Thirteen Years Among the Wild Beasts of India: Their Haunts and Habits from Personal Observations; With an Account of the Modes of Capturing and Taming Elephants*, London: W. H. Allen & Company, 1878.

Simson, F. B., *Letters on Sport in Eastern Bengal*, London: R. H. Porter, 1886.

Singh, Arjan, *The Legend of the Man-Eater*, Delhi: Ravi Dayal, 1993.

Singh, Kesri, *Hints on Tiger-Shooting (Tigers by Tiger)*, Mumbai: Jaico, 1969.

Sukumar, Raman, *Elephant Days and Nights, Ten Years with the Indian Elephant*, Delhi: Oxford University Press, 1994.

Suydam Cutting, *The Fire Ox and Other Years*, London: Collins, 1947.

Thapar, Valmik, *Tigers, The Secret Life*, London: Elm Tree Books, 1989. Reprinted as *The Secret Life Of Tigers*, Delhi: Oxford University Press, 1998.

——— *Land of the Tiger*, London: BBC Books, 1998.

Thompson, Edward, *Letter from India*, London: Faber and Faber, 1943.

Ward, G. C., *Tiger Wallahs*, New York: Harper Collins, 1993.

Whitaker, Zai, 'The Riddled Ridley', *The India Magazine*, June 1982.

Wilson, Guy Fleetwood, *Letters to Nobody, 1908–13*, London: John Murray, 1923.

Copyright Statement

The editor and publishers of this anthology are grateful to the following for permission to reprint material which they published originally, or for which they hold copyright. Copyright-holders who could not be contacted earlier because of lack of information are requested to correspond with Oxford University Press, New Delhi.

Mrs R. P. Noronha, for excerpts from *Animals and Other Animals* by R. P. Noronha; John Murray, for the excerpt from *On Hill and Plain* by Lord Hardinge of Penhurst; Banyan Books, for the excerpt from *The End of a Trail: The Cheetah in India* by Divyabhanusinh; Bombay Natural History Society, for 'Aboriginal Methods Employed in Killing and Capturing Game' by K. M. Kirkpatrick and 'A Novel Method of Destroying Man-Eaters and Cattle-lifters' by S. R. Daver; HarperCollins Publishers, for excerpts from *Nine Man-Eaters and One Rogue* and *Man-Eaters and Jungle Killers* by Kenneth Anderson.